TEACHING BIG HISTORY

Lauren Guittard, *Abstract Comic* (*Big Bang and Thresholds 1–4*), 2012.
(Photo: Lynn Sondag)

Edited by
Richard B. Simon
Mojgan Behmand
and Thomas Burke • TEACHING BIG
HISTORY

UNIVERSITY OF CALIFORNIA PRESS

University of California Press, one of the most distinguished university presses in the United States, enriches lives around the world by advancing scholarship in the humanities, social sciences, and natural sciences. Its activities are supported by the UC Press Foundation and by philanthropic contributions from individuals and institutions. For more information, visit www.ucpress.edu.

University of California Press
Oakland, California

Library of Congress Cataloging-in-Publication Data

Teaching big history / edited by Richard B. Simon, Mojgan Behmand, and Thomas Burke.
    pages   cm
  Includes bibliographical references and index.
  ISBN 978-0-520-28354-1 (cloth)
  ISBN 978-0-520-28355-8 (paper)
  ISBN 978-0-520-95938-5 (e-book)
  1. History—Study and teaching.   2. Physical sciences—Study and teaching.   I. Simon, Richard B., 1971– editor. II. Behmand, Mojgan, 1966– editor. III. Burke, Thomas, 1957– editor.
  D16.25.T39   2015
  001—dc23   2014013287

Manufactured in the United States of America

24   23   22   21   20   19   18   17   16   15
10   9   8   7   6   5   4   3   2   1

In keeping with a commitment to support environmentally responsible and sustainable printing practices, UC Press has printed this book on Natures Natural, a fiber that contains 30% post-consumer waste and meets the minimum requirements of ANSI/NISO Z39.48–1992 (R 1997) (*Permanence of Paper*).

For our dear friend and colleague Neal Wolfe, who continues to teach us how to situate ourselves in the cosmos.

CONTENTS

ILLUSTRATIONS

TABLES

ACKNOWLEDGMENTS

We have worked as a three-headed team, with Mojgan Behmand as producer, keeping the book in line with the vision and reality of the program; Richard B. Simon as editor-in-chief, envisioning the text's structure and wrangling faculty writers; and the gentleman Thomas Burke as the voice of reason, keeping us all grounded. Each of us is grateful to the others.

Simon and Burke are especially indebted to Professor Behmand, whose idea it was to write this book. Behmand encouraged her faculty from the start of the Big History program to be creative and rigorous. She secured funding for us to work, and to travel and present our ideas at conferences throughout the world. She recognized that we were making advances and that, "really, we should write a book."

We stand on the shoulders of the Big Historians. We are especially grateful to our dear colleague Cynthia Brown and to David Christian for their guidance and support. Simon and Burke were pleased to be invited to join in its making.

Mary Marcy, the president of Dominican University of California, who made our program a pillar of her vision for the institution's future, deserves our special gratitude.

We would also like to thank John Cox for his generous funding, which has enabled us to complete this task.

We are of course thankful to our editors Niels Hooper and Rachel Berchten, to the diligent Kim Hogeland, the laser-eyed Emily Park, and everyone at UC Press for helping us to bring this work to fruition.

The Dominican University of California faculty, staff, and administrators embraced and supported our work and we thank them, especially Jaime Castner.

We thank our families for their support and patience when this project has been intensive.

We are, collectively, most grateful for the diligent work of our colleagues, who, despite the pressures of teaching, grading papers, and publishing in their home disciplines, have distilled what we do in Big History in inspiring ways. We are still learning from one another. Publishing this book together is just the next step.

Finally, we would like to thank our students for joining us on this adventure. We are at our best when we keep learning from them.

Introduction

Six college professors and a librarian sat around a big oak table. At one side of the room, large picture windows opened onto a curated landscape of green lawn, gardens, and trees from around the world. On the room's blue walls was painted a mural of the fox hunt—horses and rifle-toting riders leaping fences—with subtle religious subtexts. The room had been a dining room in the redwood mansion that now housed a dormitory and classroom space. The gathered faculty had been tasked with developing, after two years of attempts and failures, a new general education curriculum for first-year students at their university. They were at an inflection point: a decision had to be made.

The question was whether to use Big History—in some fields it was called the "Epic of Evolution," in others "the New Story"—a meta-narrative that unified the sum of human knowledge in all disciplines, as the core of the general education program, a requirement for all students. Some of the faculty members were pushing for Big History as a radical rethinking of what a general education meant. Some were pragmatically resistant. The approach on the table was so fundamentally interdisciplinary that those who were to teach it would need to effectively go back to school themselves to master content from far outside their fields of expertise. Some members of the committee were certain that faculty would not go for it. Others felt that this was just the sort of challenge that professional intellectuals would—and should—relish: a challenge to learn, to grow, to incorporate new knowledge and understanding.

The mood in the room was electric with possibility. The opportunity not only to revise a curriculum but to innovate in ways that completely re-envision what a basic education should mean doesn't come around often. Considering this, the faculty asked: can this work? And they set about to answer that question by modeling a new curriculum that would reframe students' entire experience at the university, the whole of their education, and their understanding of the world and how it works.

One year later, in the fall of 2010, Dominican University of California began teaching Big History to all first-year students: first, in a semester-long survey of the history of the universe, our solar system, Earth, life, and humans; and second, in a more focused examination of that story through the lens of a particular discipline.

This book is the result of five years of revision and refinement of that curriculum, and of an intense focus on the development of practical pedagogical approaches for use in the Big History classroom.

We feel strongly that Big History, used as the core of a general education, prepares students for the challenges humanity faces in the twenty-first century. What's more, over the course of developing our program, we have learned some important lessons about how to bring this exciting and intellectually challenging new field to life in the classroom. These insights should be of use to anyone teaching Big History, whether at the podium in a large university lecture hall, around the table in an intimate seminar, or before the chalkboard in a community college or high school classroom. Our aim, in this book, is to share what we have learned.

WHAT IS BIG HISTORY?

Big History is a young, transdisciplinary field, in which scholars from diverse academic disciplines seek to make sense of the story told by the entirety of human knowledge. Big History bends what is considered "history" back to the beginning of the universe. It begins with cosmology and physics and moves on through astronomy, chemistry, geology, paleontology, evolutionary biology, archaeology, and anthropology. Ultimately, it recontextualizes traditional "recorded" human history as inseparable from natural history, environmental geography, and the story of the cosmos. And it is bound together with the art of storytelling that is the province of the humanities.

Big Historians seek to place the wealth of human knowledge into a framework that illuminates the patterns that recur throughout the universe and its history, and to explore the relationships among the events and phenomena therein. The goal is to understand the nature of the universe, and humans' place in it. That new understanding gives scholars new perspective in their home disciplines and allows them to understand their own work on a granular scale, within the context of the story of time and space on the grandest scale.

The throughline is a tendency over time toward increasing complexity—toward the emergence of more complex systems that consist of more and more varied components that are arranged in characteristic structures, that use more intensive flows of energy to maintain those structures, and that manifest new, emergent properties that their component parts could not. For example: the ability of atoms to form complex molecules; the ability of single-celled organisms to form complex, multicellular ones; the ability of humans to coalesce in a complex global civilization whose members can perceive, measure, and intuit the structures and nature of the universe.

The story of complexity, pioneered by astrophysicist Eric Chaisson and furthered by cultural anthropologist Fred Spier, has been divided by historians David Christian, Cynthia Brown, and Craig Benjamin into eight "thresholds," inflections in time/space in which some new form of complexity emerged and changed everything: the explosion and rapid expansion of a universe of matter and energy in plasma; the cooling out of the hydrogen atom from that plasma; the formation of galaxies of stars that churn hydrogen into heavier chemical elements; the formation of a solar system that consists of a single such star with planets comprised of these heavier elements; biological life that emerges on one of those planets and evolves into myriad forms; the evolution of one line of those life forms into *Homo sapiens; Homo sapiens'* harnessing of the sun's energy with the development of agriculture, the surpluses from which leads to revolutions in human social structures; and the marriage of fossil fuels and machines, which generates both great wealth and a potentially catastrophic impact on Earth's systems.

When we give students an understanding of the patterns that recur throughout the history of space and time, of Earth and life, and of human society and culture, we prepare them to understand the challenges and anticipate the opportunities that the future may hold, and we empower them to act accordingly. Students can prosper from this understanding. The human society of which they are members will benefit, too.

BIG HISTORY IN TWENTY-FIRST-CENTURY EDUCATION

We are contending with the fragmentation of knowledge at a time when information is overabundant and easily accessible. Our changed and ever-changing twenty-first-century context requires innovation in our educational models. The answer does not lie simply in experiential learning, flipped classrooms, or even massive online open classrooms, or MOOCs. Important innovations such as problem- or inquiry-based learning are effective, student-centered approaches; nevertheless, they are only pedagogies, new ways to help students assimilate the knowledge formerly delivered through lectures. Our times call for a greater progress, a shift not merely in method but in content. We need foundational knowledge that goes beyond the regional, disciplinary, or temporal. We can no longer study World History, Western Civilizations, or the Great Books and count these as sufficient foundation. Many of the issues that confront humanity and the planet today have a reach that transcends national borders and regional time frames. In order to solve such large-scale problems, today's citizens must develop the ability to see the issues at hand as inextricably linked within a large, complex global system.

Systems thinking is crucial for solving complex, systemic problems and avoiding unintended consequences. It must become a core competency in twenty-first-century education. A large frame of reference is an essential ingredient for intellectual flexibility, as is an interconnected, interdisciplinary approach to the study of past and present complexities in our world.

Our vision for a twenty-first-century education is an approach to learning that empowers individuals and prepares them to deal with complexity, diversity, and change. Our approach to Big History integrates the sciences and humanities; endows students with critical thinking, analytic reasoning, and communication skills; and fosters a sense of social responsibility.

We feel that this synthesis—of broad foundational knowledge, a logical structural framework for the knowledge students will continue to accrue, and practical intellectual skills that allow them to apply their knowledge to diverse challenges—gives students the cognitive resilience that life in a rapidly changing twenty-first century will continue to require.

TEACHING BIG HISTORY

In the intensive and ongoing process of learning how to teach the challenging material that constitutes a course in Big History, we have gained a bit of insight. Teaching information literacy is important for helping students navigate our complex, message-dense media environment. Awakening curiosity about global, national, and local issues is important for engaging them in the world; promoting engagement with the research process activates them as scholars, and as lifelong learners.

Students sometimes need coaxing to warm up to Big History. Its content defies disciplinary designation and is sometimes too abstract to be tangible for younger students, or may feel like too much information to process. More experienced students might initially resist the content of particular units as repetition or review of what they already know. To help students with varying experience synthesize the vast wealth of content that a course in Big History presents, the Big History teacher must consistently contextualize all topics and discussions within the Big History framework. The instructor must also establish the relevance of studying Big History to students' majors or fields of interest, and to their education as a whole. The constant reiteration of the overarching narrative and the themes of interconnection and interdependence are essential for students. And these themes are among those that resonate with them most deeply.

Big History can be taught in a rotational lecture format where the main lecturer invites expert guests to give talks in their area of expertise. It can also be taught in a seminar format with one faculty member leading students through the course. Independent of the format, no single faculty member comes to Big History an expert in all the thematic areas the course covers. Faculty need to train.

It is essential for the Big History teacher to distinguish between mastering Big History content and mastering Big History pedagogy. The Big History teacher must first become proficient in Big History; she or he must then develop means to successfully convey that content. We have found that the most important strategy for achieving such proficiency has been the creation of a faculty learning community. Our annual Big History summer institute has become a living example of *collective learning*. This central Big History concept holds that the sharing of information among groups of diverse individuals or societies leads to sharp increases in innovation. It is the fruit of that collective learning that we aim to share here.

Over the five years that we have spent developing our pedagogy and our program, we have moved away from lecture and a rigid focus on conveying content, and toward active learning. This approach makes students stewards of their own learning, as they participate in activities that they experience, rather than acquiring knowledge only through reading about the material. Our activities are designed with multiple learning styles in mind—visual, auditory, kinesthetic—and consider the range of skill levels one finds in a contemporary classroom. Our faculty learning cohort includes professors from disciplines across the university to ensure that our Big History content agrees with our best knowledge in those disciplines. We have incorporated reflective writing as a key method for engaging students with the content on a deep and personal level, and we collaborate with campus librarians to integrate information literacy into our Big History courses.

Considering the big questions that Big History raises is also an essential part of our approach and something to which we devote time in faculty development settings and in the classroom. It is in the tradition of the humanities to tend to the intellectual, moral, and spiritual implications of these questions. Ours is a secular university with a Dominican heritage and a cutting-edge science program. We feel that we are thus well placed to have such a conversation.

THE STRUCTURE OF THIS BOOK

In part 1, "The Case for Big History," we explain in a bit more detail our argument for making Big History the core component of any comprehensive educational endeavor, as well as how we developed our program, our curriculum, our faculty, and especially our pedagogy. We begin with a Big History primer, "What Is Big History?" Next, in "Big History and the Goals of Liberal Education," we discuss why we chose to make Big History the core of our general education program and how we began to construct our curriculum. In "Summer Institutes: Collective Learning as Meta-Education," we lay out the process of faculty development that led to our unique transdisciplinary approach, and to the further refinement of our curriculum and our pedagogy. In "Assessing Big History Outcomes: Or, How to Make Assessment Inspiring," we share our strategies for assessment, which have led to some interesting revelations about what our students take away from the study of Big History. Finally, in "Big History at Other Institutions," we invite some colleagues from the Netherlands and South Korea to share their approaches to teaching Big History, and we explore the Big History Project, a cyberspace-based high school curriculum.

In part 2, "A Practical Pedagogy for Teaching Big History," we lay out our approaches to teaching Big History, step by step. Because we use the thresholds of increasing complexity model, we start with a chapter on teaching complexity, then devote a chapter to each of the eight thresholds—the Big Bang, the formation of stars and galaxies, heavier chemical elements and the life cycle of stars, the formation of our solar system and Earth, the evolution of life on Earth, the rise of *Homo sapiens*, the agrarian revolution, and modernity and industrialization—as well as a chapter on teaching possible futures. In each chapter, we introduce key concepts, approaches, and best practices. We suggest student learning outcomes (SLOs) and assessment tools to gauge those SLOs. We broach potential challenges in teaching each specific unit in the classroom. And we offer a few active or experiential learning activities, designed and field-tested both in our faculty development workshops and in the classroom. Additional chapters on our approaches and practices follow: "Reflective Writing in the Big History Course" is a guide to using free-writing prompts to further engage students; "Activities for Multiple Thresholds" offers additional exercises for use across the course; and "Igniting Critical Curiosity: Fostering Information Literacy through Big History" suggests models for working with campus librarians to build information literacy into the Big History course or program.

Finally, in part 3, "Big History and Its Implications," we broach some of the bigger questions: we explain the deep roots of our program in "Big History at Dominican: An Origin Story"; we explore some of the challenges in negotiating the various issues surrounding Big History and religion in "Teaching Big History or Teaching *about* Big History?"; and we suggest approaches for engaging students in the profound implications of the Big History story in "The Case for Awe."

The book concludes with our "Annotated Bibliography of Texts and Multimedia Resources Useful in the Teaching of Big History." The annotated bibliography is an assignment we use in preparing our students to write research papers. Here, we have meta-modeled it as an extensive reference for Big History teachers; in it, our faculty summarize a broad selection of books, articles, films, and online resources relevant to Big History, and we suggest how to use each in a Big History context.

HOW TO USE THIS BOOK

Like Big History itself, and like our pedagogy, this text is rooted in storytelling. We have designed it to be both readable and useful as a reference. It is both argument and practicum.

Some readers who are eager to start planning their courses may want to skip ahead to the practical pedagogy, or even to the annotated bibliography. That's a perfectly valid way to use the book.

Those who are newer to Big History, as well as those who are interested in building a Big History program and those who are interested in how this approach may affect the broader mission of education in the twenty-first century, might want to start with part 1, in order to understand the causal chain of events that led to our pedagogy, as well as our program's grounding in the mission of liberal education.

An instructor who is busily preparing for tomorrow's class on, say, Threshold 6 might find it useful to look at the essential recommendations for teaching that threshold in our conclusion.

In our classrooms, and in our program, we have found that Big History does what a core curriculum for twenty-first-century college students needs to do: it instills or reinforces basic foundational knowledge in many disciplines, presents that knowledge in a larger context that makes sense of what students are learning in all their classes, and encourages the problem-solving and critical thinking skills that will make them stronger students and more innovative thinkers and leaders—with benefits for them as individuals, and for society. When we teach Big History as both foundational knowledge and analytic framework, we empower students and prepare them to deal with complexity, diversity, and change. We give them tools for understanding large and complex structures and systems, for solving the types of systemic problems humanity faces, and for making sense of a maelstrom of information that can otherwise obscure fundamental truths about basic reality and the interconnectedness of humans and the actual world in which we live.

A wonderful side effect is that our work in our home disciplines has become richer through being informed by our collective study, by our deep, far-ranging, and sometimes contentious discussions of Big History and how to teach it, and by our innovations therein.

Again, we are advocating for Big History as the core of a general or liberal education for college students. But we believe that our intensive and transdisciplinary work in developing the Big History pedagogy that follows has lessons for all educators using Big History in the classroom, at all levels, and in all educational contexts.

We invite you to join us.

Overleaf: FIGURE O.1   Danielle Dominguez, *Creation Myth Tunnel Book: The Dreaming* (*Threshold 1*), 2012. (Photo: Lynn Sondag)

ONE · What Is Big History?

Richard B. Simon

On my first day as a college undergraduate, I walked into my freshman seminar, "The Gaia Hypothesis," a course on James Lovelock's groundbreaking theory that the biosphere, all life, interacts with all of Earth's systems—climate, oceans, and the rocks themselves—in ways that maintain temperature and chemical homeostasis on Earth. The course changed my understanding of how the Earth works, and of how everything works. It was disruptive, and transformative, and it framed the rest of my education so that when I studied astronomy, paleontology, sociology, climate science—and even literature—it all seemed to fit together, to make sense, in ways that were profound. It made me want not only to make art about big ideas, but to teach others how to make art about big ideas.

So, when the committee revising Dominican's first-year program asked me whether I could see how Big History might work as the core of a new curriculum, my answer was an enthusiastic "yes!"

Well, first I had to look up what, exactly, Big History was.

Then I said "yes."

WHAT IS BIG HISTORY?

Big History uses humanities-based storytelling to span cosmology, physics, chemistry, astronomy, geology, evolutionary biology, anthropology, archaeology, and a traditional human history that has been reconciled with natural history and environmental geography. Its aim is to weave the vast realm of evidence-based human knowledge into a master narrative that tells the story of human beings on Earth, from the beginning of the universe to the present.

Big Historians seek to place the wealth of human knowledge into a framework that allows us to see patterns that repeat on various levels of reality, from the subatomic to the political to the macrocosmic, and to seek the relationships among those levels of reality. Ultimately, the goal is to try to understand the nature of the universe, and our place in it.

Because the Big History framework illuminates the structures that underlie the universe, it is a powerful analytic tool. Because its structure binds together content from all human disciplines, it is a powerful pedagogical tool. Finally, because the structure of the Big History narrative parallels the structures of the physical universe, even as it *tells the story of those structures*, Big History is at once narrative and meta-narrative. All this makes Big History an intuitive vehicle for critical thinking, and for rich, innovative intellectual exploration within students' and teachers' home disciplines, as well as within Big History itself.

Perhaps most importantly, a Big History understanding, in reframing all of human knowledge in a way that makes intuitive, logical sense, prepares us to consider possible futures, premised on the patterns we see in the past, and empowers us intellectually to act to shape the future.

BIG HISTORY AS HISTORY

Thinkers from disciplines other than history are telling versions of this story, too. These thinkers include cosmologists, biologists, geologists, astrophysicists, theologians, and more. Some approaches are more academically rigorous than others. While the name "Big History" connotes the attempt by academic historians to frame the story of the universe as history, it has also become shorthand for all the versions of this story (for a meta-narrative history of the field and some discussion of its various threads, see chapter 19, Cynthia Stokes Brown's "A Little Big History of Big History").

What is remarkable about Big History as *history* is that, traditionally, the field of history privileges primary sources—firsthand accounts of events. But while surviving firsthand accounts might offer us a snapshot of roughly the last five thousand years, our species, *Homo sapiens sapiens*, has existed for two hundred thousand years. That means that 97.5 percent of the history of humans has been off-limits to human historians!

Instead, the study of our species in those years—the Paleolithic or Stone Age—has been termed "prehistory," considered the realm of archaeologists, anthropologists, and paleontologists, and kept academically separate from "recorded history."

For scale, if in 2014, we were to write an analogous account of the history of the United States of America, we might begin at the inauguration of President Barack Obama. We would not only maintain that nothing that happened before January 20, 2009, counts as American history (perhaps because it had not been broadcast via digital social networks); we would also assume that nothing that happened before that date bears any relevant causal relationship to the events that have occurred since.

Of course, that's absurd. Yet our academic disciplines have long worked in isolation from one another, even as they are writing different chapters of the same story. We've been like the five blind men who find something in the forest and argue over whether it is a snake, a worm, a giant bat, a tree trunk, or the side of a barn, never understanding that, together, they have found an elephant.

Big Historians posit that because human knowledge has undergone—largely in modern, industrial times—a *chronometric revolution*, we have ways of knowing that are at least as valid, accurate, and verifiable as firsthand written accounts.

These newly readable texts are encoded in DNA, in the half-lives of the radioisotopes of certain atoms, in the time-layered sediments in rocks, and in events occurring all across the cosmos that we can observe with powerful, space-based telescopes. When we read these texts, we stretch the purview of history not only beyond "recorded" history to the dawn of humans, but across the fossil record to the dawn of life itself; to the earliest days of our planet, our star, and our galaxy; and through theoretical physics and cosmology to the first seconds in the existence of our universe. We can actually see events that occurred billions of years ago as if in real time, as the light from those far-off events finally reaches our telescopes. Of course, that makes astronomical observations firsthand accounts of the history of the universe. And our observations of other stars and galaxies across their predictable, patterned life spans allows us to extrapolate an understanding of our own sun, and of our own Milky Way galaxy.

Big Historians also include the history of the evidence—the story of science itself.

By using *all* the information at our disposal, we can begin to understand the breadth and depth of the history of the universe, and to see some of the remarkable patterns that recur throughout.

That history goes a little something like this.

THE BIG HISTORY STORY

We don't know what existed before, but around 13.8 billion years ago, a tiny point appeared, in which the entire contents of the universe already existed as energy and matter. It exploded rapidly, as if in a big bang, then expanded again in a period known as inflation, until the universe consisted of a vast amount of space filled with a superhot plasma of subatomic particles.

As the universe cooled, basic forces appeared: gravity, electromagnetism, and the nuclear forces that bind subatomic particles together and govern radioactive decay—along with protons, neutrons, electrons, photons, and neutrinos. As the universe continued to cool and expand, those particles were able to bind together electromagnetically as atoms, releasing loose photons in an enormous flash of light.

The new atoms of mostly hydrogen condensed into enormous clouds, which congealed in pockets of their own mass and gravity into galaxies and stars. Within one of those galaxies, our own star blinked awake when a critical mass of hydrogen, jostled by a nearby supernova (the explosion of a star that itself had been fusing hydrogen and helium into heavier elements, such as oxygen, silicon, iron, and gold), began to fuse into helium. A cloud of loose material, cast off by the supernova, continued to spiral around the new star under its gravitational influence—the heavier material closer to the star, the lighter, gaseous material toward the outside of this rapidly spinning disc of matter. This matter accreted into four rocky inner planets and four gaseous outer planets (and a few other objects).

One of the inner planets was our own, Earth. It orbited at the right distance from the star to be able to maintain water in all three states—gas, liquid, and solid—and maintained a gaseous atmosphere. Somehow, life emerged on this planet as single-celled bacteria. Some of these bacteria incorporated other bacteria and became more complex cells, which learned to reproduce sexually, to photosynthesize, and to process sugars for use as energy. Life evolved along with and affected changing planetary and atmospheric conditions—from single cells to plants to animals, including the vertebrate mammals that evolved into primates.

Some of those primates diverged from the great apes to evolve into bipedal hominines such as *Australopithecus afarensis* ("Lucy") and *Homo erectus*. *Homo erectus* evolved, separately, into both *Homo sapiens neanderthalensis* and, finally, *Homo sapiens sapiens* (they are commonly referred to as Neanderthals and *Homo sapiens*, respectively).

Homo sapiens, the wise human, spread out to populate the globe and learned to dominate and domesticate other species—namely, animals and plants. Because the invention of agriculture led humans to settle in one place, and then led to surpluses—which had to be stored and guarded against invasion, and which could also be traded—human society became more complex, and grew to include centralized yet sprawling communities, social hierarchies, and networks of exchange.

With the discovery of fossil fuels and the invention of heat-driven machines, ever-increasing flows of energy (which had originated as solar energy) pulsed through these agricultural civilizations until they became industrialized. As more and more people and ideas (from far-flung and disparate civilizations) were connected through trade, their collective learning led to more and more innovation.

As of this writing, this increased innovation has resulted in a highly industrialized global civilization, connected through rapid, energy-fueled information flows as well as advanced transportation technology. This global civilization is aware of itself; it has seen itself from space. And it is even now exploring other planets and on the verge of exploring beyond the outer reaches of its own solar system. At the same time, it faces several global problems that are existential crises, including an ever-expanding human population, resultant scarcities of energy, food, and clean freshwater resources, and an increasingly unpredictable global climate system, the cause of which is the by-product of the very consumption of energy that has led to civilization's current level of complexity. There we remain, perched on the verge of an age of wondrous new technologies and potential ecological catastrophe.

And that's the story so far. It's a cliffhanger.

A FRAMEWORK FOR HUMAN KNOWLEDGE

The framework that underpins the Big History narrative, as laid out by David Christian, Cynthia Stokes Brown, and Craig Benjamin in *Big History: Between Nothing and Everything,* is that of thresholds of increasing complexity. In this model, what binds the story together is that as the universe has progressed across time from the Big Bang to advanced industrial human civilization, it has tended toward greater complexity.

According to this model of complexity, which is based on the work of astrophysicist Eric Chaisson and that of cultural anthropologist Fred Spier, a form of complexity—for example, the universe, a galaxy, a solar system, an organism, or an agricultural or industrial civilization—is comprised of four elements:

diverse components, or *different* types of parts, arrayed in . . .

specific arrangements, or characteristic structures, such as an atom with a nucleus and orbiting electrons; a cell with a nucleus and energy-processing organelles, surrounded by a membrane; a solar system with a star at its center, orbited by planets and two concentric layers of loose debris; or a civilization with a city at its center and agricultural production at its periphery, connected by trade routes to other similar cities. Those arrangements of those components are held together by . . .

flows of energy, typically energy emitted by the fusion within stars that is being used in ways characteristic to that form of complexity and resulting in new . . .

emergent properties, or new properties that exist only in this new form of complexity—things that the whole can do that the parts could not.

The greater the number and variety of parts, the more numerous and varied the connections among them, and the higher the flows of energy, the more complex an entity is said to be. And the more complex an entity is, the more interesting (to us!) its reality-altering emergent properties are—because they eventually result in us.

Each new emergence is marked by an increase in the amount of energy that is used by that new form of complexity to maintain its structural continuity. So, a form of complexity connotes a complex system that maintains its own energy flows and thus its structure. But the amount of energy that flows through any form of complexity must remain within limits that Fred Spier calls "Goldilocks conditions." Too little energy and the complexity collapses; too much energy and it burns out.

Christian, Brown, and Benjamin use increasing complexity as the throughline that binds the entire story together. They divide the story by eight "thresholds"—transitions in time / space across which a new form of complexity "emerged" from what existed before, and so changed everything—leading toward us, contemporary *Homo sapiens.*

We think of it like this:

Threshold 1: The Big Bang

Threshold 2: The Formation of Stars and Galaxies

Threshold 3: Heavier Chemical Elements and the Life Cycle of Stars

Threshold 4: The Formation of Our Solar System and Earth

Threshold 5: The Evolution of Life on Earth

Threshold 6: The Rise of *Homo sapiens*

Threshold 7: The Agrarian Revolution

Threshold 8: Modernity and Industrialization

Threshold 9? Possible Futures

Unless the universe continues in its present form until its eventual demise, without generating any new forms of complexity that are perceptible by humans on Earth—or, from an even more anthropocentric point of view, unless human civilization does not evolve into yet another higher order of complexity—there will be a Threshold 9. It will be marked by increasing complexity—new and more numerous components, connected in many ways into complex structures, with vast increases in energy flows and remarkable emergent properties. But "Threshold 9" is not synonymous with "the future."

A PEDAGOGICAL FRAMEWORK

Because each of the eight thresholds is primarily the domain of a few disciplines of study (Threshold 1: cosmology, physics; Threshold 2: astronomy; Threshold 3: astronomy, chemistry; Threshold 4: astronomy, geology; Threshold 5: evolutionary biology; Threshold 6: paleontology, anthropology; Threshold 7: archaeology, geography, history; Threshold 8: geography, history, sociology), the Big History narrative is quite valuable as a pedagogical framework.

Big History as a field is zoomable (the ChronoZoom project, developed by legendary geologist Walter Alvarez and technologist Roland Saekow, working with Microsoft Research, demonstrates this visually), in that we can view the long story of the universe and then focus in on what is happening in a distant nebula (astronomy), or in a particular terrestrial ecosystem (ecology), or in a period of human history. We can use our understanding of the larger picture to make sense of the specific detail. Likewise, the specific detail lends granularity to the big picture. And so we fit new knowledge into both micro- and macro-level understandings of how the world works.

Students who are taught the broad story of the universe, and who understand how the knowledge yielded by the various disciplines of human endeavor fits together, have a sound and logical scaffold in which to place the specific knowledge they attain in their own eventual disciplines. It's one thing to learn trigonometry by formula in a classroom, and quite another to understand that by *using* trigonometry, we *know* the distance to stars as measurable fact, and that because our ancestors understood this (on a more rudimentary level), they were able to navigate the seas using the stars as relatively fixed points, and thus to populate every continent, to drive trade, and to spread ideas, religions, markets, diseases, genetic material, and conflict around the globe. What discipline that students might study is *not* affected by this sort of understanding?

The students in our nursing program, for example, now understand why the vaccines they use *work*, because they are learning in Big History about the forces that drive natural selection. They also know that numerous human civilizations have met their demise at the flagella of enterprising microorganisms. So they know what's at stake when such organisms—such as methicillin-resistant *Staphylococcus aureus*, or MRSA—select around the abilities of our antibiotics to contain them. They understand on both the macro and the micro scale what this means for their work in combating infection in the hospital setting. And that understanding empowers them to innovate.

META-NARRATIVE AND META-EDUCATION

Because the Big History narrative tells the story of time/space in a chronological fashion, the story itself is structured chronospatially. It also grows more complex as the universe grows more complex. That leads to interesting opportunities for critical thinking about ways in which humans structure information. In this case, the structure of the story is synecdoche, a fractal representation of the structure of the universe. And each chapter is a scale model of the whole story.

The concept of **collective learning**—the idea that diverse knowledge shared among diverse individuals and societies, and across generations, leads to innovation—is demonstrated in its own discussion in a dynamic Big History classroom. Students working and thinking and discussing Big History together solve problems together, raise new questions, and innovate—and underscore the concept of collective learning in doing so.

Likewise, a Big History faculty, working together to hone teachers' understanding of Big History, innovates in ways that lone teachers reading Big History and lecturing might not (see chapter 3, "Summer Institutes: Collective Learning as Meta-Education").

Such meta-educational surprises await every Big History teacher.

SOME CHALLENGES FOR BIG HISTORY

Certainly, Big History, as a young field, faces challenges and questions. Some scholars find the complexity model problematic. The thresholds are not bright-line boundaries in time/space but transitions chosen as the most significant events in the development of the universe on the way to us, contemporary *Homo sapiens sapiens*. In this respect, Big History is necessarily anthropocentric—it tells the stories of humans from a human point of view. But that can make the story seem as if humans on Earth are the end point of the story of the universe, rather than one strand among perhaps trillions. Big History also poses challenges for traditional religions; in telling a story that explains how the universe came to be, it does some of the important things that have long been their purview. It may meet resistance where scientists making world-shifting discoveries have often found it. We will address some of these issues later in this book.

But we should ask such questions. They are among the most profound for students, and among those they are most eager to discuss. What is the nature of the universe? How do we know? Why are we here? These are the big questions, the timeless big questions. Asking them engages students, deeply, in the world of ideas.

BIG HISTORY AND TWENTY-FIRST-CENTURY SCHOLARS

As Big History matures, it is likely that new schools will emerge within the field that focus on telling the stories of other planets, or other species, and so the larger story will be filled in. It's also likely that, like all other attempts by humans to understand our own place in the universe, Big History itself will be challenged or supplanted by some new understanding that incorporates a perspective that we could not yet begin to comprehend, just as Paleolithic humans could not build their creation narratives to incorporate observation of radio waves from billions-of-years-old supernovae.

That is another way in which Big History is subversive. Not only does its study disrupt current ways of organizing human knowledge and reframe all received and new knowledge with a cosmic understanding; it also contains the seeds of its own destruction. To understand Big History is to know that it, too, will be replaced by some future way of knowing. After all, that's another one of the key repetitions.

These are all issues with which Big History students may wrestle. The exercise helps build the critical thinking skills that will make them adaptive and resilient thinkers who will be well prepared to solve humanity's current problems. We want to challenge them and what they think they already know, to prepare them to incorporate new information, so that they may make their own individual ways through a world that they increasingly understand. We want to challenge their beliefs—but not to destroy them. Rather, we hope that students will develop the skills that they need to adapt to a twenty-first century that demands that the sum of observed human knowledge be somehow reconciled with other ways of knowing, and applied to solving very big problems.

Our ecological crises have become existential and planetary: warming and acidifying oceans; melting glaciers and permafrost; desertifying farmlands; devastating storms, droughts, and wildfires; expanding disease vectors; and mass species extinctions driven by all these problems—all caused, like so many human crises over the last 200,000 years, by our overuse of resources whose initial abundance astounded us. A Big History understanding suggests that ethical regimes developed during the age of agrarian civilizations are ill equipped to confront late Industrial Age problems.

The next generations will have to find the way.

We believe that Big History will prepare them to do it.

TWO · Big History and the Goals of
Liberal Education

Mojgan Behmand

At the age of twelve, I was certain that I would change the world; at the age of thirty-two, life in academia had made me certain that I would have little effect on it; and at the ripe age of forty-two, I became a believer and a dreamer again. The reason? Big History and liberal education.

August 2007 saw the beginning of my tenure at Dominican University of California, a 123-year-old secular institution of Catholic heritage twelve miles north of the Golden Gate Bridge. The institution had until 2000 been a liberal arts college before transforming itself into a comprehensive university.[1] This historical tidbit meant little to me at the time; I simply enjoyed my classes and the collegiality. Then, in early 2009, I was elected to be a member of a committee engaged in curricular revision. It is a truth universally acknowledged in academia that an educator in possession of an inquisitive mind is perpetually in want of the perfected curriculum. I joined the group with an odd mix of enthusiasm and wariness, as I hoped to contribute but feared outing myself as a novice in curricular development. Soon, I was assigned to the subgroup examining Dominican's first-year programming.

Collectively, we developed a unique First-Year Experience program based on Big History. My esteemed colleague Phil Novak proposed Big History as foundational content; Big Historian Cynthia Brown, professor emerita at Dominican, returned to campus to promote the idea; and the committee supported the developed concept and eventually won faculty and administrative approval. By the end of 2009, I had been tasked with developing both the detailed curriculum and organizing the requisite faculty development. Energized, my colleagues and I gathered in the summer of 2010 and collaboratively designed the courses and learning outcomes while studying Big History. The excitement of innovation and the prevalent camaraderie fed us and, in my case, rendered me impervious to the manifold and diverse assumptions we were making about our educational goals.

Incorporating Big History was relatively easy at Dominican, as the first-year programming is a place within general education that is not traditionally *owned* by a department. Knowing that first-year seminars are among the high-impact practices lauded by the Association of American Colleges and Universities (AAC&U), which advocates for liberal education, we created the program First-Year Experience "Big History" as a requirement for our incoming class of first-year students.[2] Our program became the first of its kind in the world: a one-year Big History program made up of a sequence of two courses and numerous co-curricular activities. The first-semester Big History survey course is taught in small, twenty-person seminars. The second-semester seminar courses offer a choice among discipline-based courses that reiterate the narrative of Big History and its major concepts in dialogue with a specific discipline or field of inquiry. Those courses include, "Visual Art through the Lens of Big History," "Myth and Rituals through the Lens of Big History," "Trade through the Lens of Big History," "Religion through the Lens of Big History," "Health and Healing through the Lens of Big History," and "Creative Writing and Big History."

Our endeavor had been incredibly successful as far as cross-campus collaboration went; however, it took me almost a year to articulate two important realizations: first, being proficient in Big History is different from mastering Big History pedagogy; second, adopting Big History necessitates setting goals for the program itself and not merely learning outcomes for the course(s).

The issue of Big History pedagogy was relatively easy to address: we did so by focusing extensive portions of our annual Big History summer institutes on the sharing and teaching of best practices, and we facilitated semesterly retreats and weekly pedagogy lunches. Ultimately, this book itself is our response to the need for extensive faculty development and resources in Big History pedagogy. The issue of setting goals for a Big History program is more complex. My recommendation is that any institution adopting Big History—be it at the elementary, high school, or higher education level—set goals aligned with the objectives of the larger program that houses the course and the institution's educational mission.[3]

Why this insistence on aligned goal setting? To put it bluntly, Big History is a vast, almost unruly, field. An instructor might decide to bring it to the classroom via engaging activities and assignments, because students would benefit from this knowledge, yet Big History, with its plethora of thresholds, themes, facts, and dates, can prove untamable. It needs a framing approach that will harness its energy and point the instructor and the learner in the same direction. The survey course and its offshoots have immense multidisciplinary potential, yet they need a clear underlying intention to succeed. Is our goal literacy in the sciences? Is it comprehension of natural and human history? Is it analysis of global interconnectivity and interdependence? Is it agency in shaping our planetary future? And do our goals for the course and/or program align with the educational goals and mission of the institution?

At Dominican University, liberal education proved to be our savior. Interestingly, the U.S. model of liberal education dates back to the Founding Fathers. In 1778, Thomas Jefferson expressed his support for it through a bill whose preamble asserted that "those persons, whom nature hath endowed with genius and virtue . . . should be rendered by liberal education worthy to receive, and able to guard the sacred deposit of the rights and liberties of their fellow citizens."[4] This vision of education as laying a moral obligation on the educated and serving to advance the good of the larger community has always resonated with Dominican educators, whose traditional ideals are study, reflection, community, and service. Notably, the course components of liberal education have been subject to reevaluation and revision over the years, but the desired outcomes of such an education have remained the same. Accordingly, Dominican embraced its history of liberal education.

It is important to note that a reenergizing of liberal education had taken place in 2005, when the AAC&U[5] launched "a national advocacy, campus action, and research initiative . . . that champions the importance of a twenty-first century liberal education—for individuals and for a nation dependent on economic creativity and democratic vitality."[6] This initiative defines a twenty-first-century liberal education as "an approach to learning that empowers individuals and prepares them to deal with complexity, diversity, and change. It provides students with broad knowledge of the wider world (e.g., science, culture, and society) as well as in-depth study in a specific area of interest. A liberal education helps students develop a sense of social responsibility, as well as strong and transferable intellectual and practical skills such as communication, analytical and problem-solving skills, and a demonstrated ability to apply knowledge and skills in real-world settings."[7] Clearly, a careful reading of this statement will accelerate the heart rate of any academic wishing to change the world: it is ambitious, idealistic, and all-encompassing.

Educators and industrialists alike have lauded this rejuvenated characterization of a twenty-first-century liberal education, as it debunks the perceived impracticality of the arts and humanities and cautions us in our focus on the professions and science, technology, engineering, and math (STEM) graduates. David Kearns, former chief executive officer of Xerox and a U.S. secretary of education, asserted in 2002: "The only education that prepares us for change is a liberal education. In periods of change, narrow specialization condemns us to inflexibility—precisely what we do not need. We need the flexible intellectual tools to be problem solvers, to be able to continue learning over time."[8] The Dominican faculty participating in our visioning session subscribed to the value of intellectual flexibility and also allowed themselves to be guided by the "Essential Learning Outcomes" of liberal education as articulated by AAC&U.

These outcomes stipulated that the student is to gain "1) Knowledge of Human Cultures and the Physical and Natural World; 2) Intellectual and Practical Skills; 3) Personal and Social Responsibility; and 4) Integrative and Applied Learning."[9] In fact, the first learning outcome in its entirety reads: "Knowledge of Human Cultures and the Physical and Natural World—Through study in the sciences and mathematics, social sciences, humanities, histories, languages, and the arts [and] *[f]ocused by engagement with big questions, both contemporary and enduring.*"[10] Big History is unique in uniting bodies of knowledge on the physical and natural world and human cultures, and its narrative arc creates room for extensive engagement with contemporary and universal human questions. Our faculty also hoped that Big History would provide learners with a vast framework for the scaffolding of knowledge to counteract the ever-increasing fragmentation of knowledge caused by growing specialization at institutions and in the workplace.

Undaunted by the challenge, we put our ideals into writing and collectively created a description of the one-year program:

> First-Year Experience "Big History" is a one-year program that takes students on an immense journey through time to witness the first moments of our universe, the birth of stars and planets, the formation of life on Earth until the dawn of human consciousness, and the ever-unfolding story of humans as Earth's dominant species. In studying the evolution of human cultures, students engage with fundamental questions regarding the nature of the universe and our momentous role in shaping possible futures for our planet.

We also determined the goals of the program:

> The program is designed to promote
> - recognition of the personal, communal, and political implications of the Big History story;

- critical and creative thinking in a manner that awakens curiosity and enhances openness to multiple perspectives; and
- development of reading, thinking, and research skills to enhance one's ability to evaluate and articulate understanding of one's place in the unfolding universe.

Writing the program description and goals taught us two valuable lessons: first, every institution must articulate its objectives as suited to its needs and mission; second, a collaborative articulation increases faculty investment in the program, and its assessment and continuous quality improvement. This collaborative approach has endowed us with a strengthened sense of purpose and potential. As one participant observed, "Big History is a wonderful curriculum that will prepare our students to succeed not only in their university pursuits but also in developing and attaining future goals. I feel a renewed sense of commitment to liberal arts education and excited about the possibilities for our next generation of students."[11]

The collaborative design of the Big History curriculum and the alignment of the program with our university's educational philosophy have ensured that we agree on the goals for student learning and are committed to thoughtful assessment and continuous quality improvement.[12] The objectives of our First-Year Experience "Big History" program are ambitious, and the vastness of the Big History narrative can render it unwieldy, but remarkably, this program's careful execution of curriculum design, faculty development, and assessment has created a unique platform for an elusive goal in academia: the integration of disciplines and collaboration across campus.

Hope has sprung alive again. Dominican University of California has not only adopted Big History and embraced its vision of twenty-first-century liberal education; it has also modeled living the outcomes of that liberal education. In our quest to change the world, we changed ourselves. So, at the ripe age of forty-two, I became a believer and a dreamer again. The reason: Big History and liberal education.

NOTES

1. A liberal arts college is defined as "a particular type of institution—often small, often residential—that facilitates close interaction between faculty and students, and whose curriculum is grounded in the liberal arts disciplines." Association of American Colleges and Universities, "What Is a 21st Century Liberal Education?" *Aacu.org*. Association of American Colleges and Universities, 2013. Web. 20 May 2013.

2. George D. Kuh, *High-Impact Educational Practices: What They Are, Who Has Access to Them, and Why They Matter*. Washington: Association of American Colleges and Universities, 2008. Print. High-impact practices include first-year seminars and experi-

ences, common intellectual experiences, learning communities, writing-intensive courses, collaborative assignments and projects, undergraduate research, diversity/global learning, service learning, community-based learning, internships, and capstone courses and projects. First-Year Experience "Big History" is a triple high-impact practice.

3. Big History—under different names—is taught at various educational levels: Montessori schools incorporate it into elementary education via the Cosmic Education curriculum; the Big History Project—an online curriculum funded by Bill Gates—has brought Big History to high schools across the globe; and numerous institutions of higher education in Australia, the Netherlands, South Korea, Canada, and the United States offer it at the university level as Big History. For more information, see Jennifer Morgan's notable children's trilogy *Born with a Bang, From Lava to Life,* and *Mammals Who Morph;* bgC3's website for the Big History Project, bighistoryproject.com; and the International Big History Association's website, ibhanet.org.

4. Thomas Jefferson, "Preamble to a Bill for the More General Diffusion of Knowledge." *The Papers of Thomas Jefferson.* Ed. Julian P. Boyd et al. Princeton: Princeton UP, 1950. 2: 526–527. Print. See also Philip B. Kurkland and Ralph Lerner, eds. *The Founders' Constitution.* U of Chicago P, 2000. Web. 28 May 2014.

5. AAC&U is the leading U.S. American association focused on undergraduate education with 1,300 domestic and international member institutions. Association of American Colleges and Universities, "Who We Are." *Aacu.org.* Association of American Colleges and Universities, 2013. Web. 20 May 2013.

6. Association of American Colleges and Universities, "Liberal Education and America's Promise (LEAP)." *Aacu.org.* Association of American Colleges and Universities, 2013. Web. 20 May 2013.

7. Association of American Colleges and Universities, "What Is a 21st Century Liberal Education?"

8. Kearns qtd. in Association of American Colleges and Universities, "What Is a 21st Century Liberal Education?"

9. Association of American Colleges and Universities, "Essential Learning Outcomes." *Aacu.org.* Association of American Colleges and Universities, 2013. Web. 20 May 2013.

10. Ibid.

11. Anonymous, Big History Summer Institute Evaluations, June 2012.

12. For more on this, see chapter 4, "Assessing Big History Outcomes: Or, How to Make Assessment Inspiring."

THREE · Summer Institutes

Collective Learning as Meta-Education

Thomas Burke

In late May our leafy campus with its quietly elegant architecture begins to slow toward its lighter summer pace. Though some classes are still in session, the student population is significantly less than at term time. Most faculty are studying, writing, researching, and engaging in the world of ideas away from campus. Administrative offices are peacefully diligent. The campus smells different, more natural without the hub and bub of a collective humanity intruding on it.

The last week of May 2010 found a ground-floor classroom in the Beaux Arts–style Guzman Hall filled with thirty professors, colleagues from across the campus. Some of us had laptops; some had notebooks. Many held coffee in reusable cups. The cool group sat in the back of the room and was noisy. The attentive sat closer to the front and busied themselves reviewing the assigned reading. Some among us were confident. Some were jittery. We were a typical first-day group of students, with one shift: the people usually at the front of the classroom were the ones at the desks.

Coming from varied disciplines, we all believed in the importance of a liberal education. As a community we had spent two years discussing and then approving a new core curriculum for the university. Big History would be the foundational first-year experience of that new curriculum. We thirty had signed on to teach Big History. For this one week we would be the students.

Oh, what a ride we embarked on that last week of May 2010.

In *Big History: Between Nothing and Everything,* David Christian, Cynthia Brown, and Craig Benjamin use the term *collective learning* to describe "the ability, unique to human beings, to share in great detail and precision what each individual learns through symbolic language."[1] In preparing ourselves, we, who became the Big History faculty at Dominican University of California, began a process that we would come to see as our own collective learning. We began with excitement, hope, and some necessary hubris.

We came from various departments and disciplines: art history, biology, business, communications, history, literature, occupational therapy, philosophy, and religion, among others. Our tasks were to teach ourselves how to deliver a survey course in Big History for every first-year student who would enter our university in late August and to develop courses for the spring semester that would follow the survey course in Big History. Thus was born the first Big History summer institute at Dominican, the originating component of a tiered process of ongoing faculty development centered on how to engage students.

The Big History faculty has ridden a narrative arc from hubris and excitement to fear, joy, and constant reappraisal. We have arrived at a place wherein we are comfortable knowing what we know and comfortable in knowing we need to continue to learn, collectively.

The faculty development program, our practical application of the theory of collective learning, has evolved from our needs, as a faculty and an institution, and interests in Big History. It has taken time—time that we as faculty do not have but somehow find. The tiers of our process, in which all Big History faculty participate, unfold through the year. They are

- annual summer institutes;
- weekly lunch meetings during the academic year;
- a daylong retreat at the end of each semester; and
- the development and implementation of quantitative, qualitative, and anecdotal assessments.

These components are outlined not to indicate a burden; rather, they illustrate the bottom-up faculty process that has grown out of a genuine excitement about what we are doing.

At our summer institutes, we use a combination of our own resources—ourselves—and lecturers invited from outside the campus. We set out a week—a full week. We read and prepare in advance. We come together. We listen. We discuss. We learn. While our content experts from outside the campus are very valuable, significant value comes from our own engagement with our colleagues here on our campus. We decide things together, based on our experiences teaching our own students.

While some of this book's content is specific to our own leafy and architecturally elegant confines, the lessons here about constant faculty engagement, transparency among colleagues, buy-in from the bottom up, our own collective learning, and our willingness to be adaptable—to experiment and to not be experts—will be instructive to most schools. Each institution will have its own needs, its own issues. While I am sometimes describing specific processes that grew out of our experience here, it might be useful for faculty and administrators at other institutions to look at what led to these processes and how those things might inform their own development. Our path can, we believe, be adapted in pragmatic ways in other academic environs.

There is much discussion here about collective learning and collective decision making. I want to make clear that while we, Dominican's Big History faculty, work together, we also retain our individuality and our academic freedom. Our approach is collegial and contains common components, and we all teach to our own strengths. No two Big History classrooms or teachers at Dominican are exactly the same.

I would be remiss if I did not tell readers that, while there has been true collegiality and collective learning and decision making, we have also benefited from visionary and hardworking leadership. Professor Mojgan Behmand has cultivated an environment in which our work can flourish. In addition to leading, she is also one of us. Along with directing, Mojgan teaches in the program; she is in the game with us every step of the way.

SUMMER INSTITUTE 1: LEARNING BIG HISTORY

During the first summer institute the focus was on content. Among the thirty faculty gathered, only three had taught any form of Big History. We were fortunate to have Cynthia Stokes Brown, author of *Big History: From the Big Bang to the Present* and a coauthor of *Big History: Between Nothing and Everything*, as a longtime Dominican faculty member. Brown, with biology professor James Cunningham and philosophy professor Philip Novak, had taught a set of linked classes in Big History content (this linked set was taught to upper-division students as an elective component of a prior iteration of our core curriculum). The rest of us were newcomers to Big History.

In advance of that first summer institute we read Cynthia Brown's book. We read David Christian's *Maps of Time*. On a fine morning in mid-May, while going to retrieve my academic regalia and take a much-deserved break from grading final student portfolios, I ran into colleagues Bill Phillips from psychology and Neal Wolfe from art history. Both, I knew, had signed up for the first summer institute.

"Have you done your reading for the summer institute?" Bill asked, a smile on his face.

Clearly he was enjoying our homework.

"I am really enjoying the books," I said.

"Me too," said Neal. "I have some questions about how to use them with first-year students, but I imagine we'll get through that next week."

Neal was also thinking about the sheer volume of content. This thought would inform our experience greatly over the next few years.

When I returned to my office, Bill had forwarded me two Big History–related articles. As the week progressed, other colleagues began to do the same. From the beginning we were all making the interdisciplinary connections that would become so much a part of our approach to teaching. We all found connections to Big History in the literature of our own fields, and we began to look at the popular press as a storehouse of Big History content (note how often new discoveries about the universe, solar system, the origins of humans, and so on appear in the news).[2]

As the institute began, a graduate school–like atmosphere emerged. We were taught the content of Big History by our own "experts" and had lectures on content from off-campus "experts." The lesson here is that we were focused on content over pedagogy.

Cynthia Brown led the on-campus "experts" (well, she is a bona fide expert in Big History) with a general introduction to Big History and lectures on such topics as teaching evolution. Our biology colleague James Cunningham lectured on various aspects of planet Earth. Phillip Novak from philosophy lectured on science and religion, followed by Richard Simon in English, who presented on science and religion in the classroom. Mojgan Behmand and I gave presentations on topics related to the research and writing components of our course as well as on developing a common syllabus. These were just a few of the topics covered.

With our colleagues from the university we debated, frequently at a high level, the nuance of ideas. In a decidedly collegial atmosphere we shouted and overspoke. We questioned each other and finished one another's sentences. Content—oh, the content!—engaged us. In short, we had a grand time.

So focused were we all on the content, and on our elevated discussions of it, that we neglected the very practical aspect of a pedagogy appropriate for a survey course that was to be required of all incoming first-year students. The focus on content was necessary. We came from different disciplines. We did need to learn the content, but we also needed to think about how to teach that content. Is my message sinking in? Don't sacrifice content for pedagogy; don't emphasize detail over guiding concepts.

Lest one think that a group of professional educators had completely ignored pedagogy in the first summer institute, let me add that we were attentive to aspects of it. Among the practical pedagogical outcomes of the institute was a common syllabus for our first-semester survey course. Additionally, the second-semester "Through the Lens of Big History" courses were developed. In semester one, all students would share a common intellectual experience and have the same foundational course; in the second semester, students would be free to choose from a menu of Big History–related courses. This approach, of looking at Big History as a whole in one semester and then through a focused lens in the second semester, is no small pedagogical achievement.

Such basic course ideas as literature, politics, and business through the lens of Big History were our starting point. These were not to be survey courses; rather, the topics would be explored using the Big History narrative and concepts (such as thresholds and complexity). Our group of thirty, already in that first week using methods of collective learning, devised a plan to divide into groups roughly based on disciplines. As we are a small campus, our groups sometimes included more than one discipline. For example, philosophy and religion would work together, as would gender studies and art history (some groupings were based on faculty members teaching in more than one discipline).

Each small group worked to come up with course titles and descriptions. The literature group's course became "Myth and Metaphor through the Lens of Big History." Gender studies entitled its course "Sex and Gender through the Lens of Big History." Business and history worked together to produce "Trade through the Lens of Big History." In addition, each group developed student learning outcomes for the courses and assessments for each of those outcomes. Once these documents were drafted, the small groups came back together and their work was vetted by the larger group.

All of this work and vetting led to a transparency and bottom-up buy-in that defines our work to the present. The faculty are deeply engaged in this program. One colleague described the process in this way: "From that first year I have never felt like anyone forced me, or us, to do things in ways that don't make sense. We, as a group, look at the issues at hand, talk about them, and come to a solution. Maybe the solution is not exactly 'my' solution, but I know I had a voice and the ultimate solution devised is usually better than what 'my' solution, arrived at alone, would have been."

YEAR 1, WEEKLY MEETINGS WHILE TEACHING: COLLECTIVE AND ENGAGED

In the fall, following that first summer institute, when we began to teach our first sections of Big History, the faculty met weekly to assess the program and support one another. The weekly meeting was rarely missed by any of the faculty teaching in the program. As we came in and professed both enthusiasm for and frustration with trying to communicate the content and ideas of Big History to our first-year students, one or two among us would offer a creative solution to opening up that week's content. Colleagues would share approaches they had developed to engage the students in the material, which can at times seem overwhelming in its magnitude. Early in the course, art history professor Neal Wolfe shared an activity he had developed in which students would position themselves, on one of the large campus lawns, as the planets in our solar system were positioned on that given day. They could look up into the day sky and talk about what they could and could not see then and what they might or might not see in the night sky. This kinesthetic activity brought together pages of reading—reading that had narrated processes of thousands of years.

Those weekly meetings became a real place of collective learning. Two or three faculty members from different disciplines might work together in a given week. We all shared the creative and engaged activities and ways of thinking about a unit that perhaps only such a diverse group of faculty might develop. As seen above, art historians came up with visual approaches to the Big History narrative. The literature faculty held on to the overarching story, the narrative, of Big History and looked for explicating metaphors. The scientists reminded us of the scientific method. Our colleagues in philosophy and religion continually prompted us with good questions about process, intent, and meaning. The diversity of our teaching methods and approaches was a strength.

The first summer institute was lofty, the focus on content. We were trying to make ourselves instant experts in Big History. Lofty was good to have us understand content, but the rough and tumble of classrooms full of first-year college students brought home the need to focus on developing pedagogy appropriate for this population. The first-year students were not in the course by choice; Big History is required of all. The weekly meetings during the first year dealt largely with the practical issues of engaging eighteen-year-olds in the ideas of Big History.

One meeting might have a biology professor and a literature professor down on all fours trying to simulate the progression from quadrupedalism to bipedalism.

"Oh," one professor said, "and I don't have hands so I need to pick things up with my mouth."

"Yes," the other said, "and look how much farther I can see when I stand up, even just a little bit. We could do this and then have the students take notes on the various stances and connect it to this week's reading and pose questions."

"A field observation, of sorts."

At one lunch meeting a communications professor, Mairi Pileggi, who had recently completed a course in confectionary arts, brought in the ingredients for baking bread and developed a physical metaphor for Big History's concept of complexity using those ingredients. This was developed by Pileggi and others into a classroom activity that could be used by any of us. Rather than actually bake the bread, we use the separate ingredients (flour, salt, yeast, and so on) and have bread ready for the students to eat after handling the ingredients and talking about how they come together into a complex new arrangement. This activity, early in the course, was much discussed throughout the semester by the students.

Activities like this were shared and critiqued at our weekly meetings. They would be further refined and critiqued at the next summer institute. The feedback from students became a part of our collective learning.

SUMMER INSTITUTE 2: REFINING AND DEVELOPING THE ENGAGED LEARNING PEDAGOGY

As we looked toward the second summer institute, we realized that, in addition to training new faculty members (natural attrition meant that we would need new teachers in the coming year) in the content of Big History, we had something else to teach. Our collective learning from the weekly meetings throughout the year had produced a body of engaged pedagogical approaches. Engaged learning moves beyond a lecture format in which students are passive receivers of information imparted by a teacher. Engaged learning seeks to have students participate in activities and projects connected to the content of a subject—a subject students have presumably read about in assigned readings.

More comfortable after teaching a year of Big History, we realized that we had perhaps packed too much information, too much content, into our first iteration of the course. We looked for ways to unpack the syllabus. Though lectures and PowerPoint slides have their place in learning, they can overwhelm. After year one we no longer felt the need, as Richard Simon from English said, "to prove to ourselves our mastery of the material and overteach out of fear."

Our university has made an institution-wide commitment to engaged learning. In the context of Big History at Dominican, engaged learning assumes that students have read, viewed, or listened to a text (a book, article, film, or podcast) and individually and formally reflected on that text (e.g., through a short reflective piece of writing; see chapter 16, "Reflective Writing in the Big History Classroom"). Once together in the class, students participate in an engaged activity—a discussion, an exercise, a project—related to what they have read and reflected on. For example, the text might describe the communication of both ideas and disease around the Mediterranean or the Silk Road. A class activity would lead the students through a simulation of travelers throughout those regions and the travelers' experience of plagues. At the end of the activity some students would be "living" and some would be "dead." Further reflection takes place after the engaged activity.

As we had in the previous year, we prepared for the second summer institute. Some colleagues from the first year did not return. In some cases people self-selected out. One colleague from among our part-time faculty was offered a full-time position at another university. For various reasons, we needed to add to our ranks as well as to continue with the evolving education of our now-veteran faculty.

Again, we read in advance of gathering. We again shared articles and information. In addition to looking at content, we presented some of the engaged activities developed throughout the prior year. We were looking at ways of teaching, ways of engaging students in the material. A history colleague demonstrated his use of student-led debates. Mairi Pileggi further refined and presented her ideas on complexity and the related activity on the making of bread. Jaime Castner, then our graduate assistant, led us through the use of reflection prompts. From philosophy, Lindsey Dean had us think about play and human consciousness. Each element had begun in one person's classroom. We critiqued, gave notes, and discussed the group's collective learning about what worked and what did not work. This informed how we would use these activities in the coming year. Yes, dare we say it: we were evolving in our approaches to teaching Big History.

In addition to our own Dominican colleagues, Craig Benjamin, coauthor (with Cynthia Brown and David Christian) of the text we would be piloting in a pre-publication edition in the coming year, *Big History: Between Nothing and Everything*, came and shared with us his experience with student-led discussions at his university in Michigan.

The large group critiqued one another's ideas, further refining the activities and exercises. In presenting our ideas to the institute we received feedback on how these could be used by any teacher. A template was developed. Using the template, any of us could write a description of an activity we had used. This formalized way of thinking about an activity served as a useful prompt, both to the originator's thinking and to discussion about the activity.

We devoted a formal session to the teaching of first-year students in particular. Though this might seem like what we were doing throughout the session, it was important to focus on both the experience we had gained and the scholarship that exists on this subject.

That second summer we were balancing our newfound pedagogy with the real content in Big History. While we may have been getting comfortable with the idea that we were not "experts" in Big History, we were, and are, always mindful that such experts exist. Cynthia Brown, one expert, is here among us. Our philosophy colleague Philip Novak has thought long and deeply about the ideas in Big History. The director of our program, Professor Mojgan Behmand, has immersed herself in this newly emerging field of study; she has presented her work in Big History at conferences, both nationally and internationally. I say this not to toot our horn but to remind readers that we come from a thoughtful and informed place.

YEAR 2, WEEKLY MEETINGS WHILE TEACHING: CREATIVE, INTIMATE, AND RARE

During our second full year of implementing the program, we continued to meet weekly in term time. While this takes precious hours, the faculty who teach continued to find the collective process necessary and valuable. Both the general Big History survey and the "Through the Lens" courses had been taught for a year. We had more information, more data, to work with in our second year.

Fully conscious of the need for useful Big History pedagogies, we return to the theme that our methodology has been developed from the bottom up, from one classroom to many classrooms across our campus. At Dominican, classes are typically scheduled on Monday and Wednesday for some sections, and Tuesday and Thursday for other sections. Thus one teacher can come to the weekly meeting on Tuesday and explain an activity or approach that she or he tried, or will try, and another teacher can adapt and try the activity on Wednesday. In a single week we can get some valuable assessment of how a particular activity is working. These activities are then further assessed at the next summer institute. We were getting better at this.

Armed with the collective learning from our summer institutes, at which small teams of faculty worked on focused projects, in the second year of teaching we used these small teams of faculty to work on such practical aspects as a test question bank, common assignment descriptions, and grading rubrics for common assignments. The small teams would report back to the larger group, as we had in the summer institute, for vetting and feedback.

Additionally, in our second year we had the experience of our "Through the Lens" classes from year one as well as new and added "Through the Lens" classes for year two. In the first year most second-semester classes had been taught by faculty who had also taught the general Big History course in semester one. In the second year we added new faculty, and new ideas, in the "Through the Lens" courses. Our discussions, our proposed engaged activities, began to be informed by such colleagues (and disciplines) as Lynn Sondag in fine (applied) arts, Judith Halebsky in creative writing, and Harlan Stelmach in religion. You will see their work in other sections of this book.

Cynthia Brown has described what we do in our weekly meetings as an "intimate" and "creative" process. One person will come in and say, "I am having trouble with this topic or this section of the class," and then colleagues will work together, at and after our weekly meetings, to find solutions and return them to the group. "This is very rare," says Brown, "among university faculty."

SUMMER INSTITUTE 3: THE IBHA COMETH

The third summer institute hosted both Dominican faculty and some interested faculty from other institutions. Since we had invited guests from other places we began to think about how our ideas would work beyond the specific profile of our first-year students. How might students at other institutions react to and benefit from what we were doing? This book was born out of that thinking.

We continued to look at, try out, and critique engaged activities. At the third summer institute we knew that many of us had been accepted to present at the International Big History Association (IBHA) meeting later that summer. We were ever-mindful of the "expert in content" versus the "expert in pedagogy" divide as we looked to the association's conference. Another evolution occurred at this institute and helped to bridge this divide.

There were topics that some of our colleagues had been working on, thinking more deeply about, and finding ways of presenting that could be shared with the group. Richard Simon had continued to read on and think about the concept of complexity as it is used in Big History. He presented his findings to the group, and they are discussed further in chapter 6, "Teaching Complexity in a Big History Context." Also discussed further in other chapters of this book are the work and ideas of historians Cynthia Taylor and Martin Anderson. At the institute, Taylor gave a presentation focused on the teaching of Threshold 6, the rise of *Homo sapiens*. Anderson, working from his own perspective and that of his discipline, presented a way to look at (and thus guide our students through) the approximately two hundred pages of our then-textbook for Threshold 7, the agrarian revolution. The presentations of Simon, Taylor, and Anderson, as you can read, are not related in the main to specific engaged activities. Rather, they look at how to frame large sections or concepts within the Big History narrative—how to make them both manageable and true to Big History.

From those outside Dominican we received thoughtful feedback about what might work at other schools and what might have to be adapted, and we are grateful to them. Among the Dominican faculty we felt that the third summer institute was the most productive of all. We know more of what to ask ourselves with each succeeding year.

YEAR 3, WEEKLY MEETINGS WHILE TEACHING: TESTS, RUBRICS, AND RELIGION

Year three saw us continuing to look at and learn from one another about activities. The meetings also served as a forum to collectively examine issues of testing, being mindful of the array of the religious beliefs (and non-belief) of our students, and formal assessment.

During the two prior years our test taking had focused on content. We were asking if students had retained specific items of information. We had established a common test bank of short answer, true or false, multiple choice, and essay questions. Individual teachers could choose from among this large array. In year three we began to question the use of that kind of testing. Following much discussion, much of which was informed by the experience of philosophy and ethics faculty, we came to a new mode of testing. A new bank of essay questions was developed; these were designed to be open-book exams. Students would be asked essay questions that they would then answer and support with evidence from a list of approved texts. Initial assessments indicate that students found these open-book tests "harder" and that they were more useful indicators of what students had learned. These tests tell how students can think about the concepts of Big History and how they support their ideas with appropriate textual evidence.

In addition to looking at testing, we also spent time talking about the variety of religious beliefs young students bring to us. We had danced around this issue for two years, and we had danced well if not perfectly. We invited representatives from the campus ministry (which had also been included in the summer institutes) to talk to us. Most useful and instructive was a one-day meeting we held on the subject of religion and Big History in early December. Chapter 21, "Teaching Big History or Teaching about Big History? Big History and Religion," discusses some of these ideas more fully. What is of use to report here is that many of us walked away from that meeting knowing that we, and our students, could become comfortable with discomfort. We need not ignore the discomfort. It's there for some students, and for some faculty. We can respect it and work with it. This was a sometimes difficult, always respectful, and ultimately very fruitful process.

We also respected and worked with assessment. It has become our friend. While assessment is discussed more fully in the next chapter, "Assessing Big History Outcomes: Or, How to Make Assessment Inspiring," it is worthwhile here to put assessment into the context of our collective learning. Formal assessment processes have included designing a common essay assignment, developing a rubric for the essay, and then blind scoring the essays using the rubric. The other processes do not involve the faculty as directly until the results are published and used for improvement; these can be read about in the chapter that follows.

The common assignment and the rubric were formulated by a small group, brought to the larger group for vetting, sent back, and then returned for more vetting. The rubric enjoyed an especially high degree of discussion and vetting. Here, the group engagement and transparency served us well. Essays were scored against the rubric, connected to learning outcomes, following a norming session. In some ways more useful than the hard data that came from the scored rubrics was the discussion that occurred following the scoring of the essays.

A few days later, at an end-of-semester retreat, we looked at the data from the essay reading and discussed improvements, in addition to preparing for the coming semester.

The meetings continue to be our ongoing classroom for how to teach Big History. Yes, they take time. Yes, they are worth the time.

The Big History faculty came together to learn about content, and we came away with so much more. Out of our collective learning our four-tiered process of continuous faculty development has evolved. To reiterate, these tiers are:

- annual summer institutes;
- weekly lunch meetings during the academic year;

- a daylong retreat at the end of each semester; and
- the development and implementation of quantitative, qualitative, and anecdotal assessments.

These tiers are indicative of the faculty engagement that has grown over time.

The summer institutes have allowed for the concentrated time—time that does not exist during the academic year—necessary to concentrate on our topic. The institutes allow us to be both "experts," where appropriate, and students, when that is the right role. It is at the institutes that our transparent processes began, that the first faculty buy-in occurred.

The weekly meetings serve as teacher-education classroom, information conduit, discussion forum, and even support group. The semester can get busy. Faculty are used to toughing it out alone. At the weekly meetings we are not lone wolves. We are accountable to one another, and that is a good thing. Some may ask why, after meeting each week, we really need a daylong retreat at the end of the semester. Are we not sick of each other by that point? The answer is yes, we need the meeting, and no, we are not sick of each other. Amazing, eh? The daylong retreats serve as mini summer institutes in that they provide the luxury of time, when even a day can be a luxury. We process what happened in the past semester. That processing is the beginning of the evolution of our collective learning.

Our assessment tools give us a more formal way—a way that produces data we can look at and think about—to critique what we are doing. In some cases assessment has confirmed what we have seen anecdotally. In other cases the data has drawn us to things we had not noticed before. Assessment is also an important component in attracting institutional and extra-institutional support for a program. We ignore it at our peril.

The greatest success of the tiers is the transparency with which we work. The specifics of the tiers are not so important; rather, the importance is in how they came to be. All ideas are open for discussion. All work is vetted by the faculty as a whole. This has led to buy-in by all stakeholders. Make no mistake; we have had intelligent and tireless leadership from our director, Professor Mojgan Behmand. She has carefully yet vigorously navigated the systems of the bureaucracy that is any university. Her enduring advocacy has secured funding and institutional support for the work we do. We are very grateful to her.

I close by returning to the ideas of Cynthia Brown. In letting go of our innate and natural faculty propensity to be experts, we have come together and grown a collegial approach. Through the collective learning of all the faculty and students we progress. This process has indeed been, as Brown has said, creative and intimate.

NOTES

1. David Christian, Cynthia Brown, and Craig Benjamin. *Big History: Between Nothing and Everything*. New York: McGraw-Hill, 2014. 308. Print.

2. A cautionary note here: avoid the temptation to say, "Big History is everything." That is not true. Big History, among other things, is a narrative that uses a specific framing, the eight thresholds, and concepts such as complexity to tell the story of the universe. Saying that Big History is everything says nothing, particularly to an audience that does not yet know about it.

FOUR · Assessing Big History
Outcomes

Or, How to Make Assessment Inspiring

Mojgan Behmand

Utter the word "assessment" in an academic setting and you'll elicit a range of reactions: blank stares from uninitiated students, groans from faculty fearing another addition to their heavy workload, soliloquies from administrators waxing poetic about assessment, and threats from accrediting associations couched in terms like "excellence" and "success." Interestingly, the first documented use of the term "assessment" was in the 1540s, meaning "determination or adjustment of tax rate."[1] Now, in the twenty-first century, it is part of the established vernacular in education and enjoys the same popularity as taxes. So here's a warning: approach assessment with the utmost care.

Perhaps the disengagement from assessment is caused in part by many institutions' approach to it: as an end in itself, rooted in a "compliance mentality," and as a step aimed at accountability, not improvement.[2] Granted, we are and should be accountable at all levels: in K–12 for laying the foundation of our young's intellectual development and habits of mind, and in higher education for providing our students with the commodity they are paying for with their parents' savings and their own future earnings. That sort of accountability is best achieved if inextricably coupled with improvement. The assessment literature of the last decade has emphasized "closing the loop"—that is, interpreting the assessment data to make changes—but at Dominican, our work has also emphasized the value of "opening the loop," both in our Big History program and beyond.[3]

Without a doubt, a collaborative process increases commitment to a program and its assessment. In the case of Big History, the vast interplay among disciplines made working as a collective the most sensible approach. Many of the concepts underlying our approach are captured beautifully in Jon Wergin's 2001 article on faculty development.[4] Wergin identifies four interdependent key factors in faculty motivation: "autonomy, community, recognition and efficacy."[5] He also lists a number of motivational strategies: "Align institutional mission, roles, and rewards. . . . Engage faculty meaningfully. . . . Identify and uncover 'disorienting dilemmas.' . . . Help faculty develop 'niches.' . . . Encourage faculty experimentation, assessment, and reflection."[6] The distinguishing feature of this approach is its tone of respect for the faculty and its move away from the ineffective corporate model of purely materialistic incentives. Wergin counsels us to create "an opportunity to engage in meaningful work that we have helped design, conducted within a nurturing community that recognizes the unique contributions we make to it."[7] At Dominican, we did exactly that.

In the summers of 2010 and 2011, our "opening the loop" for our First-Year Experience program consisted of working as a Big History learning community to align our institution's mission with an understanding of our own aims, roles, and individual niches to create an innovative and meaningful program. We took two important steps: first, we collaboratively designed our program goals in accordance with the "Essential Learning Outcomes"[8] of a twenty-first-century liberal education;[9] and second, we worked in small and then large groups to design our program's individual courses, craft their learning outcomes, and designate the assessment tools. For the initial step of setting the program's goals, we knew that Big History supported a "study in the sciences and mathematics, social sciences, humanities, histories, languages, and the arts. . . . [f]ocused by engagement with big questions, both contemporary and enduring".[10] We also intended that Big History provide students with a framework for the scaffolding of knowledge and for counteracting fragmentation, so we wrote our program description and goals accordingly:

PROGRAM DESCRIPTION

First-Year Experience "Big History" is a one-year program that takes students on an immense journey through time to witness the first moments of our universe, the birth of stars and planets, the formation of life on Earth until the dawn of human consciousness, and the ever-unfolding story of humans as Earth's dominant species. In studying the evolution of human cultures, students engage with fundamental questions regarding the nature of the universe and our momentous role in shaping possible futures for our planet.

The program is designed to promote

1. recognition of the personal, communal, and political implications of the Big History story;

2. critical and creative thinking in a manner that awakens curiosity and enhances openness to multiple perspectives;

3. development of reading, thinking, and research skills to enhance one's ability to evaluate and articulate understanding of one's place in the unfolding universe.

For the second step of developing our semesterly curriculum and individual courses, we studied our program goals and employed the principles of backward design as articulated by Jay McTighe and Grant Wiggins in the 1990s. Our three steps of backward design were: (1) identifying the desired student learning outcomes; (2) determining the appropriate evidence of achievement of those learning outcomes; and (3) designing instruction and learning experiences to achieve those outcomes.[11] We created a one-year program made up of a sequence of two courses and numerous co-curricular activities to support achieving our learning outcomes. The first semester consists of a Big History survey course emphasizing global interconnectivity within the context of natural and human history. The second semester offers students a choice among discipline-based courses that reiterate the narrative of Big History in dialogue with a specific field of inquiry.

Examples of our course descriptions and learning outcomes are:

SEMESTER I: "BIG HISTORY: FROM THE BIG BANG TO THE PRESENT"

In Big History we take an immense voyage through time. We witness the first moments of our universe, the birth of stars and planets; we watch as life forms on Earth, grows and develops in complexity, until human consciousness dawns. We then trace the evolution of human cultures through geography, migration patterns, and social structures, until we finally peer over the threshold of the present into possible futures for us and for our planet.

SEMESTER I: STUDENT LEARNING OUTCOMES

Students will

1. employ major Big History concepts and the eight Big History thresholds from the Big Bang to the present in developing a perspective that emphasizes

a view of themselves as embedded in the fabric of an interconnected world (assessment: Little Big History essay);

2. demonstrate an understanding of Big History themes addressed in the course through identifying, defining, explaining, and / or analyzing them (assessment: midterm and final exams); and

3. demonstrate the ability to locate and evaluate appropriate secondary sources, and extract and synthesize research, while summarizing, paraphrasing, and quoting in accordance with the MLA, APA, or CMS documentation styles (assessment: information literacy exercises, Little Big History essay).

SEMESTER 2: "MYTH AND RITUAL THROUGH THE LENS OF BIG HISTORY"

What are the stories that shape us? The reading, discussion, and performance of myths and rituals from diverse cultures of the world—from early human to contemporary mythologies—shed light on the implications of the Big History narrative as humankind imagines the origins of the universe, seeks understanding of the present, and attempts to shape the future.

SEMESTER 2: STUDENT LEARNING OUTCOMES

Students will demonstrate the ability to

1. make connections across time and cultures to engage with myths and rituals critically through a Big History perspective. This includes identifying and analyzing the Big History narrative and its thresholds as told in myth, in addition to studying and imagining its implications (assessment: quizzes, final exam, a Big History paper); and

3. formulate a research question within the framework of a myth of personal significance to examine its origins, rituals, and implications. Locate and evaluate appropriate sources, and extract, synthesize, and apply information (assessment: an annotated bibliography in MLA documentation style, a research paper).

Table 4.1 demonstrates the alignment of our general education program with First-Year Experience "Big History" with an individual Big History course.

As discussed in chapter 2, "Big History and the Goals of Liberal Education," this process taught us two valuable lessons: first, every institution should articulate its learning objectives as suited to its needs and mission; second, the alignment of the program with our university's educational philosophy has ensured that we agree on the goals for student learning and are committed to thoughtful assessment and continuous quality improvement.

TABLE 4.1   Alignment of Goals and Student Learning Outcomes

General education goals	Big History program goals	Big History course SLOs
• Learning in the essential foundations in the main areas of human knowledge (i.e., sciences, arts, and humanities) • Development of students' awareness of the moral and spiritual dimensions of existence	• Recognition of the personal, communal, and political implications of the Big History story	• Ability to demonstrate an understanding of Big History themes and concepts addressed in the course through identifying, defining, explaining, or analyzing them
• Students' understanding of themselves as citizens of diverse communities in an ecologically imperiled world	• Critical and creative thinking in a manner that awakens curiosity and enhances openness to multiple perspectives	• Ability to employ major Big History concepts and the eight thresholds, from the Big Bang to the present, to develop a perspective that emphasizes students' view of themselves as embedded in the fabric of an interconnected world
• Persistent practice of critical thinking, persuasive writing, quantitative reasoning, creative expression, and effective research and speaking	• Development of reading, thinking, and research skills to enhance students' ability to evaluate and articulate understandings of their place in the unfolding universe	• Ability to locate and evaluate appropriate secondary sources, and to extract and synthesize research in accordance with MLA, APA, or CMS documentation styles

Having designed the Big History curriculum and launched the program, it was essential to ensure that student learning was indeed happening and that we were meeting our stated goals. This is where thoughtful assessment plays a crucial role in continuous quality improvement. Dominican University's assessment of the Big History program is multi-instrumental and simultaneously formative and summative. Much of the formative assessment happens throughout the semester in our Big History classes to improve our instructional methods as the courses are taught. At the launch of the program, our tools were surveys, writing responses, and anecdotal information shared at our weekly faculty meetings. These assessment tools have also contributed to our summative assessment, in which we also draw on exams, student papers, and student focus groups to evaluate the effectiveness of our program and the achievement of competencies at the end of the semester and/or year. However, for our purposes in this chapter, it might be most effective to distinguish the assessment processes by the tools used, as some of the measures are cognitive and others are affective. Our cognitive assessment focuses on intellectual capabilities and demonstrations of learning using student-generated artifacts; our affective assessment measures engagement and confidence using students' perceptions and self-evaluation.

For our cognitive assessment, quizzes, exams, and papers have been the most appropriate student-generated artifacts. Common questions were embedded in quizzes and exams across sections to measure outcomes—comprehension and application of the Big History narrative and its major concepts—and also to measure success rates across different sections. Papers were assessed using rubrics developed in accordance with the designated learning outcomes of the assignment, course, and program.[12] The rubrics included in tables 4.4 and 4.5 are examples of such homegrown rubrics. Thus, the assessment measures knowledge and application of Big History in addition to writing and information literacy skills.[13] Results show that students make significant gains in knowledge and skills in the first year; in our initial assessments they also showed that we needed to adjust our teaching of information literacy skills in connection with Big History. The growing gap in our millennials' skills is most apparent in their manner of formulating questions and approaching research about global and local issues.[14] We need to promote engagement with the process of research while using Big History to consistently contextualize all topics and discussions to support students in synthesizing information.

Complementing our cognitive assessment, our affective assessment aims at measuring student motivation, attitude, and engagement. To achieve that end, we have developed Big History–specific student surveys and conducted student focus groups. For the first two years, our Big History surveys were student satisfaction surveys administered midsemester to help us make quick fixes in response to perceived problems in this new program. We frequently sought information regarding the syllabus, readings, textbooks, events, teacher preparation, and general responses to the courses. During this initial period, we learned that student responses to Big History were directly related to the proficiency of the faculty in Big History pedagogy; that Big History's relevance to students' majors and education as a whole needed to be emphasized; and that students perceived a dichotomy between science and religion that we faculty did not subscribe to.

An important shift in our surveys came about through the work of Rich Blundell, a PhD candidate working with David Christian at Macquarie University. Blundell

TABLE 4.2   Question 1: Do you ever think about what
you've learned in your Big History course, or talk about
it with others outside of class (in everyday life)?

Answer	Response (no. of students)	Percentage (%)
Yes	68	80
No	17	20
Total	85	100

generously shared his research and assessment tool for Big History as a transformative experience. His work drew heavily on developments of Deweyan thinking and
suggested measuring transformative experience via the following characteristics,
defined by Kevin J. Pugh as based on John Dewey:[15]

- **motivated use:** thinking/talking about the material outside of class (in
  everyday life);
- **expansion of perception:** thinking about existing knowledge in new
  ways; and
- **experiential value:** sensing a value in what is learned.[16]

Collaborating with Blundell and modifying the questions slightly for our own
institution, we administered our new Big History survey in the fall of 2012. To
measure the characteristics of a transformative experience as indicated above, we
asked the following questions, supplementing with follow-up queries as appropriate:

1. Do you ever think about what you've learned in your Big History course,
   or talk about it with others outside of class (in everyday life)? (see table 4.2)
2. Has your Big History experience changed the way you see or understand
   aspects of the world? (see table 4.3)
3. Has your Big History course changed the way that you see your role in the
   world?

That semester, 241 students were enrolled in Big History. With 89 respondents,
our participation rate was 37 percent. By the end of one semester of Big History,
80 percent of respondents said they had thought or talked about the content of the
course outside of class; 72 percent indicated that their Big History experience had

TABLE 4.3   Question 2: Has your Big History
experience changed the way you see or understand
aspects of the world?

Answer	Response (no. of students)	Percentage (%)
Yes	60	72
No	23	28
Total	83	100

changed the way they saw or understood aspects of the world; and 48 percent expressed that Big History had changed the way they saw their role in the world.

One student reported thinking about Big History "every time I look at the sky at night or whenever I talk to my family about religion and the beginnings of the universe." Students' changed perspectives ranged from seeing "the 'bigger picture,' or how all things are complex and interconnected" to "my role in the vast universe" to "the future of Earth and / or humanity." One student wrote: "Big History brought to my attention how we got here, and what our future might look like. It also showed me my place in terms of our history. It changed the way I see my role in that I realize where I stand in relation to everything else."[17] The Dominican faculty had, of course, hoped for such results but had not expected such clear evidence of Big History as a transformative experience. We also learned that, though we had made concerted efforts to stress Big History's relevance to students' majors and education, we had more work to do, and that the perceived dichotomy between science and religion remained an issue.

By the end of 2012, we felt that we needed more in-depth information regarding students' responses to Big History. To complement the assessment data we had gathered through rubrics and surveys, we decided to conduct student-led focus groups and attended a training workshop with two student facilitators.[18] Our areas of inquiry were: (1) reasons for students' enthusiasm for Big History; (2) relevance of students' levels of academic preparation; and (3) the impact of Big History on students. In addition to our student facilitators, we trained five students as notetakers. Jennifer Lucko, our program's assessment director, and I oversaw the preparation and led the data analysis, but only the students themselves were present at the focus groups. Participants were guaranteed anonymity and confidentiality. We held seven focus groups, including the pilot in March and April of 2013, and had a total

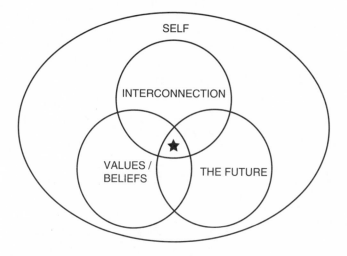

WHEN DOES STUDENT OWNERSHIP OCCUR?

SELF

INTERCONNECTION

★

VALUES /
BELIEFS

THE FUTURE

FIGURE 4.1
When does student ownership occur?

of twenty participants.[19] The participants were all first-year students enrolled in their second required semester of Big History.

The data held some surprises. It was significant that questions regarding preparation and its impact on response to the courses did not elicit the expected information. Rather, students' responses implied that the value of the course depended on the students' commitment to their education. They responded with "It works if you get interested" and "You make Big History valuable ... you have to make it interesting."[20] Ultimately, we realized through data analysis that the Big History program is successful if students taking a Big History course see their own self reflected in the course. And such inclusion of the self is best achieved through the integration of three major thematic areas, as indicated in figure 4.1 and the list below:

1. **Investigations of values and beliefs:** We learned that the issues of science, religion, faith, morality, responsibility, and obligation are not peripheral to the program or unwelcome side effects. Rather, they make up part of the core of the program. Student responses paraphrased by notetakers included: "In class she wasn't just learning facts, she was learning about herself and her own faith"; "She was able to talk to her mom about the conflicts she felt with

TABLE 4.4   First-Semester Little Big History Paper Rubric: Program Assessment Rubric

	4	3	2	1
Understanding of the Big History narrative	Conveys an above average understanding of the Big History narrative via clear and detailed explanations of each of its major thresholds.	Conveys an average understanding of the Big History narrative via adequate explanations of each of its major thresholds.	Conveys a limited understanding of the Big History narrative via limited explanations of each of its major thresholds.	Conveys a lack of understanding of the Big History narrative via unclear or insufficiently detailed explanations of each of its major thresholds.
Perspective on interconnected world	Development of the topic demonstrates an above average perspective on an interconnected world.	Development of the topic demonstrates an average perspective on an interconnected world.	Development of the topic demonstrates a limited perspective on an interconnected world.	Development of the topic demonstrates a lack of perspective on an interconnected world.
Introduction	Identifies topic with above average clarity and clearly foreshadows the conclusion.	Identifies topic with average clarity and adequately foreshadows the conclusion.	Identifies topic with limited clarity and with a limited attempt to foreshadow the conclusion.	Lacks clear identification of the topic.
Conclusion	Above average expression of resonating final thought or question that clearly connects to the introduction.	Average expression of final thought or question that adequately connects to the introduction.	Limited expression of final thought or question with a limited connection to the introduction.	Lacks expression of final thought or question.

Sources, evidence, and documentation	Uses credible and relevant sources. Above average evaluation and integration of evidence. Documentation of sources is virtually free of error.	Uses credible and relevant sources. Average evaluation and integration of evidence. Documentation of sources has only a few errors.	Uses credible and relevant sources. Limited evaluation and integration of evidence. Documentation of sources has a number of errors.	Lacks credible and relevant sources. Documentation of sources has many errors.
Control of syntax and mechanics	Above average control of syntax and mechanics for clear and virtually error-free communication.	Average control of syntax and mechanics for adequate communication with only a few errors.	Limited control of syntax and mechanics for unclear communication with a number of errors.	Lacks control of syntax and mechanics, resulting in unclear and error-plagued communication.

her Catholic faith, and Big History . . . allowed her to have more of an open mind and accept the different ideas that are in the world"; "Felt like by starting the semester with teaching the Big Bang it is challenging to people's faiths and is abrasive." We need to acknowledge the centrality of these topics.

2. **Interconnected and multidisciplinary content:** We learned that this was valued at various levels: through connecting content to the students' majors and education, through providing the students with cultural literacy in a way that boosted their self-esteem, and through interwoven and interdisciplinary areas of inquiry. Student responses were recorded as: "Big History has given her the basic tools she plans to use in her psychology major. It gives an understanding of people by looking at where people have come from and where we are going" and "Feels like s/he is a better conversationalist because she is knowledgeable about the creation of the universe." Some of our results also included partial direct quotes: "Taking the course 'made me see the big picture' in life. The student had taken many courses before but this was the one that made her stop and think. Until now she had not thought back into history as far as the Big Bang. Now she can't look at things the same way, Big History is 'at the back of my mind as a constant reminder . . . this pertains to this, this is connected to that.'"

3. **Foregrounded connections to the future:** The study of the past often leaves young students unmoved, because as interested as they might be in the past, they see little relevance to their own lives now or in the future. Simply designating the past as a general source of learning somehow applicable in the future is not sufficient. However, the study or discussion of the future in connection with the past is extremely appealing and even powerful, as it allows students to understand contingency and also to envision their own agency in shaping the future of humanity and our planet. Students respond more strongly to Big History when connections to the future are woven in throughout the narrative and the thresholds.

These three strengthened thematic emphases increase student commitment to and ownership of the Big History courses and render a vast and abstract content tangible. As it happens, these areas also accord wonderfully with our program goals and the intended outcomes of a liberal education, rendering this revision to our program effortless.

Creating a multi-instrumental assessment process for Big History has allowed a fuller development of Big History's potential through its many facets and applica-

TABLE 4.5 Second-Semester Big History Paper Rubric: Program Assessment Rubric

	4	3	2	1
Implications of the Big History narrative	Conveys an above average understanding of the personal, communal, or political implications of the Big History narrative.	Conveys an average understanding of the personal, communal, or political implications of the Big History narrative.	Conveys a limited understanding of the personal, communal, or political implications of the Big History narrative.	Conveys a lack of understanding of the personal, communal, or political implications of the Big History narrative.
Understanding of one's place in the unfolding universe	Development of the topic demonstrates an above average understanding of one's place in the unfolding universe.	Development of the topic demonstrates an average understanding of one's place in the unfolding universe.	Development of the topic demonstrates a limited understanding of one's place in the unfolding universe.	Development of the topic demonstrates a lack of understanding of one's place in the unfolding universe.
Introduction	Identifies topic with above average clarity and clearly foreshadows the conclusion.	Identifies topic with average clarity and adequately foreshadows the conclusion.	Identifies topic with limited clarity and with a limited attempt to foreshadow the conclusion.	Lacks clear identification of the topic.
Conclusion	Above average expression of resonating final thought or question that clearly connects to the introduction.	Average expression of final thought or question that adequately connects to the introduction.	Limited expression of final thought or question with a limited connection to the introduction.	Lacks expression of final thought or question.
Control of syntax and mechanics	Above average control of syntax and mechanics for clear and virtually error-free communication.	Average control of syntax and mechanics for adequate communication with only a few errors.	Limited control of syntax and mechanics for unclear communication with a number of errors.	Lacks control of syntax and mechanics, resulting in unclear and error-plagued communication.

tions. Approaching this endeavor—not with a compliance mentality but through engagement—has necessitated opening the loop, moving forward with thoughtful assessment, and closing the loop. These assessment measures and their findings have all fed constantly into the faculty learning community engaged in achieving individual, programmatic, and institutional goals. This, in turn, has promoted continuous quality improvement. We have also consistently disseminated our findings to the entire institution to shape our discussions around curriculum. Assessment may enjoy the same popularity as taxes, but seeing our tax dollars at work building roads, bridges, schools, and art centers can be a gratifying experience that leaves us uplifted. Thoughtful assessment can have the same uplifting effect and inspire us to excellence.

NOTES

1. "Assessment." *Online Etymology Dictionary*, 2013. Web. 3 June 2013.

2. Barbara Wright, vice president of the Western Association of Schools and Colleges: Accrediting Commission for Senior Colleges and Universities (WASC), uses the term in her 2008 presentation "Closing the Loop: How to Do It and Why It Matters," given at AAC&U's Summer Institute on General Education, Minneapolis, Minnesota, May 30–June 4, 2008. Available at AAC&U website, PowerPoint downloadable through Google. 3 June 2013.

3. Warmest thanks are due to Dr. Jennifer Lucko, my partner in all things assessment since 2012. Her contributions and guidance have been invaluable, and all information and data gathered in the 2012–2013 school year have been collected through our joint efforts.

4. Jon F. Wergin, "Beyond Carrots and Sticks." *Liberal Education* 87.1 (2001): 50–53. Print.

5. Ibid., 50.

6. Ibid., 51–52.

7. Ibid., 53.

8. Association of American Colleges and Universities, "Essential Learning Outcomes." *Aacu.org*. Association of American Colleges and Universities, 2013. Web. 20 May 2013.

9. Association of American Colleges and Universities, "What Is a 21st Century Liberal Education?" *Aacu.org*. Association of American Colleges and Universities, 2013. Web. 20 May 2013. For more discussion, see chapter 2, "Big History and the Goals of Liberal Education."

10. Association of American Colleges and Universities, "Essential Learning Outcomes."

11. My introduction to backward design came through an AAC&U preconference workshop. Jo Beld, "Beginning with the End in Mind: Backward Design in General Education Assessment." AAC&U Annual Meeting, Washington, D.C., 20 January 2010. Lecture.

12. Numerous resources for the construction of rubrics exist. Stevens and Levi's *Introduction to Rubrics* is a great introductory resource; AAC&U has developed sixteen VALUE (Valid Assessment of Learning in Undergraduate Education) rubrics to accord with its "Essential Learning Outcomes"; and RCampus hosts a website with an online gallery of over 280,000 rubrics sorted by level, subject, and type.

13. See tables 4.4 and 4.5 for our sample rubrics.

14. Literature on the millennials abounds, and a simple Google search elicits over a million results. That information is useful but should be used cautiously: first, because these studies have been lacking in their analysis of the *why* of a trend and have thus led to erroneous conclusions; and second, because the rate of change has accelerated exponentially. This means that the millennials are changing too quickly for the studies. Those of us in the classroom know that the millennials of two years ago differ substantially from the millennials of today.

15. Kevin J. Pugh, "Transformative Experience: An Integrative Construct in the Spirit of Deweyan Pragmatism." *Educational Psychologist* 46.2 (2011): 107–121. Print.

16. Richard Blundell, "Big History Experience Survey," e-mail message to the author, 13 June 2012.

17. All survey quotes in this paragraph are from anonymous respondents to Mojgan Behmand and Jennifer Lucko, "Dominican Big History Experience Survey," December 2012.

18. The HEDS Consortium at Wabash College offers a well-developed training workshop for academic teams, and its director, Charles Blaich, and associate director, Kathy Wise, are invaluable resources.

19. This process has also been a learning experience regarding the timing and frequency of focus groups in addition to considerations on student incentives. We were happy with our results for the first year and increased participation in 2014.

20. All survey quotes in this paragraph and below are from anonymous respondents to Mojgan Behmand and Jennifer Lucko, "Big History Focus Groups" survey, March–April 2013.

FIVE

· Big History at Other
Institutions

INTRODUCTION

Mojgan Behmand

In our first year of teaching Big History, we Dominican faculty gathered over lunch every week to discuss curriculum and pedagogy. We often found ourselves wondering how Big History was taught at other institutions. We heard about Cynthia Brown's challenges in proposing a Big History course; we listened to the story of David Christian's path from Russian history to Big History; and we generally speculated about the inception of individual courses or programs and their specific successes and challenges. In the ensuing years, we have met many of our American and international colleagues, both here at Dominican and at various conferences, and we have had the opportunity to enter into the most rich and rewarding conversations with them. We feel that the same opportunity should be afforded the readers of this volume.

This chapter contains sections on three programs external to Dominican, with each piece discussing a specific program: its history, its structure, its challenges, and its hopes. The first focuses on the Big History Project, an online curriculum developed with the support of Bill Gates, under the guidance of David Christian and Bob Bain, for students at the middle school and high school level. The program delivers a virtual textbook for teachers' classroom use, offers a curriculum created in accordance with various literacy standards, and shows evidence of enhanced student engagement. The second article describes the development of the Big History program in the Netherlands as spearheaded by Fred Spier and recounted by Esther Quaedackers. One of the most fascinating aspects of that program is its nuanced attention to structure and its use of *Big* Big History and Little Big History to recognize large patterns in Big History while creating personal and individual connections for the learner. In the third piece, Seohyung Kim, research professor at Ewha Womans University in Seoul, South Korea, describes the emergence of convergence education in South Korea, the role of the Institute of World and Global History in bringing Big History to the university and to high school students, and the program's contribution to the combined study of the sciences and humanities.

We have learned a lot from our colleagues. We hope you will, too.

THE BIG HISTORY PROJECT: BILL GATES'S FAVORITE COURSE

Mojgan Behmand

PROGRAM ORIGIN

Innovative Big History programs seem to begin with individuals' unique and amusing Big History stories. I find it enjoyable to imagine David Christian in a curriculum meeting with his colleagues at Macquarie University in the 1980s, arguing for a course that teaches the history of *everything*. His perplexed colleagues ask, "And what would such a course be called?" and Christian responds, tongue in cheek, "*Big* History." Fast forward a couple of decades to Bill Gates, one of the greatest technological pioneers of our time, watching video lectures from the Teaching Company's Great Courses series as he does his early morning cardio workout. And on that particular morning, the course is not one on science or philosophy or economics, but rather a combination of them all, "Big History: The Big Bang, Life on Earth, and the Rise of Humanity," by David Christian.[1] In that moment, the seeds of the Big History Project are sown.

The Big History Project is "a free, online course that tells the story of our Universe and humanity" with the goal of "developing a framework to help people learn about anything and everything. It's inherently interdisciplinary—combining the best of social studies, humanities, and science."[2] Of course, there is more to this story than a philanthropic technology mogul and the Big Historian he invited for a conversation. There is also Bob Bain, an associate professor of history at University of Michigan's School of Education, who is shaped by his twenty-six years of high school teaching and his concern for the "enduring and critical problems in educating American adolescents": a curriculum delivered in pieces, deficient adolescent literacy, chronically disengaged students, and what he terms the "hidden challenges." Those hidden challenges include students' "level or scales problem," which deters them from connecting the level at which they learn or live to higher levels, and students' lack of understanding of causation, which renders them unable to connect their current situations to larger structures that might have influenced them. These issues in U.S. adolescent education are in turn exacerbated by under-resourced schools.[3]

The Big History Project, a free online open curriculum, took shape at the behest of Bill Gates, with Christian's collaboration. Bain was engaged to oversee the development of the curriculum and the launch of its pilot program. The first pilot in 2011 included six U.S. schools and three Australian schools, as well as a small team led by Bain at the University of Michigan, whose job it was to collect and analyze student and teacher data for the assessment of outcomes. The pilot in its second year included sixty-five to seventy schools across the globe and analyzed data from four thousand students to determine conceptual engagement, levels of literacy, and perceptions of the course.[4] Response to feedback and continuous improvement are hallmarks of the program, as evidenced by its remodeling of the website, its virtual textbook, every year. In its third year, the pilot enrolled fifteen thousand students.

In 2013, the Big History Project launched a revised classroom curriculum. As in previous years, the classroom curriculum was available free to high schools and teachers who signed up to use it in the classroom, but this time the classroom curriculum was accompanied by a new parallel site open to the public and to more general audiences.

COURSE STRUCTURE

What does the Big History Project website actually offer, and what does it look like? Among the advantages of this curriculum being online is that it can be continuously refined and updated based on feedback from educators and students—hence, any discrepancies between what we describe here and what one may find on the website. Its classroom site is accessible in both a teacher and a student view. The visuals on the landing page lay out the outline of the 13.8-billion-year course in ten units. Teachers are also provided resources such as a console that holds unit quizzes, course tests, teacher and student surveys, teacher unit logs, and teaching resources, and a community that functions as a learning hub for the sharing of ideas, lesson plans, activities, and concerns. The current curriculum is composed of two sections and ten units, as shown below:[5]

PART 1: FORMATIONS AND EARLY LIFE

Pre-Threshold. Unit 1—What Is Big History?

Threshold 1. Unit 2—The Big Bang

Threshold 2 and 3. Unit 3—The Stars Light Up & New Chemical Elements

Threshold 4. Unit 4—Our Solar System & Earth

Threshold 5. Unit 5—Life

PART 2: HUMANS

Threshold 6. Unit 6—Early Humans

Threshold 7. Unit 7—Agriculture & Civilization and Unit 8—Expansion & Interconnection

Threshold 8. Unit 9—Acceleration

The Future. Unit 10

The course is focused by its emphasis on four major themes throughout as it attempts to "provide an overview of scientific concepts in an historic context":[6]

1. **Thresholds of increasing complexity:** Goldilocks conditions feature as a key idea here, with the concept of increasing complexity determining the organization of the course into eight major thresholds.

2. **Differing scales in time and space:** Alternation between different scales in time and space prompt students to move through the content of the course spatially and temporally, and to think at different levels and ponder causation.

3. **Claim testing:** Evaluation of a claim using the four main claim testers— authority, evidence, intuition, and logic—is a core part of the course, which attempts to engage students in the type of inquiry scholars engage in.

4. **Collective learning:** The human capacity for cooperative and shared learning that leads to the increase of knowledge over time—that is, collective learning—is a key concept in Big History, as it is an ability unique to *Homo sapiens* and the key to the human species's eventual dominance on the planet.

The content of the public parallel site mirrors that of the classroom version but offers lighter fare in a six- to eight-hour course designed primarily for parents who wish to follow along with their child's studies, or for humanities or science enthusiasts who wish to access the 13.8-billion-year history we all share. The videos feature Christian's voice but display solely animations and images intended to reinforce the information. Although the "Guest Talks" by experts are not included, the public site does draw on the original curriculum's infographics, lectures, and articles to provide an overview and offers an interactive component at the end of each unit in the form of optional quizzes to test the learner's knowledge.

COURSE LEARNING OUTCOMES

The Big History Project's course-wide outcomes are augmented by unit- and lesson-specific outcomes. As the teacher version of the curriculum establishes, by the end of the course, students will have demonstrated the ability to[7]

- use multiple perspectives, including "shifting scales" and "claim testers," to create, defend, and evaluate a cohesive narrative of universal change, from the beginning of time to the present;
- evaluate the ways historical contexts, collective learning, and scientific advancements have altered our stories about the universe, our Earth, and humankind;
- deepen an understanding of key historical and scientific concepts and facts, and the use of these in constructing explanations;
- evaluate key historical and scientific concepts from a variety of scholarly disciplines;
- locate our own place—and that of our communities and humanity as a whole—within the Big History narrative and reflect on how Big History uses the concept of "thresholds" to frame the past, present, and future;
- compare the interdisciplinary approach of Big History to other, more traditional approaches to knowledge, and use various disciplines to analyze, evaluate, and justify one's own and others' claims about the past and the present;
- conduct historical investigations by framing researchable problems, finding relevant sources of information across a range of disciplines and formats, analyzing and evaluating evidence, and constructing narratives, explanations, and arguments;
- critically read, synthesize, and analyze primary and secondary historical, scientific, and technical texts, and other resources; and

- communicate Big History ideas, evidence, narratives, explanations, and arguments to a variety of audiences through individual or shared writing, speaking, and other formats.

The achievement of these outcomes is supported by a range of diverse materials for a variety of learning styles. These materials include "videos and animations, texts, Project Based Learning (PBL) projects, infographics, comic strips, image galleries, investigations, online assessments, and turnkey lessons with a variety of activities. These components come with teaching notes and are available online via download or to print."[8] Some of these components have attractive features built in; for example, the "Main Talks" by Christian have downloadable transcripts. A number of primary texts, historical articles, and biographical essays—many penned by Big Historian Cynthia Brown—complement the lessons. "Timelines" make zooming in and out of Big History's temporal and spatial scales possible, allowing students to shift scales constantly, and, as lesson plans make clear, with intention. Investigations "invite students to take up a problem and require them to analyze, synthesize, and evaluate evidence to construct their own answer."[9] The ten featured investigations are clearly intended to engage students in solving some of the world's most pressing issues and are part of the value-added approach used in the development of the program. And the research component, the Little Big History, acquaints students with scholars' modes of inquiry while requiring the students to share their Little Big History assignments with an audience not comprised of their classmates or teacher.

HOW THE PROGRAM WORKS

Christian has crafted the course's narrative and delivered the core video content. His work has been augmented with a wide array of virtual faculty, experts whose "Guest Talks" have a personal quality as they discuss insights and ways of knowing. These materials work in tandem with real-time and face-to-face instruction by the teachers at the schools that have adopted Big History. These schools include big and small urban schools, charter schools, parochial schools, and middle schools, among others. Offering Big History in these various contexts for adolescent students becomes possible because the schools do not need resources to adopt the program, but rather use the Big History Project's open and free content. In addition, the teachers are offered professional development and community through the Big History Project website, with content that ranges from lesson plans to quizzes, exams, and chat rooms.

Bain's professional experience manifests in the program's emphasis on assessments and standards. Not only is it the first Big History course to use the latest technology to offer content and interactive components; it is also one of few existing Big History courses to be shaped in response to and in connection with standards such as the World History Standards[10] and Common Core Standards for Literacy[11]—in order to promote specific content and skill outcomes for the learner.[12]

Emphasis is placed on reading, analysis, and the use of text, and growth in literacy becomes evident through the comparison of baseline assessment results with later course assessments. For example, in the year assessed, the number of students using outside, unassigned texts to answer a posed question grew by 40 percent; the inaccuracy rate in reading dropped from 15 percent to 1 percent; and the number of students using evidence in making an argument increased by 57 percent. The perceptions inventory showed only 8 percent of students disagreeing with the statement "I like the course" and only 9 percent not recommending the course to others. For those of us well versed in program assessment numbers, these are indeed very low negatives. In short, the continuous quantitative and qualitative assessment of the program has shown that students readily draw on learned skills, engage with core concepts, and become fluent enough with the ideas and language to apply them both to other courses and to their own lives and experiences.[13]

CHALLENGES AND CONCLUSIONS

The Big History Project has faced some challenges unique to its specific tools and audiences. One of the early challenges was bandwidth, which sometimes limited the school's or teacher's access to the internet and thus their ability to stream materials during class. Downloadable content is now a core part of the program. Another technical challenge was finding ways to increase student focus on video content in order to actively engage with it. Dynamic visuals with the mixing of image genres has been a successful approach, as has been the limiting of lecture lengths. The nontechnical challenges reflect some of the issues faced by Big History educators everywhere. Teacher training is one of them, with the Big History Project undertaking teacher preparation by helping teachers ask and answer three questions: (1) "How do I teach what I don't know?"; (2) "How do I teach what I know too much of?"; and (3) "How do I teach when I'm uncertain what my students know?" The final challenge is finding or creating a place for Big History within the high school curriculum while considering issues of sequencing, connection, and territory.[14]

Ultimately, we Big History advocates believe that Big History has the ability to unify and connect knowledge across time and space, to counteract the fragmentation of a curriculum delivered "in pieces." In the Big History Project, specifically, we enter the realm where the lofty dreams of three men converge: First, Bill Gates, who intends to help solve a number of global health and education issues through investment and finds that Big History is his "favorite course of all time," because it "shows how everything is connected to everything else. It weaves together insights and evidence from so many disciplines into a single, understandable story."[15] Second, David Christian, who realizes the significance of origin stories as "they offer maps that can help us to place ourselves, our families and our communities and to navigate our world" and hopes that Big History can function as "a universal origin story that works in today's globalised societies."[16] And finally, Bob Bain, who sees Big History's potential for engaging with some of the critical issues in education and finds that, when embedded with the right goals of literacy and inquiry, the Big History Project might be just the right tool for his vision of access and equity that offers a chance at excellence to "*all* secondary students in *all* contexts."[17]

BIG HISTORY IN A SMALL COUNTRY

Esther Quaedackers

PROGRAM ORIGIN

Big History has been taught in the Netherlands for twenty years, since 1994. It is now an established course at several Dutch universities, but its beginnings were somewhat precarious. In the early 1990s, sociologist Joop Goudsblom and biochemist and cultural anthropologist Fred Spier became acquainted with the Big History course taught by David Christian at Macquarie University in Sydney and decided to plan and offer a similar course themselves at the University of Amsterdam.[18] However, they had no idea how their students, colleagues, or even the media would react to the course they were attempting to establish.

In the midst of those exciting planning days, Goudsblom and Spier's project received an unexpected boost in the form of a front-page article, headlined "Super Lectures," in *Folia*, the university's weekly. The appearance of this article, at a time when interdisciplinary programs did not yet exist in the Netherlands, attracted the attention of Jan Boorsma, the director of Teleac Radio, the main public broadcasting service in the Netherlands. Boorsma offered to produce twenty half-hour Big History segments based on live recordings of the course lectures. Goudsblom and Spier were simultaneously delighted and apprehensive. The unknowns of the course—the guest lecturers' contributions and the reception of the course by its projected thirty students—could result in negative national publicity that might undermine their Big History endeavor. Yet they found themselves acquiescing, and, in December 1994, the course launched in the electrifying atmosphere of an auditorium packed with 220 students and ample audiovisual equipment.

Fred Spier's forthright recounting of his trepidation is a good example of the anxiety experienced by all teachers of Big History: "I had only a one-year, one-day-per-week contract for organizing the course. I was especially worried that we would be shot down by critical specialists if we made any blunders in their respective fields. So I tended to be very cautious. Yet this fear has never materialized. Until today, our Big History courses have never been criticized, at least not openly, for committing errors in the historical garden."[19] This proved to be a most auspicious beginning for the Netherlands' Big History courses. Not only did it garner the course national publicity but it provided Goudsblom and Spier with additional funds, which enabled them to invite prominent scholars from around the world to join their course the following year. Big History courses in the Netherlands have remained successful ever since.

HOW THE PROGRAM WORKS

What do Big History courses in the Netherlands look like today? The persistent success of the initial Big History course attracted the attention of deans and program directors at the University of Amsterdam and other institutions, leading them to request more Big History courses. David Baker and I have joined Fred Spier—David in 2013 and I in 2006—in teaching these courses, and we are now planning to teach seven different courses, as detailed below:

- the original University of Amsterdam lecture series, which is still going strong and is regularly attended by about 300 second- and third-year undergraduate students of all majors;[20]
- a similar but shorter lecture series at the Eindhoven University of Technology that caters to about 150 engineering students;[21]
- a more interactive course at Amsterdam University College for three small groups of first-year liberal arts and science students;[22]
- a course entitled "Origins" that consists of lectures and tutorials and is required of all first-year students at Erasmus University College;[23]
- an advanced course for a small group of gifted first- and second-year students from the University of Amsterdam and the Free University; and, finally,

- two lecture series aimed at lifelong learners over the age of fifty-five: one at the Free University and the other at Utrecht University.

Our teaching of Big History courses targeting different audiences has enabled us to experiment with different course structures and teaching methods. Yet a few core ideas that reflect our understanding of the importance of Big History run through all our courses.

We think it is important for students to learn to see Big History as a framework they can use to connect different pieces of information. This seems particularly relevant in our current age, because the rise of the internet has made incredible amounts of information more accessible than ever before. Online information, however, can be rather fragmented. Big History can help to frame these fragments, which makes it less difficult to evaluate them and navigate a digital ocean of information. Big History therefore has the potential to function as a vital tool enabling people to thrive in the digital twenty-first century.

The effective use of Big History as a framework for connecting loose pieces of information has two main requirements. First, the structure of the framework must be very clear. Second, it must be equally clear in what ways information can be connected to the framework. These principles underlie our specific approach to teaching Big History in the Netherlands, where we clarify the structure of the framework with the aid of a *Big* Big History and clarify ways to connect information to the framework with the aid of a Little Big History.

So, what do these terms actually mean? *Big* Big Histories are general patterns and mechanisms that we can detect in the history of everything and that help us explain Big History in simpler ways.[24] The *Big* Big History we use in all our courses is the theory—based on the work of Eric Chaisson and developed further by Fred Spier—that throughout Big History, complexity evolves when energy flows through matter under certain Goldilocks circumstances.[25] This may sound a bit abstract, but it simply means that the availability of sufficient energy and the right conditions have been important for the development of inanimate objects like stars and planets, but also for the development of life and human societies. This idea can help us summarize Big History and provides students a compact overview of the grand narrative.

On the other hand, Little Big Histories are studies that analyze specific, often relatively small-scale, subjects from the perspective of Big History. And this is where romantic Paris plays a role in my own Big History story. In the winter of 2006, my husband, Marcel, and I were sitting in a Parisian café when he—down-to-earth engineer that he is—challenged me to explain how abstract Big

History concepts could help him understand his everyday environment better. We began talking about the Parisian street plan as seen from a Big History perspective, and I began fantasizing about all kinds of possible connections between this street plan and several Big History topics. Soon, I realized that uncovering such connections could help students to relate *Big* Big History concepts to their daily lives, making these concepts seem more relevant to them. Once back in Amsterdam, I transformed my thoughts into an assignment. In a way, the Little Big History is my response to Fred Spier's *Big* Big History approach.

In this assignment, I asked my students to pick a subject that was important to them and connect that subject to an aspect of every Big History lecture they had heard in class. In practice, this meant linking subjects ranging from beer and iPhones to students' favorite works of art or religions to topics such as the cosmic background radiation, DNA, and agriculture, with the aim of discovering some general mechanism or pattern that could connect these seemingly disparate things. Once they moved past the initial confusion (*"Do you seriously want us to do this?"*), students usually had a lot of fun and often came up with a number of creative ideas, which I in turn asked them to support with solid evidence from peer-reviewed journals or other reliable sources. The resulting Little Big Histories made *Big* Big History more relevant to our students, and helped them understand that anything can be connected to Big History, and *through* Big History to everything else. In other words, it taught our students how they could connect information with the aid of the Big History framework.

Today, *Big* Big History and Little Big Histories are permanent features of all our courses. In these courses students familiarize themselves with the Big History framework by reading Fred Spier's *Big* Big History book *Big History and the Future of Humanity* before class. And they work on their Little Big History assignment after class. During class, we discuss topics that are more specific than the general patterns that govern Big History but broader and less personal than Little Big History subjects. This approach, which sometimes requires bringing in guest lecturers to speak to students about their own field of study and its relation to Big History, shows our students how these subjects fit into the big picture. At other times, we discuss certain processes that occurred in Big History and that students find particularly interesting, such as the Cambrian explosion or the development of various philosophies during the transition to the Iron Age.

CHALLENGES AND CONCLUSIONS

The task of fitting subjects into the big picture has been a challenge, because Fred, David, and I are not specialists in those areas. Sometimes this leads students to assume that we will discuss these details in only superficial ways. Yet they quickly learn that, though we may not be specialists in the body plans of Cambrian trilobites or the ins and outs of Mohism, we have over the years become specialists in contextualizing Big History details and connecting them to one another. For instance, in discussing the transition to the Iron Age, I usually start with stellar nucleosynthesis, telling my students that because iron is the most stable chemical element, it is produced in large quantities in many stars—which makes iron a fairly abundant chemical element. And because iron is so abundant, it could be turned into relatively cheap tools and weapons once humans figured out how to extract iron from ores. The production of relatively cheap weapons eventually enabled many individuals to buy such weapons and subsequently challenge state monopolies on the legitimate use of physical force and taxation. This led to the weakening of a number of states during the transition to the Iron Age. These weakened state monopolies may have allowed for greater freedom in the development of philosophies in the Middle East, India, China, and Europe, both because social unrest created a greater demand for new ideas and because it became more difficult for rulers to suppress such ideas. As is evident, contextualizing Big History details such as these allows for a new kind of depth that arises from the connections that are made rather than from the details themselves.

Challenges in the classroom are not the only ones we have had to overcome. Many of our colleagues also assume that Big History teaching and research must, by necessity, be superficial. As a result, they are skeptical about Big History. This skepticism dates back to 1994, when Joop Goudsblom and Fred Spier first started teaching Big History in the Netherlands, and some are still dubious today. Dealing with such skeptical colleagues has been more difficult than dealing with skeptical students, perhaps because there are fewer opportunities to convince such colleagues. Yet accumulating research publications and course descriptions that emphasize integrating the more specialized disciplines in new ways with the aid of *Big* Big History and Little Big Histories have allowed us to slowly carve out a niche for ourselves within our academic environment.

All of this has led me to believe that becoming seen as integration specialists may be of great importance for Big Historians. This may be done through working on a *Big* Big History theory that revolves around energy, complexity, and Goldilocks conditions or through developing a Little Big History approach. These are just two approaches among many, and I suspect new ones will be invented by Big Historians in the near future. Developing such specialized pedagogical and theoretical methods for integrating Big History knowledge will lead to greater credibility, both for Big History teachers and researchers and for the field as a whole.

BIG HISTORY AND CONVERGENCE
EDUCATION IN SOUTH KOREA

Seohyung Kim

PROGRAM ORIGIN

Convergence education requires that every subject in middle school and high school be combined within a larger framework or context, and it emphasizes the interconnection between the natural sciences and the humanities. The Korean government first tried convergence education in 1992. The need for a combined science education had been raised and, in response, a national common science curriculum was established as a required subject in high school, divided into the disciplines of earth science, physics, chemistry, and biology.

Over time, awareness of the need for and importance of interdisciplinary studies grew, and the idea of convergence education was revised. As a result, science education became something more than a combination of two or more subjects in the sciences.

In 2011, for the first time, the Big Bang was included in a Korean science textbook, *Science*. This science textbook is composed of two parts: "Universe and Life," followed by "Science and Civilization."[26] The first part of the textbook includes the origins of the universe, the solar system, and the Earth, as well as the evolution of life; the second part contains information about communication, human health, scientific technology, energy, and the environment. Still, this science textbook is far from incorporating a Big History perspective.

Both the new science textbook and Big History begin with the Big Bang 13.8 billion years ago, but the science textbook treats these subjects without fully interlinking them, while Big History converges the natural sciences and the humanities through direct connection. This originally elusive difference was in evidence in our Big History program, in which twenty-three out of twenty-four students—that is, 96 percent—responded that they found Big History to be different from other ordinary subjects. Furthermore, 25 percent of the students expressed that the Big History perspective was totally different from anything they had experienced before.

Since 2009, the Institute of World and Global History at Ewha Womans University has taught Big History, in the form of a new course entitled "A History of Everything after the Big Bang."[27] Currently, this is the only university-level Big History course in Korea. In addition, in 2011, the Institute of World and Global History launched the first program in true convergence education for particularly talented middle school and high school students. This program is similar to the Big History Project in the United States and Australia. Since September 2011, the Big History Education Program for Talented Students has met five times at Ewha Womans University. Support from the Korea Foundation for the Advancement of Science & Creativity (KOFAC) enabled the institute to invite David Christian, one of the founders of Big History, to participate and to conduct collaborative research.

PROGRAM GOALS

The goals of this program are twofold. The first goal is to help students understand today's global society through the convergence of many different scientific perspectives. Unfortunately, since the nineteenth century, most disciplines have become so specialized that the natural sciences and humanities developed as completely separate and different disciplines. However, linking diverse disciplines is essential in order to interpret and understand our world and modern global society. In this context, the Big History perspective—which doesn't limit our understanding of history to the history of human beings or to the period since the invention of writing, but reaches back to the Big Bang and the origins of the universe—can help students move beyond the limited perspectives offered within specialized disciplines. Moreover, Big History can offer an important foundation for considering the meaning of history, looking for new intellectual synergies in the relationship between the natural sciences and the humanities, and developing new levels of creativity.

The second goal of this program is to cultivate a sense of the unity of our modern global society and the importance of global citizenship through Big History education, which unites diverse perspectives and viewpoints. Today, there are many global agendas and problems, but we struggle to find active solutions or alternatives. Instead, the contradictory interests of different nations block the finding of global solutions. It seems plausible that solutions will be imposed by the powerful nations that have the necessary political, economic, and even military might. However, if we learn to consider all humans as members of a single global community, rather than of a national or regional group—a perspective that Big History emphasizes— we may able to understand global problems more precisely and comprehensively, analyze them within new frameworks, and develop better solutions.

PROGRAM STRUCTURE

Our Big History program consists of five units: (1) the Big Bang and the origin of the universe; (2) life on Earth and evolution; (3) the era of foragers; (4) the agricultural era; and (5) the global era. This division of time and space from the Big

Bang to the present creates five units that show an increase of complexity and the development of technology and innovation.

In the first unit, the Big Bang and the origin of the universe, we look at the creation of the universe and its historical meaning. We try to understand the place of human beings in the universal context of time and space. Through studying the formation of stars, galaxies, and Earth, we understand Earth as the habitat of the human race. We illustrate common themes in the natural sciences—such as physics, chemistry, and Earth science—and in human history. And we examine why we need knowledge of the natural sciences to understand the nature and characteristics of human beings and human society.

In the second unit, life on Earth and evolution, the aim is to understand the process of the emergence of various kinds of life on Earth and their significance. It is especially important to discuss commonalities between the history of the Earth and the history of humans and to reveal the appropriate scale for historical analysis by contextualizing human history within the evolution of Earth and the evolution of life. We can see an increase in complexity through specific examples of evolution and the development of life.

In the third unit, the era of foragers, we begin to examine the appearance of human beings. And we study the significance of Africa, often excluded in traditional historical research and world history, by looking at the migration of the first human beings within and then out of Africa. We study the life patterns and characteristics of foragers and examine the complexity of forager society through comparison with today's world. This approach requires interdisciplinary study of archaeology, anthropology, history, geography, and geology.

In the fourth unit, the agricultural era, we focus on the communication and exchange of knowledge between the diverse disciplines of history, sociology, political science, economics, and genetics. The comparison of different regions showing evidence of the domestication of plants and animals allows us to identify both differences and commonalities. By studying the process of domestication, we recognize the vital role of agriculture in the development and growth of human societies. In this section, we also investigate the diverse conditions under which varied civilizations evolved and their distinctive historical features.

Finally, in the fifth unit, the global era, we discuss the development of today's global networks, which formed at many scales throughout human history. This requires a multicentric approach to global history, excluding narrower forms of centrism that privilege particular regions. Global networks have different features according to time and space, so if we analyze global networks from a Big History perspective, we can better understand global phenomena. It is especially important to note that comprehending the rapidly growing role of humans since the nineteenth century allows us to appreciate the huge impact of humanity on the biosphere. When we understand and reestablish the relationship between humans and the environment, it is possible to better understand the role of human beings on Earth and in the universe. This is an essential educational foundation for solving common problems that humans face.

OUTCOMES

Students who participated in the first Korean Big History education program were talented students with special ability in mathematics, physics, or chemistry. They were second and third graders in middle school (equivalent to eighth and ninth graders in the United States). Thirty-three students were selected for the Big History education program, and they took five Big History classes at Ewha Womans University.[28] When the first class began, few students knew anything about the Big History perspective or methodology, and they didn't understand the need for convergence between the natural sciences and the humanities. However, a survey of the students' responses and changes in their views at the end of the program provides evidence for the necessity of Big History education in Korea.

For the most part, students showed great interest in the Big History perspective itself and in its emphasis on convergence.[29] They agreed that though the natural sciences and humanities are completely dissimilar disciplines, they can be converged through Big History. Interestingly, the majority of the students found the unit on the Big Bang and the origin of the universe most interesting, but 75 percent of students developed an interest in other subjects, such as history, sociology, or geography, as a result of the program. So, the Big History perspective and methodology can catalyze talented students with interests in the sciences or mathematics to develop interest in the humanities as well.

In addition, more than 50 percent of the students responded that Big History explained their preexisting knowledge in new and different ways. Big History helped them see the changes in the histories of humans, the Earth, and the universe in a new light by reorganizing knowledge from varied disciplines at the largest possible scale of time and space. Thus, Big History is an essential methodology for a new mapping of knowledge and for bringing together diverse perspectives in our global era to solve global problems.

Looking at students' responses to the Big History class more closely, we find that 71 percent of students thought that the natural sciences and humanities have similarities and interconnections in Big History. For example, one girl responded that she could see the natural sciences and humanities converge and was acquiring new perspectives about specific problems or phenomena. One boy replied that he had had an interest in the sciences but, after taking Big History, had developed an interest in other subjects and wished to study humanities from a Big History perspective. Another boy replied that he had a new interest in and passion for knowing how everything is connected in Big History. These responses show that Big History helps students to organize and understand knowledge in new ways, to develop the ability to analyze within different perspectives, and to develop new kinds of intellectual creativity.

CHALLENGES AND CONCLUSIONS

This was the first time Big History was taught in Korea at the middle school level, so there was naturally room for improvement. One issue is that our students were a preselected cohort: they were all talented students with special ability in the natural sciences. We need to expand our Big History program to include other types of students and to learn about their responses to convergence education through Big History. In spite of this limitation, most students showed an affinity for convergence education between the natural sciences and humanities within the Big History perspective.

A Big History program is necessary if we are to realize the potential of interdisciplinary convergence education as a way of helping us solve the problems faced by our global society. To truly meet the desired outcomes of convergence and interdisciplinary education, all time and space must be considered to enable students to think on a scale large enough to embrace the universe and its human societies. Thus, the Big History perspective, which tries to analyze the many relationships between human beings and the natural environment—and the value or meaning of creativity within such interactions—can expand and promote new kinds of convergence education, like STEAM (science, technology, engineering, art, and mathematics) or STS (science, technology, and society), in a global society.[30]

In order to creatively tackle global issues, such as global warming or nuclear crises, convergence education is essential. It unites our understanding of the histories of Earth, the human race, and the universe. Today's world is multicentric. Poverty, energy shortages, and infectious diseases are no longer problems that one nation or country can resolve on its own. In this context, the Big History perspective, which admits differences and varieties while seeking universality, is required to find solutions that allow human beings to coexist.

NOTES

1. David Christian. "Big History: The Big Bang, Life on Earth, and the Rise of Humanity." Video lecture series. *The Great Courses*. The Teaching Company, 2008. DVD.

2. "A Story for Everyone." *Big History Project*. bgC3. 2011. Web. 3 November 2013.

3. Bob Bain, "Big Questions about Big History in U.S. Schools." Metanexus Symposium on Teaching Big History at the Harvard Club of New York, 21 September 2013. *Metanexus*. Web. Accessed 3 Nov. 2013.

4. Ibid.

5. *Big History Project*. bgC3. 2011. Web. 3 November 2013.

6. "Teaching Big History." *Big History Project*. bgC3. 2011. Web. 3 November 2013.

7. Ibid.

8. Ibid.

9. Ibid.

10. National Center for History in the Schools. "World History Content Standards." National Center for History in the Schools, UCLA. 1996. Web. Accessed 3 Nov. 2013.

11. National Governors Association Center for Best Practices, Council of Chief State School Officers. "English Language Arts Standards." *Common Core State Standards*. National Governors Association Center for Best Practices, Council of Chief State School Officers, Washington D.C., 2010. Web. Accessed 3 Nov. 2013.

12. Bain, "Big Questions."

13. Ibid.

14. Ibid.

15. Blair Hanley Frank, "Big History: Bill Gates Wants You to Take His 'Favorite Course of All Time' For Free Online." *Geekwire*, 1 November 2013. Web. 3 November 2013.

16. David Christian, "Why We Need to Teach the Modern Origin Story." *The Conversation*, 7 November 2012. Web. 4 Nov. 2013.

17. Bain, "Big Questions."

18. Fred Spier, "The Small History of the Big History Course at the University of Amsterdam." *World History Connected* 2.2 (2005). Web. 14 October 2013.

19. Ibid.

20. Fred Spier and Esther Quaedackers, "Big History." University of Amsterdam, 2013. Web. 14 October 2013.

21. Esther Quaedackers, "Universiteitscollege Big History." Eindhoven University of Technology, 2013. Web. Accessed October 2013.

22. Fred Spier, "Big Questions in History." Amsterdam University College, 2013. Web. 14 October 2013.

23. "Erasmus University College: Academics Year 1." Erasmus Universiteit Rotterdam, 2013. Web. 14 October 2013.

24. Fred Spier, "Big History Research." *Evolution*. Ed. L. E. Grinin, A. V. Korotayev, and B. H. Rodrigue. Volvograd: Uchitel Publishing House, 2011. 26–36. Print.

25. Eric Chaisson, *Cosmic Evolution*. Cambridge, Mass.: Harvard UP, 2002. Print. Fred Spier, *Big History and the Future of Humanity*. Hoboken: Wiley-Blackwell, 2011. Print.

26. Wanho Jeong and Hyunduk Ko, eds., *Science*. Seoul: Kyohaksa, 2011. 443. Print.

27. "A History of Everything after the Big Bang" was created by the World Class University Project and supported by the National Research Foundation of Korea.

28. The class comprised twenty-three boys and ten girls. Eight students were in the second year and the others were in the third year of middle school.

29. Twenty-four students responded to the survey that was conducted after the last Big History class, on 26 November 2011.

30. STEAM or STS is an attempt to develop creativity by seeking common ground in the convergence between the natural sciences and other disciplines and reorganizing knowledge in new ways. Big History has the largest possible scale to accommodate the convergence of the natural sciences and the humanities, starting with the Big Bang 13.8 billion years ago.

PART TWO · A PRACTICAL PEDAGOGY FOR TEACHING BIG HISTORY

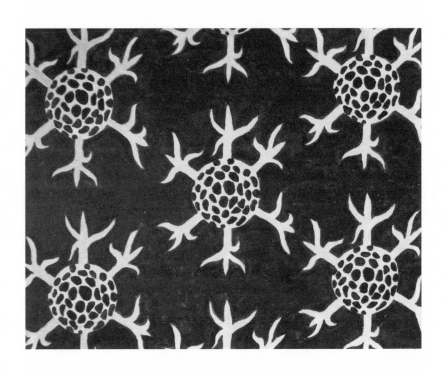

Overleaf: FIGURE 6.1 Sara Pimentel, *Design Principles in Microorganisms: Spumellaria (Threshold 5)*, 2012. (Photo: Lynn Sondag)

SIX · Teaching Complexity in a Big
History Context

Richard B. Simon

I was preparing to teach our first-semester course, and studying the preliminary edition of David Christian, Cynthia Brown, and Craig Benjamin's textbook, *Big History: Between Nothing and Everything*, when I was asked to teach a unit on industrialization for Dominican's second Big History summer institute.[1] As I compiled my materials and sketched out an outline, I realized that while I had a lot to say about industrialization—probably way *too* much to say in a brief hour—I didn't yet have a way to focus, to narrow and crystallize everything important about the age of industrialization in a way that would be pedagogically useful for colleagues.

The night before my presentation, I sat down to watch Charlie Chaplin's 1936 film *Modern Times*, thinking that it might have something to offer as an exemplar of the industrial age (in fact, we had shown it to students on a "Big History Movie Night" in the first semester of our new program).[2] The film begins with a ticking clock— regimented time, marked by a machine. Then, a flock of sheep fills the screen—with one black sheep running among them. It's an image from the age of agriculture, and a symbol of nonconformity against a growing conformity that is, itself, emblematic of the industrial age. The camera cuts slyly from the streaming sheep to men streaming into a heroic-looking brick factory—a flow of potential energy, breakfast sugars packed into human bodies, piling into the factory to work. Inside the factory, the Big Boss, president of the Electrosteel Company, is sitting at his desk, reading the comics and doing the crossword puzzle at his leisure. He pops a pill, glances up at a screen that allows him to surveil the factory, speaks into an intercom, and orders an underling—"control"—to speed up the production line. The underling, a giant, shirtless, barefoot man—an icon of 1930s New Deal industrial labor—runs across the floor and pulls an enormous lever that sends electricity from a mass of turbines into the factory's machinery. The energy flow is thus increased by a command passed down a power hierarchy. When the machines speed up, the humans who operate them must, too—and we see quickly that the humans in the factory have become cogs in the machine. The Little Tramp rebels—he punches out to go to the bathroom, sneaks a cigarette, and is scolded by the Big Boss on the screen: "Back to work!"

Flashbulbs went off. The film was a perfect illustration of the four features of complexity manifested in Threshold 8, modernity and industrialization. Again, those features, adapted just a bit from a synthesis of Christian, Brown, and Benjamin, with Eric Chaisson and Fred Spier, are

> **diverse components**—the most basic units or building blocks that comprise the particular form of complexity;
>
> **specific arrangements**—the precise structures into which those components or building blocks are arranged or organized, and the connections among them;
>
> **flows of energy**—the ways that energy flows through those diverse components in those precise arrangements, including the amounts of energy flowing through them over a set period of time; and
>
> **emergent properties**—the new properties characteristic only of such a form of complexity, with just such flows of energy (at Goldilocks levels) through those particular components in those precise arrangements. These emergent properties may lead to new forms (or higher orders) of complexity.

Through a simple thought experiment, which colleagues—and, later, students—could do, we were able to reveal a lot about the text (the film), about the period in which it was made (1936, as the United States was transitioning from an agrarian civilization to an industrialized one), and about the larger patterns in the story of increasing complexity from the Big Bang to humans' current situation.

We tried it in summer institute the next day. First, we recapped the four features of complexity. Then we watched, together, the first fifteen minutes of *Modern Times*.

After his smoke break, the Little Tramp returns to work. At lunchtime, an "automatic feeding machine" is tested on him. The machine goes haywire, and pummels his face with food and dishes and devices, like a motor that spins an ear of corn like a lathe. Bullied by the machine, Chaplin's character goes back to his post and falls onto the production line conveyor belt—he is chewed up in the machine and spat out. This leads, eventually, to a nervous breakdown—and to the protagonist's rebellion against his employer, his colleagues, and the machine of which he has become a part. The police are called, and the Little Tramp is subdued and taken away to a mental hospital. On his release, he picks up a red flag that falls off the back of a truck and is mistaken for a communist agitator. A riot ensues, complete with police violence. The Little Tramp is arrested.

We stopped the film and began the thought experiment: think about how those four features of complexity manifest in the film. We worked collectively, using the complexity model as an analytic tool, to tease meaning out of Chaplin's text and onto the whiteboard.

It went a little something like this:

Diverse components: Workers. Machines. Big Boss. Foreman.

Specific arrangements: Social hierarchy with Big Boss on top, handing orders down to the controller, who pulls the levers that control the energy flows. At the bottom, line workers such as Chaplin's character interact with and operate machines.

Flows of energy: Energy almost always flows from the sun. Here, it flows from the sun to photosynthesizing plants and the livestock that eat them—both of which comprise the food that workers consume to power them. In addition, such plants that lived hundreds of millions of years ago and were buried in anoxic swamps became coal, which, when used to heat water into steam to drive turbines, is the likely source of the electricity that powers the factory's machinery. The humans, working in synchrony with machines, do a lot of work—and generate great surpluses of products and, therefore, wealth.

One useful way to think about energy flows is to diagram them in simple flowcharts.

sun → plants → animals → humans →

                                → products → wealth $$$

sun → plants → fossil fuels → turbines → machines →

Emergent properties: Industrial capitalism. Wealth. Leisure time. Wealth disparity. Regimented time. Rebellion. Repetitive stress disorder. Mental illness. Communism. State violence as social control.

PhDs might take this kind of activity further (and farther afield) than students. But our students did an admirable job, too. In both cases, the exercise revealed that *Modern Times* was a key text for understanding the industrial period. Chaplin, the creator of this work of art—of fiction—captured the spirit of his time so well that the work is representative not only of the interwar industrialization of the United States but also of the process of industrialization more generally—of the transition from an agrarian civilization to an industrialized one, and of the human response to that transition. And because Chaplin's film so clearly illustrates the four features of complexity, it is not only an important Big History text but also a useful tool for teaching both industrialization and complexity. The exercise reveals what *makes* a text useful to a Big History perspective. It also demonstrates that the complexity framework, itself, is an incisive analytic tool.

ILLUSTRATING COMPLEXITY IN THE CLASSROOM

Fred Spier purposefully distinguishes between a system and a regime, because to him, as a social scientist, the word "system" connotes stability—where "there are no forms of complexity that are completely stable over time."[3] He suggests "regime" as a catchall for both **complex adaptive systems** (those that adapt in response to changing conditions, such as organisms or civilizations) and **complex nonadaptive systems** (those that do not adapt in response to changing conditions, such as stars or galaxies). Still, we can use what we think of as systems to illustrate what we mean by complexity, and how it functions. And while students can begin to get their heads around the concepts with any form of complexity, the easiest complex adaptive system for college students to visualize is the one in which *they* are the key components—their campus.

Start by introducing the four features of complexity. Define them. Explain them at, perhaps, the atomic level. Start with a hydrogen (H) atom:

Diverse components: One proton, one neutron (in some cases), one electron.

Specific arrangements: The proton and neutron stick together in the nucleus; the electron orbits (uncertainly!).

Flows of energy: Electromagnetic energy attracts electron to proton. Strong nuclear force binds neutron to proton. The electron orbiting the nucleus is not only an energy flow; it's the one with which we are most familiar: electricity.

electron $\rightarrow$ (electromagnetism) $\leftarrow$ proton

proton $\rightarrow$ (strong nuclear force) $\leftarrow$ neutron

Emergent properties: These components in these arrangements with these energy
flows and electromagnetic affinities make the H atom apt to bond with other
atoms—and form water and other molecules. Or, in a star, to undergo fusion
and become helium—and all other elements. So, the key emergent property,
which changes everything (and gives rise to new complexity) is the ability to
form other elements and compounds. In the early universe, this allowed for
stars and galaxies to light up. But because stars continually burn through their
core elements and create new, more complex elements (see chapter 9, "Teaching Threshold 3: Heavier Chemical Elements and the Life Cycle of Stars"), it
also led to all the other known elements in the universe. The ability to become
anything and everything else? That's some emergent property!

Once students begin to get their heads around the key features, the basic concepts,
then you can ask them—as a whole class, or in groups—to think about the four
features of complexity for the campus. What are its diverse components? What are
the precise arrangements in which those components are placed? How does energy
flow through the campus? And what new emergent properties arise as a result?

Think about it yourself, for a moment.

Like much in the Big History classroom, it's a thought experiment.

So . . .

Diverse components: Students. Teachers. Staff. Buildings. (Can you be more
specific?) The cafeteria. Dorms. Classroom buildings. The gym.

Specific arrangements: On Dominican's campus, the cafeteria lies near the
library—and both are located centrally, between the classroom buildings
and the dorms. A city road separates the gym, sports fields, and a student
parking lot from the rest of campus. Only three quiet streets run through
campus. Significantly, in human societies, we have not only physical structures but societal ones as well. Within the classrooms, we have the professor,
in the front of the room, and students, either at desks, facing the professor,
or seated around a large table, facing one another. Administrators are above
faculty in the power pyramid, with the college president at the top. Then,
above her, is the board of trustees—who control significant flows of money
into the campus. Speaking of which . . .

Flows of energy: Much like Chaplin's factory, energy enters the campus in the form of food, which is delivered from farms where it is grown (or via food distributors) or raised, on energy from our star, the sun. This food is photosynthesized into sugars usable by biological cells, such as those in students' or teachers' bodies. In addition, electrical energy flows into the campus on wires, typically from hydrothermal or gas-fired power plants—or, increasingly, from solar electricity producers. That energy is used to light classrooms and to power technology. And let's posit that a third type of energy, information (at least, information is what *organizes* energy), is flowing through that technology, as well as through the teachers and through texts. Finally, fossil fuels are being expended to transport teachers, in particular, to campus. (And, of course, money, which is essential to maintaining all of the above, must flow through as well.)

sun →   plants →   (animals) →   students and faculty →

sun → plants → (animals) → fossil fuels → automobiles → students and faculty → classrooms

sun →   plants →   (animals) →   fossil fuels →   electricity →   technology →

And also:

sun → plants → (animals) → humans → products → wealth → grants, gifts → administration → faculty → information → student

                                                ↘ staff → facilities/operations → student

When you have students and faculty, fueled on food and maybe coffee (sun → photosynthetic coffee plant → human → brain)—to say nothing of (ahem) "*energy drinks*"—all transported to campus using fuel consumed in motors, coming together in classrooms lit with electricity and bristling with electric-powered digital information–conveying technology, a few new properties arise . . .

Emergent Properties: Education. Knowledge. Understanding. Job preparedness. Potential for wealth. Self-awareness. Community. Shared identity. School spirit. "The university."

Once students get a few concrete examples of what complexity is, how its features manifest in familiar ways, they'll have an easier time visualizing—and understanding—the complexity framework that underpins the Big History narrative. And when they see those recurring patterns at every level of reality, at every new threshold or in each regime, they'll begin to understand that similar processes are at work throughout the universe, across time. It will certainly allow them to understand that the intellectual and material connections among what have long been treated as separate, unrelated, siloed academic disciplines are real, organic, and quite logical.

At the very least, we'll be able to begin our story.

STUDENT LEARNING OUTCOMES AND POSSIBLE ASSESSMENTS

By the end of this unit, students should be able to

1. define the four features of complexity (assessment: quiz or test);
2. apply the four features of complexity to analyze any complexity or system (assessment: brief in-class or at-home writing assignment);
3. demonstrate understanding of the greater concept of complexity (assessment: class discussion or group work with written assignment); and
4. demonstrate understanding of the key Big History concept of thresholds of complexity (assessment: at-home reflective writing assignment; see chapter 17, "Activities for Multiple Thresholds").

CHALLENGES IN TEACHING COMPLEXITY

Some students—and some teachers—may resist the concept of complexity, and the framework of increasing complexity altogether. In a way, it seems like a device. But (as Cynthia Brown has asked) is it? Is it a trick to make the somewhat arbitrary thresholds hang together long enough to seem like a unified narrative? Or is it the actual, material, physical reality? Each of us will have to wrestle with that question.

For students developing critical thinking skills in an expansive, interdisciplinary course on the history of everything from the beginning of time, it's not a bad wrestling match to have.

COMPLEXITY IN PEDAGOGY

Understanding the importance of the four features of complexity as a teaching tool was a turning point in our faculty's understanding of our own Big History pedagogy. In year one, we had focused on content—on making sure that our students got as much information as possible (enough to persuade ourselves that we had mastered the content!) so that the story would emerge from the details—like in the film *The Matrix*, in which a green rain of ones and zeroes suddenly materializes into a palpable, full-color virtual reality.[4] But we got bogged down in the nitty-gritty. We had students reading long assignments from two challenging Big History texts, and we were spending class hours projecting PowerPoints to try to make sure that we didn't leave out anything important—and quizzing and quizzing to make sure that the facts were sticking. But we didn't leave as much time as we could have to allow our students some breathing and thinking room, some space in which to *make sense* of all the new foundational knowledge we were cramming into them. And that ability to make sense of this great wealth of human knowledge is one of our essential learning outcomes.

Typical of our summer institutes, this was a meta-cognitive lesson for the faculty. In year one, we had all crammed in order to wrap our heads around all this foundational knowledge, so much of which was new for many of us. And in the second year, once we felt a bit more comfortable with the content, we were able to sit back and say, hey, wait a minute, we're missing something important here. We shifted the focus of our summer institute from learning the material to developing our pedagogy. And we unpacked the syllabus we had developed collectively for our first-semester Big History course.

Over the next year, we learned that changing our focus a little bit, from information overload to a more elegant balance of foundational knowledge and conceptual throughline, using the complexity model more centrally, freed up class time and allowed our students to process all that information, to see the framework into which it fit, and which did underpin the entire story.

We shifted some of our focus from lecturing to learning activities that were fun for students (and for us teachers) and yielded intellectual fruit, including a growing recognition of the patterns that recur from threshold to threshold: that the universe, as we move from the Big Bang to us—from hydrogen atoms to students with smartphones drinking coffee in classrooms—tends toward greater complexity (even as that complexity adheres to very simple recurrent patterns); that in recounting that story, certain key emergences stand out as having changed everything, in ways that are profound for our species; and that we might call those emergences "thresholds," doorways into increasingly complex regimes, that matter and energy crossed on their way to forming us.

We suggest starting with the four features of complexity at the beginning of the Big History course, and then checking back in with them as we cross each threshold (or at the beginning and end of each unit). That way, as we move from talking about, say, the formation of our sun and solar system and Earth to studying the origins of life on Earth, we can illuminate how the particular conditions that exist on earth—"Goldilocks" temperatures and distance from the star; the chemical composition and structure of the planet; the availability of water in liquid, solid, and gaseous forms—which are functions of the components, arrangements, and energy flows in our solar system—allow for organic, carbon-based life as we know it to emerge.

We will continue to follow that pattern as we lay out our practical pedagogy in the chapters that follow. Each chapter contains a look at the key concepts in that threshold, a discussion of the four features of complexity as that new threshold emerges, some suggested learning outcomes and assessments, and a few of the challenges teachers might face in teaching that particular threshold. We also include a few learning activities that we've developed for each threshold, which you might use or adapt, or which might serve as inspiration for your own approach.

ACTIVITIES AND EXERCISES FOR TEACHING THE FOUR FEATURES OF COMPLEXITY

BAKING WITH COMPLEXITY

Mairi Pileggi

Threshold Related to the Activity: Pre-thresholds

Category of Activity: Brief lecture, learning activity

Objective: Christian, Brown, and Benjamin use the concept of increasing complexity to frame the Big History narrative in their text, *Big History: Between Nothing and Everything*. Through this short activity, students will

- comprehend and retain the definition of increasing complexity; and
- be prepared to discuss how it relates to the Big History thresholds.

Overview of Activity: Students participate in an interactive demonstration of the concept of the four features of complexity. The entire activity, with brief lecture, should take approximately thirty minutes.

Faculty Preparation for Activity:

- Prepare a brief lecture on the concept of complexity and its four features as they relate to Big History's eight thresholds. You might also wish to use the first five minutes of David Christian's TED video as an introduction.
- Bring a sampling of ingredients required for baking bread: flour, water, dry yeast, and salt. Place each ingredient in an unmarked jar or bag. Also, plan on having a loaf of fresh bread to show the product (fresh is better so students may eat it!).

Cost: About $10–$15 for all ingredients

Student Preparation for Activity: Students should have read the definition of complexity.

In-Class Sequence of Activity:

A. Begin with a brief lecture in order to
  1. emphasize the importance of increasing complexity as a *frame* for our Big History narrative. A *frame* helps us to organize a large scope of information and decide on the events that make it into our narrative. In

this case, the only elements that we will include in the Big History narrative are those that can be verified by science.

2. articulate to students the goal of today's activity, as stated above: to understand what increasing complexity is and how it relates to the Big History thresholds.

3. introduce or review the four features of complexity. Complex things
   a. consist of multiple diverse components,
   b. are arranged within a precise structure,
   c. are held together by flows of energy, and
   d. have new emergent properties.[5]

   You might use this phrase to help students relate these principles: ***Emergent properties*** **arise from a selection of** *diverse components structurally related* **in a** *specific* **way and activated through** *energy flow.*

B. Now, begin the activity by handing out your contained and anonymous ingredients to students in the class. Give them a few minutes to analyze the substance by examining it. For example, they should open the jar of liquid (water) and smell it, and pour a bit of the powders onto a paper towel and taste them. You want them to get a physical sense of the ingredients they are examining; then, ask them to share with the class the identity of their mystery ingredients.

C. Discuss the following aspects of baking to demonstrate the tenets of complexity:

1. Ingredients represent multiple diverse components because baked goods generally require basic, independent elements in order to create the desired whole.

2. Measurements represent precise structure because too much yeast or too little flour can dramatically change the final product.

3. Flows of energy (heat) added to the previous two features result in . . .

4. Emergent properties, which are the qualities and wholeness of the final product: a delicious loaf of bread!

D. Finally, unveil a loaf of freshly baked bread you have kept hidden up to this point and share with the class.

Assessment of Learning: If you have time, ask your students to identify the four features of complexity in a casual discussion or a pop quiz.

Origin of Activity: This is an original activity.

Source Consulted: Christian, David, Cynthia Brown, and Craig Benjamin. *Big History: Between Nothing and Everything.* New York: McGraw-Hill, 2014. Print.

Richard B. Simon

Threshold Related to the Activity: Pre-thresholds / all thresholds
Category of Activity: Learning activity
Objectives:

- To illustrate complexity through modeling.

- To help students understand visually, spatially, and kinesthetically the four features of complexity and recurrent patterns across the thresholds of Big History.

- To help students understand that each type of complexity is the key building block in the next higher level of complexity (that is, in the next threshold).

Overview of Activity: Students break into teams of three or four. They use modeling materials to make model representations of given types of complexity and consider components, specific arrangements or structures, energy flows, and emergent properties, as well as patterns that might recur, and the connections from one type of complexity to the next. You might try this at the beginning of the semester or at the end—or perhaps both.

Faculty Preparation for Activity: You'll need a modeling system; for example, children's interlocking building blocks, large sheets of paper, and colored markers. (The concept is inspired by old chemistry models that used colored wooden balls, metal connecting rods, and springs to allow students to recreate different atoms and molecules; you could use marshmallows and colored toothpicks or pipe cleaners to model in three dimensions.)

Cost: About $50 for blocks, markers, and paper

Student Preparation for Activity: If the activity is done at the beginning of the semester, students should have read about complexity and been led through the thresholds in brief. If the activity is done nearer to semester's end, students may need a refresher on the four features of complexity.

In-Class Sequence of Activity:

1. Break students into groups of three or four. Give each group at least ten blocks (they can be all the same size and shape, or perhaps include a few of varying shapes); two markers (one warm color, such as red, orange,

or yellow, and one cool color, such as black, blue, or purple); and several sheets of paper.

2. Have students, all together, create a model of a hydrogen atom. Think about what parts constitute a hydrogen atom (one proton and one electron.) Lay the model atom out on the paper. Use the cool-colored marker to draw in the structure (the proton as the nucleus; the electron orbiting the electron, indicated by a circular orbit). Use the warm-colored marker to draw in arrows or vectors that represent energy flows (in this case, the electron orbiting the proton *is* a flow of electrical energy, so a red vector redoubles the orbit); also, the proton and electron are attracted to each other by electromagnetism, a two-way arrow between the two components). Finally, have students remove the blocks and draw in the proton and electron in their place.

3. Have students take a new sheet of paper and create a helium atom (two protons, two neutrons, two electrons), repeating the process from step 2. Think about the differences between the two—for one thing, the helium molecule has more parts (double the protons, double the electrons). It also has more types of parts (it has two neutrons). It has a similar structure. But a new force is involved—the strong nuclear force, which binds protons and neutrons in the nucleus. Because the helium atom has more parts, more types of parts, and more connections among those parts in slightly different structures, it is more complex than the hydrogen atom.

4. Now have students repeat the process for a few more types of complexity that correspond with the thresholds of Big History: A biological cell. A human body. A hunter-gatherer village. An agrarian civilization. An industrial-age city—or a college campus. Wander the room and help the groups that might be having a hard time visualizing each level of complexity, or what its parts might be.

5. Have students take the sheets of paper that now represent models or diagrams of complexity across several thresholds and lay them out to view them in sequence. Let the smaller groups discuss for five minutes or so what they see.

6. Have the class re-form for groups to report out and for all to discuss their findings. Make sure to reiterate that each complexity forms the basic building block for the next higher-order complexity (atoms make up cells, cells make up human bodies, humans make up villages, and villages make up civilizations).

Assessment of Learning: Class discussion. Possible additional assessment: at-home reflective writing assignment, in which students think about the activity in relation to the course reading. If you try this activity both when you first introduce complexity and nearer semester's end, you will be able to gauge students' acquisition of complexity understanding, and thus their achievement of complexity learning outcomes.

Origin of Activity: This is an original activity.

Sources Consulted:

Christian, David, Cynthia Brown, Craig Benjamin. *Big History: Between Nothing and Everything*. New York: McGraw-Hill, 2014. Print.

Spier, Fred. *Big History and the Future of Humanity*. Chichester, West Sussex, U.K.: Wiley-Blackwell, 2011. Print.

NOTES

1. David Christian, Cynthia Brown, and Craig Benjamin, *Big History: Between Nothing and Everything*. New York: McGraw-Hill, 2014. Print.

2. *Modern Times*. Dir. Charlie Chaplin. Perf. Chaplin and Paulette Goddard. United Artists, 1936. The Chaplin Collection. Warner Home Video and MK2 SA. DVD. 2003.

3. Fred Spier, *Big History and the Future of Humanity*. Chichester, West Sussex, U.K.: Wiley-Blackwell, 2011. 21. Print.

4. *The Matrix*. Dir. Andy Wachowsky and Larry Wachowsky. Perf. Keanu Reeves, Laurence Fishburne, Carrie-Anne Moss, Hugo Weaving, and Joe Pantoliano. Warner Bros., 1999. Film.

5. Christian, Brown, and Benjamin, *Big History*, 6.

SEVEN

· Teaching Threshold 1

The Big Bang

Richard B. Simon

> When God began to create heaven and earth, the earth
> was then without form, and void, and darkness was over
> the deep, and God's breath hovering over the waters.
> And God said, "Let there be light." And there was
> light. And God saw the light, that it was good, and God
> divided the light from the darkness. And God called the
> light day, and the darkness he called night. And it was
> evening and it was morning. A first day.
>
> *The Book of Genesis,* ILLUSTRATED BY R. CRUMB

In the early 1990s, the American university was in the process of shifting from one approach to general education—the Western civilization model—to another, the world cultures model. At the school I was attending as an undergraduate, GenEd 101 was still "The Roots of Western Civilization." My section was taught by a tweed-jacketed, golden-curled classics professor of British origin. We read a verse translation of Homer's *Odyssey,* and we studied Chartres Cathedral, and in between we read the Bible, as literature. That is to say, we read it without the underlying assumption that the story related in the text was true and factual and had been handed down by a supreme being. For a young person brought up in a traditional faith, the effect of considering seriously, for the first time, the idea that the Adam and Eve story was something other than God-given literal fact was disorienting.

We are at an interesting moment in human history, when the previously dominant cosmological paradigm is yielding to new understandings of how the universe works. These new understandings are the result of modernity and industrialization, and of the ensuing **chronometric revolution** that Christian, Brown, and Benjamin describe— the ability to measure time, space, and the age of rocks, fossils, bones, and historical artifacts with atomic precision. People whose belief systems are entwined with prior ways of knowing—prior cosmological epistemologies—may feel that their core beliefs are threatened by the contemporary way of knowing about the universe.

In *Big History: Between Nothing and Everything,* Christian, Brown, and Benjamin place Big Bang cosmology alongside other cosmologies—Hopi, Chinese, the Hindu

Rig Veda, an Islamic-Somali origin story, and the Book of Genesis. They set all cosmologies on the same level, as ways that humans tell their own story and explain and understand their place in the universe. But they suggest, firmly, that Big Bang cosmology is based on evidence, rather than on faith—and set it aside from other cosmologies on that premise.

"The modern origin story," Christian and his coauthors write, differs from other origin stories "in important ways." They continue:

> Above all, it offers a literal account of the origin of everything. It expects to be taken seriously as a description of what actually happened beginning about 13.8 billion years ago. It is not simply a poetic attempt to make up for ignorance. It claims to offer an accurate account of the very beginnings of history because it is based on a huge amount of evidence, generated through numerous measurements over several centuries, and based on rigorous and carefully tested scientific theories. It is the only origin story accepted by scientists throughout the world. But because it is based on evidence, and new evidence can always turn up, the same scientists also know that many of its details will change in coming years. It is not a fixed or absolute story and does not claim to be perfect.[1]

At Dominican, too, we begin our journey into this story by comparing it to other cosmologies. In an exercise developed by Cynthia Brown and our Big History faculty, we give our students brief summaries of six or seven origin stories from cultures around the world, including both the Book of Genesis and the Big Bang.

Students break into small groups, each of which must study, quickly adapt, and perform an interpretive version of one of the origin stories. It's an icebreaker. Students must let their guards down, bond with classmates, and be creative—if not just a little bit competitive. After the performances, students compare several key features among all the origin stories—each version's depiction of the creator of the world; the material from which the world was formed; *how* Earth was formed; the age of the Earth; what the first life-forms were; how humans formed; and what the relationship is among humans, other living things, and the natural world as a whole.

At the end of the exercise, students see that all origin stories fulfill common functions—to explain where the world came from, where people came from, and what our relationship is with plants and animals. They also see that the Big History

story does a few things that other creation myths might not (such as to quantify time).

Ultimately, Big History sets itself alongside previous explanations of the nature of the universe as the latest attempt by humans to tell our own story (it's just that now, we've got space-based telescopes that can see billions of years back in time.)

This is the frame through which we begin our explanation of how modern cosmologists believe the universe came to be.

KEY CONCEPTS IN THRESHOLD 1

THE BIG BANG

Big Bang cosmology holds that the universe began some 13.8 billion years ago, when what Christian, Brown, and Benjamin describe as a tiny, atom-sized space, filled with all the matter and energy in today's universe (inconceivably hot and shifting back and forth between those two states), appeared out of nothingness and began very, very rapidly to expand—creating both time and space—and to cool, creating distinct matter and energy, as well as gravity, electromagnetism, and the strong and weak nuclear forces that bind atoms together internally and govern radioactive decay. Both atomic (or baryonic) matter and dark matter appeared.[2] This happened within a fraction of a second.

Now, still within the first second of its existence, the universe expanded (in a process known as "inflation") to what may have been the size of a galaxy. Matter and energy stabilized into a plasma of protons, electrons, neutrons, and photons. "Most of the matter in the universe," Christian, Brown, and Benjamin write, would have been "crackling with electricity and constantly buffeted by intense electromagnetic energy."[3] The universe existed in this state for about 380,000 years, after which time it cooled enough (to a temperature similar to that of the surface of Earth's sun) that previously heat-jostled protons, neutrons, and electrons could succumb to electromagnetism and nuclear forces and begin to bind together and form atoms of hydrogen and helium. The resulting net neutral charge released the photons from the other particles' electromagnetic grip, resulting in a "huge flash" of light—as newly freed photons shot off in all directions.

And the universe continued to expand.

The Evidence   Christian, Brown, and Benjamin explain that five main pieces of evidence support the Big Bang theory, and they provide a history of the scientific discovery and the development of the theory.

1. **The cosmic background radiation:** This is a distinct electromagnetic signature, a static fuzz, detectable everywhere in the universe (in every direction one might point a radio telescope from Earth). Discovered in 1964, it aligns with the theory of the "huge flash" of newly freed photons.

2. **Hubble redshift:** Using spectrography, we can analyze the chemical compositions of stars by breaking their light into diffraction spectra. A dark band or shadow that appears in the distinct rainbow cast by each star's diffracted light indicates the presence within that star of a particular element, which we know absorbs light at that wavelength. So stars of a certain composition have recognizable patterns of "absorption lines" in their spectra. That said, sometimes these patterns appear "shifted" to the red or to the blue end of the spectrum. That's due to the Doppler effect: objects that are moving toward us appear shifted to the blue (as light "bunches up" ahead of them) and objects that are moving away from us appear shifted to the red (as light stretches out behind them). Astronomer Edwin Hubble observed, in the 1920s, that just about everything in the universe appears to be expanding out from a central point. That would be the Big Bang.

3. **Age:** We can discern the age of stars from their temperature and chemical composition. According to Christian, Brown, and Benjamin, "No object in the universe appears to be older than about 13 billion years."[4]

4. **Change over time:** The universe has a history—which is to say that it changes over time. "The most powerful modern telescopes can detect objects billions of light years from Earth," Christian, Brown, and Benjamin write. "By doing so they are, in effect, looking at those objects as they were billions of years ago, because it has taken the light they emit billions of years to reach Earth." Our most powerful telescopes can actually see the early universe, and it was "more crowded, and it contained objects . . . that are very rare in today's universe."[5]

5. **Hydrogen and helium:** Because the universe cooled so quickly, only the simplest elements—hydrogen (H) and helium (He)—should have had time to form, early Big Bang theorists believed. Again, with spectrography, we can determine the chemistry of much of what we see out there. And it's mostly hydrogen (about 75 percent) and helium.

We can tell they are there by their effects on everything around them—but we don't know what dark matter and dark energy are. Dark matter may have aided (gravitationally) in the formation of galaxies and stars. Dark energy may be driving the universe to accelerate in its expansion. But science doesn't know what this stuff is—and it constitutes perhaps 96 percent of the universe!

That's awfully troubling. This uncertainty underscores that, while we can tell a story based on the best of what we do know, there's an awful lot we don't know—and that will certainly always be the case. Believe it or not, these enormous question marks in the story the science tells can provide a bit of comfort for students amid the tremendous cognitive dissonance of having their religious beliefs challenged by scientific evidence.

A big lesson here is that our understanding of all this is always changing—and that's the nature of science. The more we learn through observation and testing and retesting those observations using the scientific method, the more the story changes. The "discovery" of the Higgs boson (the particle believed to give matter its mass) in the summer of 2012 is still so recent that Big Historians are not quite sure how it changes our understanding of the early universe.

It's also only in the last few years that we have learned that not only is the universe expanding, its expansion is accelerating. Scientists believe that's due to dark energy.

Again, wrestling with these concepts and learning to continually reevaluate what one thinks in the face of new information, are very useful in developing critical thinking skills.

And those skills are the basis for new discovery in all disciplines.

THE FOUR FEATURES OF COMPLEXITY IN THRESHOLD 1

You might think about the four key features of complexity as shifting across the first threshold as the universe itself moves through a few different phases. At first, we have that tiny point that emerges from nothing:

Diverse components: Matter / energy

Specific arrangements: Everything massed together in an atom-sized space

Flows of energy: Phase shifts from energy to matter and back again, or

$$energy \rightarrow \leftarrow matter$$

Emergent properties: The potential to expand into the entire universe

Then, after a second or so, we have the plasma universe:

Diverse components: Protons, neutrons, electrons, photons, neutrinos

Specific arrangements: A homogenously arrayed, superhot plasma, expanding

Flows of energy: High temperatures from massed energy keep particles agitated and prevent electromagnetic bonding; electromagnetic energy ripples through the plasma

Emergent properties: The potential to form atoms

Finally, about four hundred thousand years later (give or take a few tens of thousands of years!), the expanding plasma universe cools sufficiently so that we have a universe of loose hydrogen and helium, which Fred Spier describes as electromagnetically neutral and newly "transparent" because it is cool enough for the plasma to settle out into matter.[6]

Diverse components: Hydrogen and helium (and dark matter)

Specific arrangements: A cooling, expanding, "transparent" universe of hydrogen and helium (and dark matter), arranged homogeneously—but beginning to break up into molecular clouds

Flows of energy: Nuclear forces bind protons with neutrons. Electrons orbit proton and neutron nuclei. Photons, freed, shoot off into the void as light. Matter exerts gravity.

Emergent properties: Gravity begins to pull atoms of hydrogen and helium together into molecular clouds, with the potential to congeal into galaxies of stars. Light.

It's a universe in which galaxies and stars may begin to form. And so the key emergent property of the Big Bang, atoms—specifically, atoms of hydrogen and helium—yield the diverse components of Threshold 2, in which gravity begins to pull all that hydrogen and helium together into stars and galaxies.

As a reading of the underground cartoonist Robert Crumb's quite literal take on the Book of Genesis reveals, the two stories are not so different, after all. Is this because our scientific understanding of the universe is filtered through brains fluent in Abrahamic cosmology? Or is it that ancient cosmologists' thought processes were not so different from those of modern cosmologists?

STUDENT LEARNING OUTCOMES AND
POSSIBLE ASSESSMENTS

By the end of this unit, students will demonstrate the ability to

1. define cosmology (assessment: paper);

2. compare and contrast Big Bang cosmology with prior cosmologies (assessment: origin stories exercise);

3. describe key concepts and events in Big Bang cosmology (assessment: quiz or test); and

4. summarize and / or evaluate the key evidence that supports the Big Bang theory (assessment: quiz, test, or written response).

CHALLENGES IN TEACHING
THRESHOLD 1

Some students will find it difficult to conceive of the physics that underlies these early universe events. It helps to bring in solid visuals to explain spectrography and redshift. It also helps to use sound and diagrams to explain the Doppler effect—so that we can understand what redshift is (for detailed instructions, see the "Redshift Demo" activity at the end of this chapter). One must also be somewhat familiar with basic chemistry—a teacher can draw or display images of hydrogen and helium atoms, so that students can see how loose particles become our most basic elements. This part of the Big History story is particularly difficult to fathom—again, visual aids, especially for students with visual learning styles, are key.

Christian, Brown, and Benjamin, historians all, spend a lot of time laying out the history of the evidence. That part of the reading may be difficult for students. But their reaction is often, wow, I didn't realize that there *was* evidence for the Big Bang. Let alone that we can take photographs of the universe as it was billions of years ago (because the light from stars billions of light years away is reaching us only now).

Students may also rebel against the deprivileging of their own cultures' origin stories.

The most significant potential roadblock is probably the challenge of confronting students' faith traditions with an alternative explanation of the origins of the universe. While some Big History texts may have an atheistic or an agnostic materialist bent, others ascribe spiritual meaning specifically to the events that science details. At our university, we bridge a Dominican heritage (and a diversely spiritual faculty) and a contemporary, secular, materialistic approach to education in the arts and sciences. We aim to challenge and test our students' beliefs, but not to damage them.

CONCLUSION

One student in my first-year Big History course found the experience of having her belief system challenged not only disorienting, but also undermining. I ask my students to write freestyle responses to each week's assigned reading, to respond honestly to the text. This young woman's responses grew ever more perplexed, even angry. Why, she asked, was this textbook, this course, trying to convince her that God did not exist? In class, her facial expressions betrayed some heavy-duty cognitive dissonance. Her engagement with the course material had been endangered by what she perceived as a challenge to her faith.

I drew on my own experience to try to guide her. When I studied astronomy and geology and cosmology in college, I found that what I was learning aligned quite readily with a faith tradition from which I had already grown distant. One of the most important prayers in Judaism is the Shema:

Shema Yisrael, Adonai Elohaynu, Adonai Echad!
Hear, O Israel, the Lord our God, the Lord is One!

If the universe just before the Big Bang is essentially a single point that contains all matter and energy, and the universe at its end point is a homogeneous space filled with reasonably evenly distributed matter; if everything in the universe is energy, sometimes arranged into different forms of matter, in different configurations—atoms or stars or cars outside bars—then what was the difference between that homogeneity and the "One" of the Shema? That was a life-altering epiphany.

Don't worry so much, I wrote to my student. You're focusing on what's difficult and challenging, and on your religious objections (and you're entitled to them!). But don't lose sight of what is beautiful and amazing and wondrous about how science describes the universe. There's still plenty of room for awe in this story. (Neal Wolfe elaborates on this approach in chapter 22 of this book, "The Case for Awe.")

Big History does pose a challenge to traditional religions, because, like they do, it attempts to place humans in the universe. It explains what happened over the last 13.8 billion years to get us from that pinpoint of matter and energy to where we are right now, and how it happened.

But it can't tell us why.

ACTIVITIES AND EXERCISES FOR TEACHING THRESHOLD 1

CREATION MYTH CREATIVE WRITING WORKSHOP

Judith Halebsky

Threshold Related to the Activity: Threshold 1

Category of Activity: Learning activity, writing activity

Objective: For students to triangulate the Big History story with creation myths and their own beliefs. Students will become familiar with Big Bang cosmology and creation myths from multiple traditions. The activity asks students to reflect on the cultural, social, and religious influences that have shaped their own beliefs.

Overview of Activity: This activity requires one to three class meeting periods and includes a homework assignment. Ideally, two class periods cover the Big Bang and creation myths. A third class period allows students to share the creation myths they devised with their peers in a creative writing workshop environment.

Faculty Preparation for Activity: Teachers should prepare a lecture on the Big Bang and mythology that highlights areas of scientific knowledge and cultural belief. This lecture should cover the transmission of oral knowledge and draw attention to how knowledge (both scientific and cultural) has changed over time. The "Creative Writing and Big History" course also connects these knowledge areas with modern literary concepts of fiction and nonfiction. (This can be adapted for a general research skills activity to highlight academic and popular sources of information.)

Cost: Students are asked to bring one print copy of their creation myth for each student in the class (i.e., twenty copies for a class of twenty students). The instructor may choose to take on that labor and expense.

Student Preparation for Activity:

> In class: In the class session prior to beginning the activity, students reflect on their beliefs about the origins of the universe in their journals.

> Homework: Next, students read the chapters on Thresholds 1–3 in their Big History textbook and creation myths (as a handout excerpted from, for example, J. F. Bierlein's *Parallel Myths*). Students journal on how their beliefs connect with or differ from those presented in the assigned readings.

In-Class Sequence of Activity:
Class 1: Creation Myth

In class: Students break into small groups and choose a creation myth from the assigned readings. Make sure to have available a larger number of creation myths than the number of student groups. This will allow groups to choose which creation myth they will share with the class. Each group then narrates or performs their selected creation myth for the class. This allows students to engage with the myths and to learn a particular creation myth in detail.

Class 2: Writing a Creation Myth

In class or as homework: Students read Simon Rich's "Center of the Universe" for an example of how a creation myth might be used in creative writing.
Homework: Students write their own creation myths.

Class 3: Creation Myth Creative Writing Workshop

In class: Students bring typed copies of their creation myths to class for a writing workshop. Each student reads her or his creation myth aloud to the class and then the writing is discussed. The sharing of student work and the discussion on that work is an important aspect of student learning in this assignment.

Make sure to establish a supportive atmosphere for the sharing of student writing. The ability to embrace and constructively employ critical feedback is a skill that takes time to develop. At lower levels, it is most useful to focus brief discussions of student writing on strengths and questions. As a general ground rule, comments should begin with positive feedback. Have students sit in a circle and invite the two students sitting beside the author to begin the discussion. After each of those two students has spoken, invite other comments. Four comments is generally sufficient for a good discussion. The author should not respond to questions until all the comments have been made.

Assessment of Learning: This activity requires class discussion, reflective writing, and the students' written creation myths.
Origin of Activity: This is an original activity.
Sources Consulted:
Bierlein, J. F. *Parallel Myths*. New York: Ballantine Books, 1994. Print.
Rich, Simon. "Center of the Universe." *New Yorker* 9 Jan. 2012. Web. 12 May 2014.

Lynn Sondag

Threshold Related to the Activity: Threshold 1

Category of Activity: Studio art activity / project

Objective: Students illustrate the pivotal events of a creation myth, exploring how cultures narrate and visualize unknown events, using both imagination and observation. When each student in the class adopts a different creation myth, we are able to draw comparisons among them, noting their similarities and differences.

Overview of Activity: This project takes three to four class periods. The tunnel book comprises four main scenes of a creation myth in a three-dimensional book structure so they can be viewed simultaneously. The stories are edited down to pivotal events describing the beginning, middle, and end stages. The visual rendering of the story also incorporates the artistic style of the culture or region where the story originated. (Page 9 of this book features a fine example of finished student work.)

Faculty Preparation for Activity:

- Create a prototype of the book structure.
- Gather the following materials: black card stock, X-Acto knives and scissors, rulers, glue sticks and PVC glue, colored pencils and markers, and Sharpies.
- Prepare four 7.5″ × 9″ card stock pieces and six 3″ × 9″ card stock sides for each student.

Cost: About $20 for card stock, colored pencils, glue sticks, X-Acto knives, PVC glue, and rulers (but total cost depends on the number of students and their ability to share materials)

Student Preparation for Activity:

1. Students read a variety of creation myths and select one for their book.
2. Students edit the narrative down to four main scenes describing the beginning, middle, and end.
3. Students sketch out a visual description of these events using the dimensions of the book, cutting out windows in each of the first three panels so that each image can be seen from the front, even when the pages are stacked in front of one another (see figure 7.1).

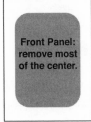

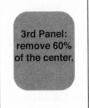

FIGURE 7.1
Cutting out the four panels for the tunnel book. (Illustration:
Lynn Sondag)

4. Students note the cultural or ethnic heritage of their chosen myth's origins
   and research the artistic "style" of this culture. Given that the culture will
   most likely have several varieties of artistic styles, students will need to restrict their search to a particular period or region within that cultural style.

In-Class Sequence of Activity:
Class 1 and 2:

1. Pass out four 7.5″ × 9″ card stock panels to each student and have them
   draw ¾″ borders around the perimeter of each panel.
2. Students draw each scene and transfer their drawings—either directly, by
   pasting, or by redrawing onto each panel. Using an X-Acto knife, students
   remove the excess negative space around their drawings on the first three
   panels. The last panel is the background and needs to be a full sheet.
   Students must make sure that they can view all four illustrations when the
   panels are stacked one in front of the other (again, see figure 7.1).

Class 3:

3. Figure 7.2 illustrates how to assemble the tunnel book. Pass out six 3″ ×
   9″ card stock sides to each student. Have each student fold each card stock
   side in half lengthwise to form a *V*, then fold each side of the *V* in half,
   away from the center, to meet the middle fold of the *V* along the outside,
   creating an *M*. These are the spacers that will be glued along the long sides
   of each panel to connect one to the other.

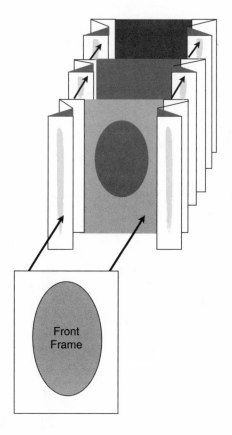

FIGURE 7.2
Assembling the tunnel book. (Illustration: Lynn Sondag)

4. Students can now assemble their books. Beginning with the front panel, have students glue one spacer on both the right and left side behind the panel. When this is dry, have them glue the next panel to the spacers on the front. Have students repeat this step to attach the third panel to the second, and the final panel to the third. The final panel does not have spacers glued on the back.

Assessment of Learning: Students share as a group and engage in a critique conversation. The artist of each piece describes his or her story, the four main scenes, and how she or he appropriated the cultural style. At the end of the presentations, the students can identify the main features and characteristics that are similar and different in each of the myths. Table 7.1 is a rubric for grading the tunnel book art project itself.[7]

TABLE 7.1 Assignment Rubric

Criteria	100–90%	89–80%	79–70%	69–60%
Use of disciplinary conventions: applying the formal visual elements of an artistic style	Composition demonstrates detailed attention and successful execution of a specific artistic style.	Demonstrates consistent and successful execution of a specific artistic style.	Meets adequate expectations for effectively applying and organizing a specific artistic style.	Attempts to use a system that effectively applies and organizes a specific artistic style.
Connecting, synthesizing, and transforming Big History content with the assigned art form	Makes choices that transform ideas or solutions that connect and synthesize content into an innovative and complex art form.	Connects ideas or solutions for merging form and content in novel ways.	Synthesizes ideas of form and content with a solution representing a coherent whole.	Recognizes existing connections among ideas or solutions for merging form and content.
Problem solving	Not only develops a logical, consistent plan to solve problems, but recognizes consequences of solution and can articulate reason for choosing solution.	Having selected from among alternatives, develops a logical, consistent plan to solve the problem.	Considers and rejects less acceptable approaches to problem solving.	Only a single approach is considered and used to solve the problem.

Origin of Activity: This is an original activity.

Source Consulted: Golden, Alisa. *Creating Handmade Books.* New York: Sterling Publishing, 1998. Print.

REDSHIFT DEMO

Robin Cunningham

Threshold Related to the Activity: Threshold 1

Category of Activity: Brief lecture that can include a spectroscope activity

Objective: To explain redshift and its role as primary evidence that the universe is expanding and has been for some time

Overview of Activity: Through this activity, students will come to understand redshift as the Doppler effect for light waves. Observing the universe with this knowledge enables us to conclude that the universe was formerly much smaller than it is today. Indeed, it appears, in the best judgment of physicists, that the universe had a beginning.

Faculty Preparation for Activity: Though not entirely necessary, it is advised that the instructor have a spectroscope available so that students can see the many colors that exist in "white" light.

Cost: About $10 for an inexpensive spectroscope for classroom use, or $55–$150 for a more sophisticated model.

Student Preparation for Activity: Students should read the sections of the textbook that refer to the Big Bang and early expansion of the universe.

In-Class Sequence of Activity: A summary of the lecture follows.

1. Why do scientists think the universe had a beginning? (This was not always the case; even Einstein got it wrong.) It turns out that one observation, along with a little logic, makes this a simple conclusion to reach.

2. Using NASCAR as a reference, note that cars always sound like they are moving faster as they approach than they do as they move away from the observer. Draw from the students (or remind or inform them) that this is due to the Doppler effect for sound.

3. Explain the Doppler effect for sound. Each sound wavelength produces a different pitch; thus, we hear sound wavelengths. Figure 7.3 provides a useful diagram to help students understand the Doppler effect for sound.

4. The key point is that, even with our eyes closed, we can often determine whether a car moving at constant velocity is approaching or leaving, based on the pitch (or wavelength) of the sound of its engine.

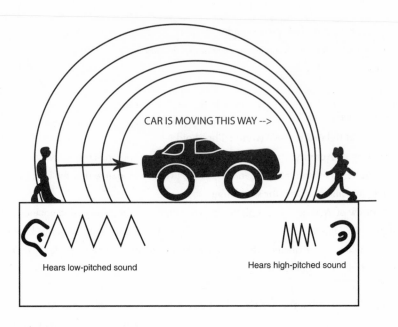

CAR IS MOVING THIS WAY -->

Hears low-pitched sound

Hears high-pitched sound

FIGURE 7.3
The Doppler effect for sound. (Illustration: Richard B. Simon)

5. Moving on to light, explain to students that, just as we hear the wavelengths of sound (as pitch), we see the wavelengths of light (as color). Longer wavelengths are toward the red end of the color spectrum, and shorter wavelengths are toward the blue end of the spectrum. This is a good moment to show students the spectrum contained in white light by having them look at a source of relatively white light using a spectroscope. (Note: Do not use spectroscopes to look at the sun.)

6. A star or galaxy typically emits a wide color spectrum of light. When we look at the vast majority of stars and galaxies, this spectrum is shifted toward the red direction. This is because the wavelengths of light are longer than they would be if the star or galaxy were in a fixed position relative to us. This is the Doppler effect for light waves. This observed **redshift** enables us to conclude that

   a. the vast majority of galaxies are moving away from ours; and
   b. the universe is expanding.

7. It is important to emphasize the following three points to students:

a. Bright students may wonder, "If the entire spectrum shifts in the red direction, how do we notice, since we still see all the colors?" This is an excellent question! Although stars emit a wide color spectrum, they emit less light at wavelengths that are absorbed by hydrogen and helium (which most stars contain in abundance). As a result, there are dim spots in the spectrum emitted by a star, and it is actually these dim spots—or **absorption lines**—that we see shift in the red direction.

b. Even galaxies very far from ours show redshift. Since light from these galaxies has taken billions of years to get to us, this shows that the universe has been expanding for a very long time. The universe was once much, much smaller than it is now. In fact, cosmologists believe that the universe started as an object so small that it at one time had no dimension whatsoever. That is, it proceeded forth from a singularity—*there was a beginning*. Around 13.8 billion years ago, the universe occupied (or consisted of) a very small space. (See the "Inflation Balloons" activity in chapter 8.)

c. We have described only one source of redshift for the purpose of explaining what it is. The most important redshift that we observe is actually due to the expansion of space between the star emitting the light and the observer of that light during the billions of years it takes the light to reach us. As the light travels through expanding space, its wavelength increases, along with the space itself.

8. The universe is currently expanding at roughly 7 percent per billion years.

9. **Questions to Ponder:**

a. It is a fact that the stars of the Andromeda galaxy (the nearest galaxy to the Milky Way) appear bluer than their chemical compositions would predict. What alarming conclusion should we draw? If students can answer this question, they really understand the day. (Answer: The Andromeda galaxy is moving toward the Milky Way and they will collide in roughly one billion years.)

b. If the day comes that we stop observing redshift, what will that suggest?

c. If twenty billion years from now, astronomers observe that most light from galaxies throughout the universe is shifted in the blue direction, what will be the logical conclusion?

Assessment of Learning: Class discussion
Origin of Activity: This is an original activity.

NOTES

1. David Christian, Cynthia Brown, and Craig Benjamin, *Big History: Between Nothing and Everything*. New York: McGraw-Hill, 2014. 14. Print.

2. Ibid, 19.

3. Ibid, 19.

4. Ibid, 21.

5. Ibid, 21.

6. Fred Spier, *Big History and the Future of Humanity*. Chichester, West Sussex, U.K.: Wiley-Blackwell, 2011. 50. Print.

7. Terrel L. Rhodes. *Assessing Outcomes and Improving Achievement: Tips and Tools for Using Rubrics*. Washington: Association of American Colleges and Universities, 2010.

EIGHT · # Teaching Threshold 2

The Formation of Stars and Galaxies

Kiowa Bower

> Doubt thou the stars are fire,
> Doubt that the sun doth move,
> Doubt truth to be a liar,
> But never doubt I love.
> O dear Ophelia, I am ill at these numbers. I have not
> art to reckon my groans; but that I love thee best, O
> most best, believe it.
>
> WILLIAM SHAKESPEARE, *Hamlet*, Act II, Scene ii

Toward the beginning of our Big History survey course, I invite my students to sit in a circle outside on the grass and share whatever they would like to about the class. In one class, a student expressed that the early material scared her, forcing her to think about things outside her comfort zone. The immense scale of the universe was overwhelming, and she was completely lost trying to grasp the early universe described by the Big Bang theory. She felt lost in the particle soup and the unintuitive concepts of the early cosmos. While there was still some anxiety associated with Threshold 2, especially because there was this "fusion bomb" of a star nearby, she was beginning to enjoy the class more. "It's nice," she said, "to be learning about things that I've at least seen a picture of before."

Threshold 2 can be an opportunity to reconnect with students who feel lost or overwhelmed by Threshold 1. Some are lost because they have trouble with the abstract physics, while for others the more prominent issue is how the Big Bang conflicts with their own origin beliefs. Regardless, the tangible nature of what we study in Threshold 2 and the intuitive nature of gravity-driven processes is an opportunity to engage with those students not yet fully on board for the journey.

KEY CONCEPTS IN THRESHOLD 2

GRAVITY

Gravity is key in the evolution of complexity because it facilitated the first structures in the universe and was essential in creating energy gradients. About 380,000 years after the Big Bang, the universe was an expanding cloud of matter with only tiny variations in density.[1] In areas of higher density the gravity was slightly more powerful and the inward forces pulled these regions together, leaving other regions more empty. As density increased further, so did the force of gravity, and the process accelerated. After a few hundred million years, matter had clumped into protogalaxies, and within these formed the first stars. Gravity had transformed the universe into a much more interesting place.

With so much discussion of the collapsing effects of gravity in Threshold 2, it is important to remind students that the universe is still expanding on the largest scales. Gravity can hold galaxies together, but galaxy superclusters are all moving away from one another; this is the expansion that was observed by Edwin Hubble.

DIFFERENCES IN DENSITY

Density differences among different regions of the early universe were critical in facilitating the further evolution of structure and complexity. The cosmic microwave background radiation (CMB or CMBR, sometimes referred to as the CBR, or cosmic background radiation) data is a key piece of evidence in our understanding of how the universe developed its first structure. The heterogeneity of the CMB is thought to have arisen from quantum fluctuations in the very early universe.[2]

In our unit on Threshold 1, we learn that the CMB is some of the strongest evidence for the Big Bang. The theory predicts the release of this radiation during the recombination epoch, when the universe first became transparent to light and photons were set free in every direction.[3] At this time the universe had cooled to about 3,000 kelvins (K), but since then, space has expanded a thousandfold and the photons within it have stretched out, along with the space through which they are traveling. As a result, the photons' wavelength has expanded, and they have cooled to about 3 K.[4] Though the radiation is now much cooler, we can still detect the relative temperature differences in the early universe in the CMB data. Thus, the CMB data is essentially an image of the young universe.

You can show students the WMAP image in figure 8.1 and explain how the small differences in temperature correspond to differences in density within the early universe. (NASA's "Universe Evolution" animation[5] effectively illustrates the connection between WMAP data and early matter clustering.) It was these differences in the distribution of matter that allowed gravity to pull more strongly in some areas and dictated the evolution of the large-scale structure of the universe (see figures 8.2 and 8.3). This filamentous structure has been explored by studies such as the Sloan Digital Sky Survey, which mapped almost one million galaxies in three dimensions.

FIGURE 8.1
The cosmic background radiation, or CBR. Nine-year WMAP
image of background cosmic radiation. (Image: National
Aeronautic and Space Administration. Goddard Space Flight
Center. *Wilkinson Microwave Anisotropy Probe.* "Nine Year
Microwave Sky." NASA / WMAP Science Team. NASA, 14 April
2014. Web. 6 June 2014.)

GALAXY FORMATION

So, we have already observed that small differences in density existed at the time of
the CMB. As the universe cooled further, these slightly denser areas developed into
clumps of dark matter, which in turn pulled together huge clouds of gas. These
clouds collapsed to become small protogalaxies, and within these, the further collapse of even smaller gas clouds led to the formation of the first stars. Evidence for
small early galaxies of about one million stars includes recent observations by the
Atacama Large Millimeter / submillimeter Array (ALMA) and by the Hubble Space
Telescope. Over a few billion years, the early galaxies eventually joined together to
become the much larger spiral and elliptical galaxies of one hundred billion stars
that we see today. One strong piece of evidence for why we believe large galaxies
took this long to form is because we do not observe any such structures when we
look out at galaxies that existed during the first few billion years of the universe.

STAR FORMATION

The first light of our universe was that from the Big Bang. It was a powerful ema-
nation of energy, but its light is destined to slowly fade as the universe continues to
expand and cool. Aside from the dim glow of the CMB, the universe was dark for

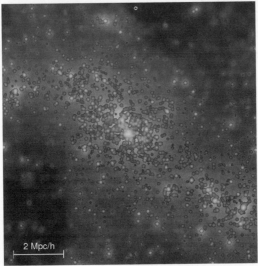

2 Mpc/h

FIGURE 8.2 AND FIGURE 8.3

Suggested media for students: the large-scale structure of the
universe, from the Millennium Simulation Project. Figure 8.2 is
about one billion light years across. Figure 8.3 zooms in on a galaxy
cluster. Each spot represents a galaxy. (From Volker Springel et al.
"Simulations of the Formation, Evolution and Clustering of
Galaxies and Quasars." *Nature* 435.7042 [2005]: 629–636. Images
courtesy of Volker Springel and the Virgo Consortium.)

about two hundred million years after the Big Bang.[6] It was only after those two hundred million years that the first stars lit up the cold emptiness of the cosmos.

Stars are formed by clouds of gas that are sufficiently dense to undergo gravitational collapse. As matter falls inward, the atoms smash together more and more violently and the gas begins to heat up (gravitational potential energy is converted into thermal energy). At temperatures between 2,000 and 10,000 K, the protons of hydrogen are stripped of their electrons and the gas becomes a plasma. (It is worth reminding students that this is the same state of matter that existed in the universe *before* the release of the cosmic background radiation.) Collapse continues, and as density in the core increases, so does the temperature of the plasma, until it reaches ten million degrees. At this temperature the protons are moving so fast that they can overcome the repulsive force of their positive charges when they collide, and they begin to fuse together.[7]

During the fusion reaction a small amount of mass is converted into a relatively huge quantity of energy (remind students of Albert Einstein's famous equation, $E = mc^2$). This fusion of hydrogen into helium is the same process that occurs during the detonation of a hydrogen bomb. Mentioning this to students can help them conceptualize the energy available from fusion. The fusion reactions in the core of the newly forming star begin to exert an outward pressure that balances the inward pressure from gravity. The opposing forces reach equilibrium in the star, and it forms a stable sphere of glowing plasma. This object is now stable and will radiate enormous quantities of energy into the cold surrounding space for millions, or even billions, of years.

We can observe star formation still occurring in stellar nurseries. Molecular clouds such as the Orion nebula can be up to one hundred light years wide. Look online for an image of this colorful celestial object.

THE FOUR FEATURES OF COMPLEXITY IN THRESHOLD 2

DIVERSE COMPONENTS

Hydrogen, helium, protons, neutrons, and electrons.

SPECIFIC ARRANGEMENTS

Early galaxies are small and globular but collect and evolve into sophisticated elliptical and spiral structures. Characteristics include a dense and active core region with a supermassive black hole at the center, and "Goldilocks regions" toward the outer edges, with dark matter throughout and probably also surrounding the galaxy.

When we view the universe at the large scale we can see that it is composed of filaments and sheets of galaxies separated by large voids in between.

Stars have a spherical, layered structure with a fusion core, conductive zone, radiative zone, chromosphere, and photosphere.

FLOWS OF ENERGY

gravitational potential energy → heat energy to start fusion
hydrogen + hydrogen → helium + energy

(Note: It takes the input of four protons to make helium. The true reaction scheme is complex, but it is not quite correct to say that two hydrogen atoms make one helium atom.)

energy of star's mass → photons into space

EMERGENT PROPERTIES

1. Energy gradients, essential for the further development of complexity
2. Matter under conditions of extreme heat and pressure
3. Many phenomena that are new in the universe and associated with stars: localized sources of energy; high intensity radiation of infrared, visible, and ultraviolet light; macroscopic gravitational and magnetic fields; convection currents within stars; and so on

HOW EMERGENT PROPERTIES GIVE RISE TO THRESHOLD 3

With the formation of stars in the universe emerged conditions of continuous extreme heat and pressure ideal for the creation of greater chemical complexity. While these extreme conditions were present in Threshold 1, they existed only for a few minutes, which was not enough time to begin fusion of heavier elements.[8] The conditions in the center of massive stars, however, were stable for a period on the order of millions of years. In addition, the intense gravitational pressures kept the star structure intact until the fusion potential of the star had run out.

The physics governing the deaths of large stars makes them an ideal implement not only for making elements but also for spreading newly minted elements into the universe. Rather than fizzling out and collapsing into themselves, large stars end their lives in spectacular supernova explosions. This not only creates all the rest of the elements that could not be fused during the star's life but facilitates the dispersal of all these elements into the galaxy so that they have the opportunity to form second-generation stars.

An essential aspect of this lesson is the relevance of stars to the generation of chemical elements. However, students should also understand the significance of the immense energy sources now scattered throughout the universe. This fact is of central importance for the development of further levels of complexity. Specifically, the Goldilocks conditions near these hot spots supported the evolution of life and human culture on Earth.

STUDENT LEARNING OUTCOMES AND POSSIBLE ASSESSMENTS

By the end of this unit, students will demonstrate the ability to

1. describe how tiny variations in density evolved into composite structures on multiple scales in the universe (assessment: quiz or written response);

2. explain the physical principles of star formation and fusion (assessment: quiz or short essay);

3. discuss the importance of hydrogen and helium as the building blocks of all matter in the universe (assessment: quiz or written response); and

4. summarize and explain cosmological scales of time and space (assessment: in-class response or summary).

CHALLENGES IN TEACHING THRESHOLD 2

One challenge in teaching Threshold 2 is communicating the enormous scales of the universe. It is often difficult for students to fully understand how large the solar system is, let alone how far away the next galaxy supercluster is. Visuals and animations are very helpful in illustrating some of these concepts.

Graphics are particularly important for illustrating the relationship between the CBR and the distribution of matter in the universe. You can point out that a location with more matter in a particular region of the early universe would produce more photons at the time of the CBR release.

Another challenge, especially for nonscientists, is grasping the physics in this lesson. The conversion of matter into energy is not intuitive. Try to get students to see the link with the earliest moments of the Big Bang, when matter and energy were one and the same. Other parallels can be found between Thresholds 1 and 2, because matter in Threshold 2 is returning to hotter, more condensed states, similar to that which existed in the early universe. For example, matter returns to plasma as it heats during star formation.

Students sometimes have trouble understanding how scientists could know so much, especially about things that are so far away. Some students feel that what they are being taught is just something that scientists are guessing about. This can be addressed by describing the evidence for the presented information whenever possible. There is so much material in this course, however, that one must be selective about what to bring into the discussion. The challenge is often how to make convincing arguments but still keep the class engaged and not lose students with abstract scientific theories.

CONCLUSION

I particularly enjoy the opportunity to instill a sense of awe when teaching the first two thresholds of Big History. Because we teach the class to first-semester freshmen, we end up exposing students to ideas well outside anything most of them have ever contemplated.

Students frequently comment that learning about the cosmos has drastically shifted their perspective. Many students are enthusiastic about this experience, but as happened with the student described in the beginning of this chapter, such shifts in perspective are not always welcomed.

The feelings that students have toward scientific cosmology vary widely, but a strong reaction is usually a good place to start, even when that reaction amounts to cognitive discomfort. An emotional response, even a negative one, means that the student is engaged. The trick is to make sure that the student stays engaged and to direct that engagement into a positive learning process in which the student is able to think critically about the material.

ACTIVITIES AND EXERCISES FOR TEACHING THRESHOLD 2

INFLATION BALLOONS

Robin Cunningham

Threshold Related to the Activity: Threshold 2

Category of Activity: Learning activity

Objective: To give students a model for an expanding universe and its impact on the placement of galaxies, on redshift, and on the CBR that we observe today

Overview of Activity: Students use balloon models of an expanding two-dimensional universe to discuss redshift, CBR, and what an expanding universe looks like to those within it.

Faculty Preparation for Activity: You will need quite a few balloons, preferably soft blue or another light color, as well as colorful Sharpie markers. Clever students might add just a bit of red ink to each galaxy as a metaphor for redshift.

You should be familiar with redshift and the origin of cosmic background radiation, when the universe cooled enough to form atoms and release the light now known as the CBR.

Cost: About $20.00 for balloons and Sharpies

Student Preparation for Activity: Have students read the sections of the textbook that refer to the Big Bang and early expansion of the universe.

In-Class Sequence of Activity:

1. Divide students into as many small groups as the class will accommodate. Give the groups a linear order (1, 2, 3, . . ., *n*).

2. Give each group a balloon and colored Sharpies that they can use to create a complete universe. Ask the first group to create a model of the universe in its earliest (smallest) stage. They may choose to show the galaxies as distinct objects or as part of a lumpy soup. Each successive group should produce a larger universe than the group before, complete with six galaxies. The last group will have the largest universe.

3. Place the balloons in linear order somewhere in the room and then review several key points:

   a. Each universe is two-dimensional—that is, it exists only in the material of the balloon itself, NOT in the interior space inside the balloon. All lines of sight are within the balloon. Each galaxy can see five other galaxies.

   b. As the models of the universe expand, it is space itself that is expanding; the traveling of the galaxies is part of that expansion. Our history is not

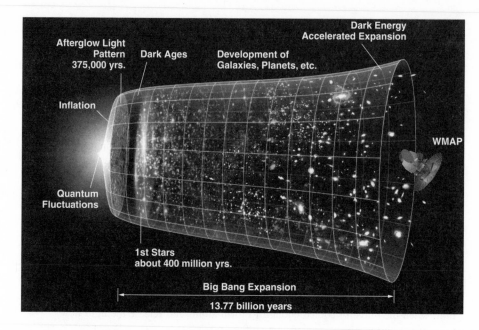

FIGURE 8.4

NASA's *Time Line of the Universe*. (Image: National Aeronautic
and Space Administration. Goddard Space Flight Center.
Wilkinson Microwave Anisotropy Probe. "Timeline of the
Universe." NASA/WMAP Science Team. NASA, 21 December
2012. Web. 6 June 2014.)

    one of galaxies pushing out into existing space but of space expanding
and taking galaxies with it.

c. Although the space between galaxies is expanding, the galaxies themselves should not be expanding. Why do the galaxies not expand if space
is expanding? (Answer: gravity.) It is also gravity that formed the galaxies in the first place.

d. Note that everyone sees redshift in the other galaxies.

e. If the universe grows a certain percent per billion years, note that the
distances between faraway galaxies are increasing at a faster rate (not a
higher percent) per year. Thus, where should redshift be the greatest?

f. Around the time of a small universe, CBR was released, and we know
what wavelength it had based on the process that formed it. This is
now called **the Big Flash,** but it is the event that gave the Big Bang its
name long ago. The light from this event has been traveling around the

universe since then, but its wavelength has increased exactly as much as space has expanded. Thus, by looking at the wavelength of CBR today and comparing it to the wavelength when it was formed, we can tell how much the universe has expanded since the CBR was formed.

g. We see CBR pretty evenly in every direction. This tells us that it occurred everywhere in the universe at roughly the same time.

h. Using the balloons, students have to make a decision about the size of galaxies relative to the space between them. They may be interested in the true relationship: if galaxies are the size of dimes, the space between them should, on average, be about a meter.

Figure 8.4 shows another great timeline of the universe from NASA. The time labeled "Afterglow Light Pattern" represents the time of the Big Flash that is the source of CBR today. The artist made a little joke by using the WMAP image of how the CBR looks today to depict the Big Flash. The Dark Ages are between the Big Flash and the formation of the first stars four hundred million years later.

Assessment of Learning: Class discussion, paper describing how the universe has developed from the primordial soup and what evidence we have for its current and past growth

Origin of Activity: This is an original activity.

NOTES

1. David Christian, Cynthia Brown, and Craig Benjamin, *Big History: Between Nothing and Everything*. New York: McGraw-Hill, 2014. 19, 23. Print.

2. Sang Pyo Kim, "Quantum Fluctuations in the Inflationary Universe." *Modern Physics Letters A*, 22.25–28 (2007): 1921–1928. Print.

3. Christian, Brown, and Benjamin, *Big History*, 19–20.

4. D. J. Fixen, "The Temperature of the Cosmic Microwave Background." *The Astrophysical Journal* 707.916 (2009): 916. Print.

5. National Aeronautic and Space Administration. Goddard Space Flight Center. *Wilkinson Microwave Anisotropy Probe*. "Universe Evolution." NASA / WMAP Science Team. NASA, 31 August 2012. Web. 6 June 2014. Animation.

6. Christian, Brown, and Benjamin, *Big History*, 24.

7. Ibid., 23.

8. Adam Frank, "How the Big Bang Forged the First Elements." *Astronomy* 35.10 (2007): 32–37. Print.

NINE · Teaching Threshold 3

Heavier Chemical Elements and the Life Cycle of Stars

Richard B. Simon

Dark star crashes
pouring its light into ashes
Reason tatters
the forces tear loose from the axis
Searchlight casting
for faults in the clouds of delusion
Shall we go, you and I while we can?
Through the transitive nightfall of diamonds

ROBERT HUNTER, "Dark Star"

The culminating project in our first-semester course is a Little Big History. Each student picks a subject to examine across Big History using a simplified version of the form pioneered by Esther Quaedackers. The subject can be an object, a person, or even a concept. Each student writes a thesis-driven final paper that traces her or his subject through each threshold, either chronologically or counter-chronologically (see the "Little Big History Essay" activity in chapter 17, "Activities for Multiple Thresholds," and chapter 19, Cynthia Brown's "A Little Big History of Big History.") It's a causal argument, really. This is a challenging assignment, for students and faculty alike, in that there are many ways one might trace the history of a given subject across the cosmic river of time. One of my students was stumped. She couldn't come up with a subject she wanted to investigate. I talked her through some possibilities—including a soccer ball (she was a student athlete) or the game of soccer itself. She looked down at her own hands and the light came on. "How about jewelry?" she said.

David Christian reminds readers, in *Maps of Time*, that "the gold or silver ring you may be wearing was made in a supernova."[1] (One reminder that the details of this story are ever changing: a July 2013 study by the Harvard-Smithsonian Center for Astrophysics suggests that the universe's gold may have been formed by colliding neutron stars, which themselves are the remnants of supernovae.[2])

The student was wearing silver rings and bracelets, laced with turquoise, worked by Native American silversmiths in a tradition that would have dated back thousands of years and required the use of fire, harnessed by Paleolithic humans. Then there was the concept of adornment, of aesthetic. When and how did humans begin to wear jewelry? And why? What evolutionary advantage might such adornment confer? Why would that have developed in *Homo sapiens*? And then, how did the rock form? The silver?

It was a perfect topic: simple to trace through the ages—and particularly illuminative of what may be the most challenging threshold, Threshold 3, the formation of heavier chemical elements and the life cycle of stars. Because it is in the stars that nearly all the elements we know are forged.

This type of direct, palpable connection—between something a student can hold in her hand and the events that unfold inside distant stars we'll never visit, which look to us like mere lights in the vault—is a useful and unforgettable way to engage students in the material.

KEY CONCEPTS IN THRESHOLD 3

STELLAR NUCLEOSYNTHESIS

By now, we have read how, in the early universe, pockets of higher mass (and therefore gravity) within vast clouds of hydrogen and helium pulled more and more matter and energy to themselves until, under high pressure and therefore temperature, they ignited the fires of the universe—becoming galaxies of stars.

It's important in teaching Threshold 3 to refresh students' understanding of basic chemistry. An atom is the smallest possible unit of an element, a single particle that contains, at the least, a nucleus of one positively charged **proton**, orbited by a negatively charged **electron.** This would be an atom of hydrogen—atomic number one on the periodic table of elements, for its one proton. The nuclei of some isotopes (atoms of a particular element with slightly different numbers of neutrons from the most common atom) of hydrogen contain a **neutron**, as well, which has no charge.

We know that in our star, the sun, hydrogen is being burned through **fusion** into helium. Helium has *two* protons, *two* neutrons, and *two* electrons—and thus has an atomic number of two and is number two on the periodic table, for its two protons.

Our sun, a large yellow star, is about halfway through its life. It is a second- or third-generation star,[3] meaning that it is not one of the first stars to have lit up in the early universe, but was rather a star born of debris, in a collapsing field of the remnants of long-dead stars, brought together through gravity and maybe jostled by another nearby supernova, and lit up (reaching a critical mass of hydrogen and helium at high pressure and, therefore, temperature) much more recently—some 4.6 billion years ago.

A star is essentially a mass of burning plasma caught in a tug-of-war between gravity and the ongoing explosion of its core. As we learned in our unit on Threshold 2, when a molecular cloud collapses into its own gravity, the atoms of hydrogen are slammed together with such force that a nuclear fusion explosion is triggered. Under the pressure of all that violent force, the collapsing material is hot enough (ten million degrees Celsius) to counteract the repulsive electromagnetism between protons in the nuclei of, say, two hydrogen atoms. This allows them to fuse together into helium. This reaction releases **photons,** free packets of pure energy, which travel slowly to the star's surface and are then freed into space, as light.

The ongoing fusion explosion inside a star should be enough to blow the star apart—but for gravity, which is so intense that its pull counteracts the force of the fusion explosion perfectly. So, a star is spherical—and of a consistent size and shape—because the enormous force of its core's explosion is in perfect equilibrium with the enormous force of gravity pulling all its matter in toward its core.

But when the star exhausts enough of its supply of hydrogen that fusion can no longer stand up against gravity—and its core is thus full of helium—the star collapses. When this happens, of course, such high pressure and temperature are generated in the core that the mass of helium begins to fuse into carbon—and the cycle begins anew (even as what's left of the hydrogen can continue to fuse into helium in a new layer just outside the core). This process, the manufacture of new elements inside stars, is called **stellar nucleosynthesis.**

This cycle continues, on up the periodic table, fusing bigger and bigger atoms together until, in the largest stars, the core fills with newly minted iron. In such stars, several fusion reactions can continue in concentric shells, yielding an onion-like structure through which photons must travel to get to the surface and escape to space.[4]

$$^{12}C + {}^{4}He \rightarrow {}^{16}O$$
$$^{12}C + {}^{16}O \rightarrow {}^{28}Si$$
$$^{16}O + {}^{16}O \rightarrow {}^{28}Si + {}^{4}He$$
$$^{28}Si + {}^{28}Si \rightarrow {}^{56}Fe[5]$$

As Fred Spier explains, the largest stars can create even larger elements—including copper, zinc, silver, and gold—through neutron capture, the capture of new neutrons by existing nuclei.[6] But when the star's fuel runs out, it collapses and explodes one last time, as a **supernova**. If the star is eight to thirty times the size of our sun, it may become a **neutron star,** an ultra-dense clump of neutrons left when protons and electrons fuse together, spinning unfathomably rapidly. If the initial star is larger than thirty suns, it might collapse into a **black hole,** the most gravitationally dense object in the known universe.[7] Scientists believe that many galaxies, including our own Milky Way, have enormous black holes at their cores. If the star is larger than sixty suns, it may break apart, perhaps into two stars.

Most importantly for our story, the supernova is so hot that it creates not only neutrons, but all the rest of the naturally occurring elements up to uranium (with ninety-two protons). And in the force of the explosion, all these new elements are blasted off into space, someday perhaps to accrete into new stars, or planets of rock or gas or ice, or all of the above.

The Evidence   How do we know all this? To students, as to many of us faculty who are not (yet) astronomers, much of this may seem like a bit of a leap of faith. That is, it does until we learn about the evidence. So, as Christian, Brown, and Benjamin emphasize, it's key to teach some of that, as well.

To understand how we think we know so much about stars, students should know how we know a star's **chemical composition,** its **distance** from the Earth, its **luminosity,** its **mass,** and its **size:**

1. **Spectrography.** We can identify a star's chemical composition through **spectrography.** A spectroscope is an instrument, invented by Joseph Fraunhofer in 1814, that breaks a star's light into a distinct rainbow pattern—like the triangular prisms students will likely have used to break the light of our own star. When we use a spectroscope to look at the light from a star, we notice a few things. One is that a few black bands or shadows appear throughout the prism. These correspond to particular chemical elements that are present in the star, and that absorb light at particular wavelengths. So, if a black band—an **absorption line**—appears at the wavelength at which carbon absorbs light, then we know that carbon is present in the star.

   As we learned in our unit on Threshold 1, if a familiar pattern of absorption lines appears to be shifted to the red end of the spectrum, we know the star is moving away from us (and we can determine how far away it is). Likewise, if the pattern of absorption lines is shifted to the blue, we know it's moving toward us. This is called redshift or blueshift.

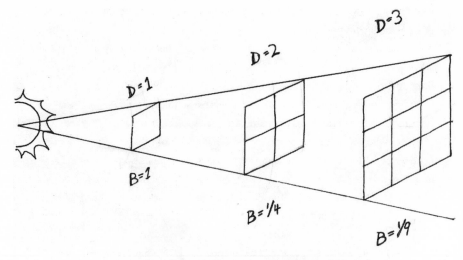

THE INVERSE-SQUARE LAW

FIGURE 9.1
The inverse square law: twice the distance, one-fourth the
luminosity; three times the distance, one-ninth the luminosity.
(Illustration: Richard B. Simon)

2. **Parallax.** We calculate the distance to a star using trigonometry. If we have
   a triangle, and if we know the size of two of the angles and the length
   of the side that connects them, then we can easily calculate the length of
   the other two sides (and the third angle). So if we stand at two observatories that are a thousand miles apart and calculate the angle between an
   imagined straight line that runs from observatory to observatory and a
   line that connects each observatory to the star, we have the dimensions of
   the entire triangle—and the distance to the star. (This difference between
   what the two observatories see as the star's position is called **parallax**.) In
   the same way, we can measure the distance to faraway stars by taking our
   measurements at opposite ends of the Earth's orbit of the sun (Earth orbit
   parallax), which gives us the distance measure **parsec**, from *par*allax and
   arc-*sec*ond, a unit of length of the Earth's orbit.[8]

3. **Luminosity.** If we measure a star's **apparent brightness**, or **luminosity**, and
   we know how far away that star is, we can use the inverse-square law (see
   figure 9.1) to calculate its **actual luminosity**, or the amount of energy it is

emitting. The light from a star weakens with distance in a way that is in proportion to the distance squared (multiplied by itself)—so that if a star is (to oversimplify) two light years away, it is one-fourth as luminous as if it were right in front of us, and if it were three light years away, its light would be one-ninth as strong. So if we consider the apparent luminosity in accordance with the distance the star's light has traveled to reach us, we can calculate its actual luminosity.

From a star's actual luminosity, we can gauge its mass—because stars with greater mass are hotter, and appear brighter.[9] And mass dictates size.

We can also gauge a star's **temperature** by the wavelengths of the light it emits—by its color (redder stars are cooler, while bluer stars are hotter). And we can determine the amount of energy a star is emitting based on its temperature.[10]

4. **The Hertzsprung-Russell diagram.** Such relationships among chemical composition, distance, luminosity, and mass are charted on a **Hertzsprung-Russell diagram,** named for the two scientists who, in 1910, plotted luminosity against temperature and discovered that most stars seem to fall on a clear curve, which they called the "main sequence," from high luminosity and high temperature (big and blue, in the top left of the graph shown in figure 9.2) to low luminosity and low temperature (small and red, in the bottom right). That meant that most stars were quite similar—they were still burning hydrogen—and that the relationship between luminosity and temperature could also indicate a star's size and mass. Astronomers have read this to mean that main sequence stars are all at a similar time in their life cycles but are of different sizes and therefore have different masses, temperatures, and luminosities—and different life spans.

Because stars of different masses burn up at different rates (more massive stars burn their fuel more quickly), when we compare various stars within the same star clusters—all of which must be roughly the same age—and gauge where individual stars are in their life cycles, we can get a sense of the age of the oldest star—and therefore of the cluster. As Peter B. Stetson, senior research officer at the Dominion Astrophysical Observatory in Victoria, British Columbia, explains in *Scientific American:*

The duration of the stable, or "main sequence," phase depends on a star's mass. A star 10 times as massive as the sun contains, clearly, 10 times as much

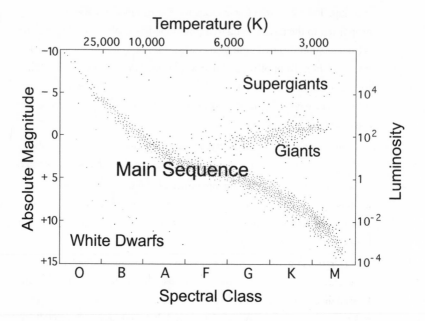

FIGURE 9.2
The Hertzsprung-Russell diagram. (Image adapted from
National Aeronautic and Space Administration. Goddard Space
Flight Center. "Imagine the Universe!: Stars." NASA, December
2010. Web. 2 February 2014.)

fuel. It consumes that fuel roughly *10,000* times faster than the sun, however.
As a result, it has a total lifetime 1,000 times shorter than that of our sun. . . .

All we have to do is look at the cluster and determine how hot and how
massive is the hottest, bluest, most massive star that has not yet entered the late,
unstable period of its life. The star's mass tells us how much fuel the star had
when it was born, and the star's brightness tells us how fast it is burning that
fuel. . . . The ratio of how much fuel the star had in the beginning to how fast
it has been burning that fuel tells us how long the star has been alive.[11]

Since 1910, astronomers have gathered this data on millions of stars—at all different stages in their development. Some stars are weirdos—big, cool red giants
and small, hot white dwarfs. These behave differently from main sequence stars—
because they are dying. Red giants are running out of fuel, and gravity from their
dwindling masses doesn't hold as well against their inner fires. They don't have what

it takes (enough mass) to keep burning up the periodic table, and their outer envelopes expand as their fusion fires cool. White dwarfs are the remnants of supernovae, which have blown off their outer layers and remain hot cores under great gravitational pressure, slowly cooling.[12] They, too, have moved off the main sequence.

And this is the fate of our sun. It will burn hydrogen into helium, then helium into carbon, and when it runs out of carbon, it will become a red giant, engulfing Mercury, Venus, Earth, and Mars, then blow off its outer layers in one final collapse and end its life as a white dwarf—cooling until it becomes a black dwarf. A cinder.

THE FOUR FEATURES OF COMPLEXITY IN THRESHOLD 3

We do want to connect what we're learning about how the life cycle of stars creates chemical complexity to our larger story about increasing complexity as we move from the Big Bang toward us. And the complex life cycle of stars yields some important building blocks for use across the next threshold, in which our solar system accretes.

DIVERSE COMPONENTS

Protons, neutrons, electrons. Atoms of hydrogen, helium, carbon, and every naturally occurring element, on up to uranium. Photons. Neutrinos, too.

SPECIFIC ARRANGEMENTS

A spherical ball of plasma, containing a core in which the product of fusion is accumulating; one or more shell layers in which fusion reactions are taking place; and a cooling, convectant outer envelope, which emits the energy loosed in those reactions into space. Size is ordained by the amount of matter that comprises the star (its mass), shape by the tension between the energy emitted by the fusion explosions within and the force of gravity, which is dependent on the star's mass.

FLOWS OF ENERGY

hydrogen + hydrogen → helium + energy (photons emitted to space)

or

fusion reactant + fusion reactant → fusion product + energy (photons emitted to space)

photon from core → outer layers → space

fusion →←− gravity

EMERGENT PROPERTIES

New chemical elements that can be gathered up by younger stars and that may accrete into planets (solid or gaseous), moons, and other bodies. Radiant energy as freed photons that can create the conditions under which new forms of complexity (such as life) may emerge.

HOW THESE EMERGENT PROPERTIES GIVE RISE TO THRESHOLD 4

Our sun pulled itself together under its own gravity, from leftover hydrogen and helium, about 4.6 billion years ago—along with everything else in our solar system. As inside a more massive star, the denser, more solid elements accumulated nearer the star itself, while the gaseous elements gathered farther away on the sun's **accretion disc,** which itself is the result of the spinning and flattening out of the field of matter that became the sun. This gravity- and mass-determined distribution of different types of matter, under the influence of centrifugal and centripetal forces, led to the clumping together of loose matter—debris from other stars' supernovae—into rocky inner planets and gaseous outer planets. As we will see later, that had a few important effects. Our own planet is a rocky planet just far enough from the sun that radiation is low enough to be shielded by Earth's atmosphere and rendered less harmful to fragile forms of new complexity—life. Yet Earth is still close enough to the sun that all those photons can warm and cool the atmosphere and, eventually, oceans—and can be turned by green plants, through photosynthesis, into usable energy. Not to mention that *everything* on Earth is composed of elements forged in the hearts of stars. Including us.

STUDENT LEARNING OUTCOMES AND POSSIBLE ASSESSMENTS

By the end of this unit, students will demonstrate the ability to

1. define what a star is and describe the forces at work inside it (assessment: short essay quiz or test);

2. explain basic chemistry and how it fits into the broader story of Big History (assessment: quiz or written response);

3. explain the basic processes of stellar nucleosynthesis (assessment: question-prompted in-class response or summary); and

4. report and describe the key evidence we have of these processes (assessment: short essay quiz or test).

CHALLENGES IN TEACHING
THRESHOLD 3

The first big challenge in teaching Threshold 3, especially for interdisciplinary instructors, is becoming familiar enough with this complicated material to relate the concepts to students without overwhelming them. Boiling down the processes at play to essential concepts—using metaphors and comparisons where appropriate—will make them easier for students to comprehend. Using visuals is also important in this unit—especially of spectra with absorption lines, the Hertzsprung-Russell diagram, the structure of a star, and the inverse-square law. Students may have widely varying experience with chemistry—for some, it may have been more recent (those students may roll their eyes at the review of the basics), while others may not have taken a chemistry class at all. Likely, few of them will have studied astronomy.

CONCLUSION

The friendlier you can make this material, the more likely students, especially non-scientists, will be to retain it. Once a student understands that—and how—the silver on her finger (not to mention the calcium in her bones, and the iron in her blood) was forged in a star, it's hard to forget. What's important in teaching this heady stuff is to contextualize it in a way that is meaningful for students. Some activities follow, which might help students internalize some of the basic concepts experientially.

Of course, when we're talking about stars, there's nothing like a night out under them, perhaps with a telescope. While writing this chapter, I took a walk around the small lake in the park by our home and looked up at Orion in the winter sky. There in the hunter's shoulder was a star whose redness I hadn't really noticed before—Betelgeuse. It's a red supergiant, off the main sequence, big and cool, running out of fuel, and fixin' to die, supernova and the resultant finger hardware to follow. By the end of this unit, students may well look up at those glowing pinpricks and see them for what they truly are—not dead, inert lights, but vast, dynamic systems, churning out new forms of complexity: the building blocks of planets, life, and us.

ACTIVITIES AND EXERCISES FOR
TEACHING THRESHOLD 3

GRAVITY VS. FUSION

Richard B. Simon

Threshold Related to the Activity: Threshold 3

Category of Activity: Learning activity

Objective: To illustrate the tension between gravity and fusion inside a star

Overview of Activity: Each student is given a large balloon, ideally with a tube mouthpiece that allows the student to inflate the balloon but also to use lung power (rather than pinched fingers) to keep the balloon from deflating. The idea is to illustrate that the fusion reaction inside a star (represented by the flow of air from the student's lungs) is what keeps the star inflated, but if that fusion reaction stops, the force of gravity (represented by the elasticity of the balloon, atmospheric pressure, and the re-inflation of the student's lungs) will cause the star to collapse.

Faculty Preparation for Activity: You'll need a large balloon for each student, and one for yourself to demonstrate. Of course, it's never a bad idea to have a few extra balloons on hand. You'll also need some short plastic tubes (½" or so in diameter) that allow students to stretch their balloons across the end to be inflated.

Cost: About $10–20 for balloons

Student Preparation for Activity: Students should have at least read about Threshold 3, so that they are familiar with what a star is, with stellar nucleosynthesis, and with the processes that are occurring inside stars.

In-Class Sequence of Activity:

1. Give each student a balloon and a mouthpiece.

2. Have each student attach her or his balloon to her or his mouthpiece.

3. Instruct them: Put the mouthpiece in your mouth and blow up your balloon until you are out of breath. Now use your lungs to hold the balloon in its inflated state. When you need to breathe again, inhale the air from the balloon.

4. Have them give it a try. The balloons will inflate, students will run out of breath, the balloons will collapse.

5. Discuss: The balloons, of course, are stars. Your body represents the fusion reaction: your body is converting oxygen into carbon dioxide, and thus producing breath. When you run out of oxygen, you are no longer able to produce breath to keep the balloon inflated, so the balloon col-

lapses. Likewise, a star like our sun is fusing hydrogen into helium, an explosive fusion reaction that keeps the star inflated. When the star runs out of hydrogen, the explosive fusion reaction stops, and the star collapses.

If you were to keep breathing only the air from the balloon, eventually the balloon would be filled entirely with carbon dioxide (and water vapor), and your body would contain no more oxygen. Don't try this, of course! That's called asphyxiation. (You should also take necessary precautions in class to prevent your students from passing out or falling down.) Then, the star would die. And so would you.

As long as you have enough oxygen to keep the balloon inflated, the balloon can maintain its lovely round shape. But the elasticity of the balloon, representing gravity, also keeps the balloon from simply exploding. Likewise, in a star, the inward force of gravity counteracts the outward force of the fusion explosion, so that the star maintains its lovely round shape. And if the force of the fusion explosion ever exceeds the force of gravity, the star goes nova, and casts its material off into space.

That's what you tell them when their balloons pop. All that CO_2, spit, and rubber can be gathered up by another nearby star to accrete into a new type of solar system.

Assessment of Learning: Laughter. In-class discussion. Possible reflective writing assignment.

Origin of Activity: This is an original activity, inspired by Robin Cunningham's "Inflation Balloons" activity (see chapter 8).

Sources Consulted:
Christian, David. *Maps of Time: An Introduction to Big History.* Berkeley: U of California P, 2005. Print.

Christian, David, Cynthia Brown, and Craig Benjamin. *Big History: Between Nothing and Everything.* New York: McGraw-Hill, 2014. Print.

"How Do Scientists Determine the Ages of Stars? Is the Technique Really Accurate Enough to Use It to Verify the Age of the Universe?" *Scientific American.* 21 Oct. 1999. Web. 6 January 2013.

National Aeronautic and Space Administration. *Goddard Space Flight Center.* "Imagine the Universe! White Dwarf Stars." NASA, Feb. 2010. Web. 6 January 2013.

Rieke, G. H. "Evolution of Stars" (lecture notes). *The University of Arizona.* The University of Arizona, n.d. Web. 7 Jan. 2013.

Spier, Fred. *Big History and the Future of Humanity.* Chichester, West Sussex, U.K.: Wiley-Blackwell, 2011. Print.

Judith Halebsky

Threshold Related to the Activity: Threshold 3

Category of Activity: Learning activity, writing activity

Objective: This activity uses creative writing to explore the concepts and vocabulary of the life cycle of stars.

Overview of Activity: This activity joins skills in creative writing with content in Big History. Students connect aspects of poetry writing with the life and death of stars.

Poets find material for poems all around them. When the material that inspires a poem is only slighted edited, the finished poem is called a "found poem." One example of a found poem is Naomi Shihab Nye's "One Boy Told Me," which was written by arranging lines spoken by her young son. Other poems use found material as a catalyst for the writing process. In this activity, students use concepts from Threshold 3 as the raw material for a poem. This activity includes in-class writing and watching a PBS video of Nye reading her found poem. Plan one class period to generate the poems and twenty minutes in a subsequent class period for students to share revised versions of their poems.

Faculty Preparation for Activity: Plan in-class writing prompts (as described below) and prepare equipment to show the short online video.

Cost: n/a

Student Preparation for Activity: Assign students to read found poems online at poets.org and the Threshold 3 section of the textbook.

In-Class Sequence of Activity:

1. Ask students to write in the first person from the perspective of a star to answer these questions: How were you born? How old are you? What does your future hold? (Students can refer to the textbook as they write.)

2. Give a brief introduction to found poems. Watch Naomi Shihab Nye read her poem "One Boy Told Me" online (at www.pbs.org/wgbh/poetryeverywhere/nye.html).

3. Ask students to write out three short quotations from the textbook that offer dynamic or surprising facts about stars or their life cycle.

4. In a creative leap, have students explain in writing how one of their chosen quotations can be a metaphor for human experience.

5. Let students share their human experience metaphor aloud with the class. (Allow students to "pass" if they are reluctant to share their work). Or, in a large class, have students share their writing with a partner.

6. Ask students to write a poem that includes two direct quotations from the textbook on the life cycle of stars. Invite students to share their work in progress with the class.

7. Ask students to revise their poems as homework.

8. At the beginning of the next class, have students share the writing they did as homework. Invite student to read their work aloud. Classmates are welcome to share responses to the writing of their peers.

Assessment of Learning: Students bring a revised poem to class. They share their work in a writing workshop format.

Origin of Activity: This is an original activity.

Source Consulted: Nye, Naomi Shihab. "One Boy Told Me." *Poetry Everywhere with Garrison Keillor.* PBS, 3 March 2011. Web. 10 June 2013.

NOTES

1. David Christian, *Maps of Time: An Introduction to Big History.* Berkeley: U of California P, 2005. 51. Print.

2. "Earth's Gold Came from Colliding Dead Stars." Harvard-Smithsonian Center for Astrophysics, 17 July 2013. Press release. Web. 14 October 2013.

3. Christian, *Maps of Time,* 52–53.

4. G. H. Riecke, "Evolution of Stars" University of Arizona, n.d. Lecture notes. Web. 7 January 2013.

5. Ibid.

6. Fred Spier, *Big History and the Future of Humanity.* Chichester, West Sussex, U.K.: Wiley-Blackwell, 2011. 60. Print.

7. Christian, *Maps of Time,* 51.

8. Eric Chaisson, *Cosmic Evolution: The Rise of Complexity in Nature.* Cambridge, Mass.: Harvard UP, 2001. 82. Print.

9. David Christian, Cynthia Brown, and Craig Benjamin, *Big History: Between Nothing and Everything.* New York: McGraw-Hill, 2014. 26. Print.

10. George O. Abel, David Morrison, and Sidney C. Wolff, *Exploration of the Universe.* Philadelphia: Saunders College Publishing, 1991. 123. Print.

11. "How Do Scientists Determine the Ages of Stars? Is the Technique Really Accurate Enough to Use It to Verify the Age of the Universe?" *Scientific American*, 21 October 1999. Web. 6 January 2013.

12. National Aeronautic and Space Administration, Goddard Space Flight Center. "NASA's Imagine the Universe!: Stars." National Aeronautic and Space Administration, Goddard Space Flight Center, December 2010. Web. 2 February 2014.

TEN · Teaching Threshold 4

*The Formation of Our Solar System
and Earth*

Neal Wolfe

Even after all this time the sun never says to the earth,
"You owe me."
Look what happens with a love like that,
It lights the whole sky.

HAFIZ (fourteenth-century Persian mystic and poet)

When I asked the freshmen in my Big History class to place the following celestial entities in order outward from Earth—the moon, the edge of the solar system, the edge of our galaxy, the sun, and the Andromeda galaxy—I was astonished that several placed the sun between Earth and the moon (not to mention placing the Andromeda galaxy inside our solar system). We can't assume that students possess even a basic understanding of what our solar system actually is, let alone of its formation. Of course, they have learned about the sun and the different planets in their earlier schooling, and while they know what the sun is in the sky, they may have treated this information like so much else that they are required to learn—as abstract facts that must be remembered for the next test.

There are many excellent resources that can help students learn about the scientific basis for our understanding of how our star and solar system formed, but there is more to teaching the formation of the solar system than making sure that students understand the science, important as that is. The solar system is not just something to know about as a set of facts, through textbook explanations and informative lectures; there is a difference between intellectually knowing the science in order to be able to demonstrate comprehension on a test or in a paper and understanding what it all *means* in the real world. To engage students, we need to do more than have them memorize the facts.

Naturally, as with every other threshold in our narrative, the amount of knowledge about the formation of our solar system and Earth is vast, and we must choose those facts that communicate the essence of this threshold in an economical but effective manner. How much detail any teacher incorporates depends upon need, amount of time available, and preference. I would like, though, to suggest a larger context within which to construct an approach to teaching the scientific content, in order to not only facilitate student engagement but also relate the information to the *reality* of what students are learning about.

I find it useful to take students outside, especially when the weather is nice. Our place on a planet in a solar system is more easily made apparent there, where one can see the sky and the sun. I draw students' attention to the sun, reminding them of what they now know about stars, thanks to their study of Thresholds 2 and 3. If we have ever wondered what a star in the night sky might look like close up, well, we live very near one. Surprisingly, many students aren't aware that our sun is a star just like those they have wished upon in the night sky. Here it is useful to remind them, or tell them for the first time, of the "facts." I encourage students to consider that the sun is ninety-three million miles away, yet we feel its warmth, uncomfortably at times, and that it takes eight minutes and nineteen seconds for its light to reach us. I tell them that, if we could imagine a line drawn from where we stand, extended straight through the sun and beyond, six months from today, we will be on that line on the other side of the sun. And that, relatively speaking, one year from today we will be back here where we are now. And that they have made such a journey however many times they are years old. From there, one can discuss solstices and equinoxes as real events to be understood in relation to our orbit around our star.

We can help students realize their location—on a planet—if we simply draw their attention to it, rather than only teach them the science. For example, I may ask them to consider that if they were to extend a line from their bodies through Earth's center to the opposite side of Earth, a person standing at the end of that line would be "upside down" relative to themselves. ("Why don't they—or we—fall off?" Ah, gravity.) This comes as quite a startling revelation to many of them. I may also ask them how far away they think that person on the opposite side of Earth would be from them—guessing can be a fun exercise. The answer? Approximately eight thousand miles.

It is important to explore the history behind the collective learning that has led to our current understanding of how the solar system formed. It is not sufficient to abandon the students (or ourselves!) to prepackaged and delivered information. The sun is ninety-three million miles away? We are in orbit around it? *How do we know these things?* Ask the students to consider how we figured out we were even living on a planet in the first place. If we had only our personal experience to go on, we wouldn't know that Earth is round, spinning, a planet orbiting a star. So how did we discover these things?

Consider telling students the fascinating story of Eratosthenes, an Egyptian Greek who lived in the third and second centuries B.C.E. He noticed that on the date of the summer solstice in his town of Syrene the sun did not cast a shadow, while in Alexandria, 480 miles distant, it did. Eratosthenes brilliantly deduced that, to account for the discrepancy, Earth must be curved. He used simple geometry to calculate the number of degrees of arc that, if Earth were spherical, would account for the difference in shadow. He divided that number—about seven degrees—into 360 (the total number of degrees in a circle), and multiplied the result by 480, the number of miles.[1] His calculation not only confirmed that Earth is spherical but also gave us its circumference.

It is useful to examine how our understandings of solar system formation, as well as other cosmological processes, came about. Students can relate to the nearly fifteen-hundred-year reign of the incorrect geocentric Ptolemaic scheme of solar system structure if they are asked to consider what assumptions they would make if they weren't the recipients of modern scientific evidence to the contrary. It is fascinating to follow the progression of discovery, beginning in the sixteenth century with Copernicus. Even though he correctly proved that the solar system is heliocentric, so strong was the influence of church teaching that he asserted that his findings were true only mathematically, not in the universe itself. Then we follow the brave souls, such as Kepler and Galileo, who, inspired by Copernican findings, continued the inexorable push to refine our understandings. And of course, this is an ongoing process, even today, as we expand our knowledge. Many things we now take for granted were relatively recently confirmed; for example, it wasn't until 1917 that the sun was removed from its assumed position at the center of the universe, and not until eight years later that, thanks to Edwin Hubble, we ascertained that the Milky Way galaxy does not comprise the entire universe.[2]

We humans are, by nature, observers. It is important for our students to understand how scientific understandings occur. For millennia, dedicated astronomers have tirelessly studied the heavens, and kept track of their findings. For example, early astronomers noticed that some stars are brighter than others, but they could not know whether that meant the brighter ones are closer to us, or are intrinsically brighter, or both, until enough information was gathered to allow an educated guess to be made. Sometimes astronomers' conclusions are incorrect, or only partially correct (as in the case of Copernicus), but through continued observation, accumulation of data, fresh approaches, and especially new technology, our understanding of what is out there—and our place in it—grows and develops. Unlike scientific discoveries that can be conducted in laboratories, astronomers are handicapped by having to make educated guesses about celestial phenomena millions of miles or light years distant.

KEY CONCEPTS IN THRESHOLD 4

SOLAR NEBULAR THEORY

According to the currently prevailing nebular theory, just over 4.5 billion years ago, a small part of a massive molecular cloud of 98.5 percent gas (mostly hydrogen and helium, along with a minimal amount of heavier elements) collapsed as a result of gravity, possibly triggered by shock waves from a nearby supernova explosion. As the cloud continued to collapse, it heated up, rotated, and formed a bulge in its center, where most of its mass collected, surrounded by a flattened proto-planetary disc. The intensely hot, spinning central bulge, over the course of about one hundred thousand years, stabilized into our sun, which contains more than 99 percent of the solar system's mass. Within another hundred million years, the remaining matter, held in orbit around the sun by gravity, formed the planets, their moons, and the asteroids, comets, and other bodies through the process of **accretion.**[3]

The scientific explanation of the formation of the solar system is critical for us to understand, but it is so much more than just abstract science to us. The nebular theory describes a profound cosmological process on a massive scale of time and space. It is the process that gave birth to our home—the sun, the solar system, and Earth itself. I am inclined to ask students what it means to them to be products of this beautiful, magnificent process. I believe it is essential to place them *within* the story.

The Evidence   As with the Big Bang and galaxy and star formation, evidence for the solar nebula theory is based on observation and data accumulated by a multitude of astronomers, astrophysicists, astrochemists, and geologists over a period of centuries, leading to continual development and refinement of our understanding. There are three main observational tools:

1. **Telescopes.** Although their ocular effectiveness is limited by Earth's atmosphere and human-caused light pollution, **land-based telescopes** have given us observational access to the wonders of the universe and allow us to accumulate a remarkable amount of knowledge about things so far away. Since we can't bring celestial objects into the laboratory to examine and test, we have had to rely upon observations from Earth to discover and examine them.

Since the 1960s, highly sophisticated **orbiting telescopes** have allowed us to view distant objects in space with a clarity not previously possible for land-based telescopes. They allow the observation of the full bandwidth of electromagnetic radiation, enabling us to "see" beyond what is within our normal visual perception. NASA's Hubble Telescope alone has transformed our knowledge of space, in particular, with its ability to look further into the universe's distant past—and to photograph it, resulting in the remarkable image of the Hubble Deep Field.

Until now, using land-based telescopes, we have been able to observe processes occurring elsewhere in the galaxy that *appear* to be similar to that of our own solar system's formation. These observations have provided compelling support for the solar nebula theory. However, the international ALMA (Atacama Large Millimeter/submillimeter Array) astronomy project, currently being constructed on Chile's remote Atacama Desert plateau at sixteen thousand feet above sea level (thus transcending light pollution), will employ fifty 12-meter antennas to provide unprecedented ability to *directly observe* the formation of planets elsewhere. This, in addition to probing the first stars and galaxies and researching the physics of the "cold" universe of dark energy and dark matter. The new knowledge ALMA is expected to yield will no doubt transform much of what we think we know about the universe, as well as clarify—if not reconstruct—our understanding of the formation of our own solar system. Maybe, suitably inspired by our unit on Threshold 4, one or more of our students will contribute to the discoveries that lie ahead.

2. **Unmanned spacecraft.** While we have landed humans on the moon, we have used unmanned spacecraft to examine other planets in the solar system. Several craft have landed on Mars, enabling us to learn much about the red planet. Venus has also been closely examined. NASA's *Voyagers* *1* and *2* have flown past the outer planets, and they continue to send back data as they leave the solar system.

3. **Radiometric dating.** In use since the 1950s, radiometric dating has allowed us to accurately determine the age not only of Earth, but of the solar system as a whole. Through our understanding of the rate of decay of nuclei in some isotopes, we have been able to calculate the age of rocks and minerals containing those isotopes. This process allows us to date the age of Earth at about 4.5 billion years. By dating moon rocks and asteroids, we have determined that the oldest objects in the solar system are approximately the same age as Earth, which suggests that Earth and the rest of the solar system were formed at roughly the same time.[4]

ACCRETION

While accretion is not yet fully understood, according to the nebular theory of solar system formation, the bits of debris in the proto-planetary disc that were not sucked into the forming sun assumed elliptical orbits and participated in a violent process of crashing into one another. Some of these objects, called **planetesimals,** disintegrated on impact, while others coalesced and formed into larger objects, which gathered more and more debris. So, through accretion, the planets, including Earth, grew in size as they formed.[5]

DIFFERENTIATION (CHEMICAL DIFFERENTIATION)

As Christian, Brown, and Benjamin describe in *Big History: Between Nothing and Everything,* "early Earth was incredibly hot because of continuing violent collisions with nebular debris, the decay of internal radioactive materials, and increasing internal pressure caused by the crushing effect of gravity."[6] As a result, the heavier metals—iron and nickel, primarily—melted and sank to the new planet's center, forming its core. Meanwhile, lighter materials rose toward the surface, so that it had a plentiful amount of silicon and aluminum, with lesser amounts of the other elements. This process, known as differentiation, accounts for Earth's layers: core, mantle, and crust.

THE HADEAN EON

The Hadean eon is the name given to the first geological period of Earth's history (4.5 to 3.8 billion years ago). The term "Hadean" was chosen because early Earth would have been a "hellish" place, teeming with volcanic activity, with no breathable oxygen, a dim sun, and a red sky due to high levels of carbon dioxide, and frequently bombarded by meteorites and comets. The carbon dioxide atmosphere created a greenhouse effect, trapping the sun's heat, with temperatures so hot that early on, no liquid water existed on the planet's surface; all water vapor was held in thick clouds. By the Late Hadean eon, though, Earth had cooled enough that the water vapor stored in clouds was released, resulting in a steady rain that persisted for millions of years, until the planet's surface was largely covered by water. (How would you like *that* weather forecast?) Two main theories account for the existence of water on the planet. One is **outgassing,** the theory that gases and water vapor escaped from Earth's interior through volcanic activity; the other posits that water arrived on the planet through hundreds of millions of years of cometary bombardment.[7]

I find that students take particular interest in learning about what early Earth was like. Most of them are surprised to find out that the planet wasn't always more or less like it is now. Especially when we are outdoors, I like to ask them to first take a good look at their natural surroundings—to situate themselves where they are—and then to imagine themselves in the same location during the Hadean eon. What do they notice? What is it like? This thought experiment is especially resonant because, even though we could not survive in the Hadean environment, the Earth that existed then is nonetheless the same planet that we live on today. Because we can imagine ourselves wherever we are on the planet's surface during any time in its history, this sort of exercise can be done during other threshold units as well. The intent is to facilitate a connection between students' experience today and what they are learning about the planet's past.

Students generally respond well to this sort of mental exercise. It takes them out of their usual orientation to course material, makes their learning experience more personal, and engages their imaginations. Sometimes a productive conversation ensues between the students and me quite naturally, while other times it helps to ask them to write a response reflecting on their experience, and then to share and discuss.

CONTINENTAL DRIFT

While there was speculation as early as the sixteenth century that Earth's continents had once fit together, the theory of continental drift first published in 1885 asserted that all continents had, millions of years earlier, formed two giant supercontinents. Although correct, the theory was largely rejected due to lack of a credible explanation for what causes the continents to move. Only in the mid-twentieth century, when the theory of plate tectonics was established, did continental drift become widely accepted.[8]

PLATE TECTONICS

The theory of plate tectonics, which confirmed and refined the theory of continental drift, was proposed by Princeton University geologist Harry Hess in the 1960s. Hess asserted that recently observed **seafloor spreading,** resulting from volcanic activity deep beneath the oceans, causes new ocean floor to be created while parts of the existing floor are driven back down into Earth's crust. It has been determined that the outer part of the planet—the lithosphere, which includes the crust and a portion of the upper mantle—is broken up into seven or eight major plates and several minor ones. Since 1968, according to Christian, Brown, and Benjamin, the theory of plate tectonics has served as "the core paradigm of earth sciences and the key to understanding most basic geological processes, from mountain formation to tectonic drift."[9]

THE FOUR FEATURES OF COMPLEXITY
IN THRESHOLD 4

DIVERSE COMPONENTS

Initially, the components are molecular hydrogen and helium, with very small amounts of heavier elements. As the new star forms and planetary accretion takes place, the components of the solar system—sun, planets, asteroids, comets—take shape. The newly forming Earth differentiates into separate layers—core, mantle, and crust.

SPECIFIC ARRANGEMENTS

Molecular components form as a vast cloud, subject to gravity. As the cloud collapses, smaller particles of matter combine chemically or physically to form larger ones, which through accretion form the bodies that make up the solar system. The components of the newly formed solar system are arranged with the four smaller terrestrial (denser) planets nearest the sun and the four larger gaseous (less dense) planets farther out. Similarly, within the newly forming Earth, intense heat causes the heavier materials to sink into the center, while the lighter materials rise to the surface.

FLOWS OF ENERGY

The molecular cloud collapses because of gravity; the gravitational collapse is possibly triggered by shock waves from a "nearby" supernova. Once the solar system forms, energy flows include electromagnetic radiation (photons) from the sun falling on the Earth (and other orbiting bodies), and from the center of the Earth, where the temperature is greatest, outward toward the surface.

sun → electromagnetic radiation (photons) → planets

Earth's core → heat → Earth's surface

EMERGENT PROPERTIES

Planets, including Earth, formed from the nebular disc as a result of accretion of the debris in orbit around the newly formed sun. With the formation of planets, a new, more complex level of development is reached—Threshold 4.

HOW EMERGENT PROPERTIES GIVE RISE TO THRESHOLD 5

With the formation of Earth and its privileged "Goldilocks" position relative to the sun—not too hot, not too cold, just right—the conditions were set for the gradual transformations of the new planet that allowed for the development of the next threshold—the emergence of life—to occur.

STUDENT LEARNING OUTCOMES AND
POSSIBLE ASSESSMENTS

By the end of this unit, students will demonstrate the ability to

1. explain the scientific foundation for the formation of the solar system— sun, Earth, other planets, asteroids, comets, and so on (assessment: quiz or test);

2. relate the formation of the solar system to the process of star formation in Threshold 2, as well as the creation of heavier elements in Threshold 3 (assessment: quiz or test);

3. explain how our knowledge of the solar system has been derived, through the history of observation and scientific inquiry that has led to discovery and refinement, which is ongoing (assessment: quiz or test);

4. summarize the development of early Earth, from its formation up to Threshold 5 and the emergence of life (assessment: quiz or test); and

5. illustrate a heightened awareness of where we actually *are*—on a planet (actually, as *products* of the planet) orbiting a star, as part of a system of planets (assessment: reflection paper or written response).

CHALLENGES IN TEACHING
THRESHOLD 4

Some students balk at accepting the scientific explanation for how the solar system formed, as with the origin of the universe itself. Learning that so much is unproven— including our understanding of solar system formation, which is technically hypothetical—some are disinclined to accept it, especially if it conflicts with their religious beliefs. However, if we educate them about the process by which we learn about our world, through which our understandings arise—that they are not merely unsubstantiated guesses—we can construct a good case for appreciating the scientific process. We see that the nebular theory, for example, is based on a tremendous amount of evidence: both observations of the solar system itself and observations of similar processes of star and planet formation taking place in other parts of the galaxy.

In addition, as with the other early thresholds, for many students who are not science-oriented, the scientific account can be intimidating or otherwise daunting. It is important to remember that our goal is not to turn out experts in astrophysics, astrochemistry, or geology. An understanding of the formation of the solar system can be taught most effectively if we keep to the major processes at play. The solar nebula theory is largely based on star formation, of which students presumably already have a basic understanding from Threshold 2. There is an endless amount of scientific detail one could go into regarding the formation of our solar system and Earth; by sticking to the basics, and engaging students in relating themselves to the story, we have a good chance of being effective facilitators of learning.

TEACHING MODELS AND EXERCISES

In this chapter I have emphasized the role of perspective—making "real" the scientific account of what the solar system is, how it got here, and how we know what we know. A number of exercises and activities have been suggested throughout. There is also ample opportunity for using in-class reflective writing, which serves to encourage critical thinking and personal connection to what students are learning. There are many excellent visual resources to aid the teaching of this unit, including websites; videos that reenact the solar system's formation, conditions on early Earth, and how plate tectonics works; and breathtaking photographs.

Also useful are constructed visual images of the solar system, showing the size differentials among the smaller inner "rocky" or terrestrial planets and the much larger outer gaseous planets. And let us not just focus on Earth and sun; while all eight planets (sorry, Pluto) and the sun are composed of the same elements, it is important to recognize that each planet contains different amounts of each element and in different forms (states), and to understand why this is.

In addition, we have developed an outdoor solar system activity ("Accretion," detailed at the end of this chapter), which invites students to reenact the process of accretion and orbiting of planets around the sun in order to make palpable the formation, structure, and movement of the solar system. Part of the activity makes more experiential how it is that we can see certain planets at certain times, and why. Ideally, students (and teachers) become more cognizant of what is going on "out there" and our relationship to it.

CONCLUSION

The unit on Threshold 4 is an opportunity to bring our attention—students and teachers alike—to a new or deeper understanding of and appreciation for this spinning, orbiting, beautiful blue sphere we call home. As with the earlier thresholds, we continue to see the ongoing movement of the development of the universe. In my view, if we wish for our students to be engaged, even excited, about learning this, we need to be excited ourselves. And that is one of the most valuable benefits of being teachers—we get to immerse ourselves in the awesome story we tell. Students notice. And we look at the world in which we live, and our own lives, with heightened awareness.

I am fulfilled by seeing my students' faces light up with recognition when, after our solar system activity, they "get" that Earth is part of a system of planets orbiting a star, and that Earth's formation occurred as part of the process of the sun's formation; when they understand why it is that at any given time, certain planets are visible from Earth and others are not; when a student proudly announces at the beginning of class that she has seen Jupiter in the night sky, even if it is after weeks of my encouraging them to look; and when they realize that the same Earth we take for granted was a very different, inhospitable place for billions of years. Their knowledge, awareness, and perspective are expanding. As teachers, we are privileged to play a part.

ACCRETION

William Phillips and Neal Wolfe

Threshold Related to the Activity: Threshold 4

Category of Activity: Learning activity

Objective: To familiarize students with the processes that formed the sun, Earth, and planets, and to see how the solar system has a life of its own

Overview of Activity: The activity takes place outside in a large open space. Students are given different colored ribbons that represent different "accretion" groups. The instructor (posing as the sun) stands in the center and the students begin to "orbit" around the instructor. As the students orbit the "sun," they link hands and "accrete" to other students with the same colored ribbon. Once all the students have "accreted," there should be eight groups of students, each of which represents one planet.

Faculty Preparation for Activity: Bring eight different colored ribbons to class (nine if you want to include Pluto); these will represent the planets. Cut the ribbons into small pieces that can be tied around students' wrists. The number of ribbons that you have of each color will roughly represent the size of the planets (colors are arbitrary); see table 10.1 for an example of how to divide a class of twenty or twenty-two students.

Cost: About $20 for ribbons

Student Preparation for Activity: n/a

In-Class Sequence of Activity:

1. Briefly review the previous lecture, emphasizing how a supernova creates all the elements of the periodic table (oxygen, iron, nitrogen, and so on). Explain how these supernovae leave behind planetary nebulae, clouds of hydrogen and other elements that become the birthplace of new stars. Review how the hydrogen gases begin to collapse under the force of gravity to form a star, and how planetesimals (large chunks of rock, iron, and ice) begin to circle around the newly formed star.

2. Take the class outside to a large open area. Hand out the ribbons and tell the students that they are all "planetesimals" that will be "orbiting" around the "sun" (either you or a student). Do not tell them that the ribbons represent planets in our solar system. As the students orbit (walk around the

TABLE 10.1   Number of Ribbons Needed for Activity

If twenty students in your class:

COLOR 1	1 piece	Mercury
COLOR 2	1 piece	Venus
COLOR 3	1 piece	Earth
COLOR 4	1 piece	Mars
COLOR 5	6 pieces	Jupiter
COLOR 6	4 pieces	Saturn
COLOR 7	3 pieces	Uranus
COLOR 8	3 pieces	Neptune

If twenty-two students in your class:

COLOR 1	1 piece	Mercury
COLOR 2	2 pieces	Venus
COLOR 3	2 pieces	Earth
COLOR 4	1 piece	Mars
COLOR 5	6 pieces	Jupiter
COLOR 6	4 pieces	Saturn
COLOR 7	3 pieces	Uranus
COLOR 8	3 pieces	Neptune

"sun"), they are to join hands and "accrete" with students that have like colored ribbons. Make sure there is plenty of space and that students are spread out around the sun before the activity begins.

3.   Once all the students have accreted (joined hands), ask them to now align themselves (by group) to represent the alignment of planets in our solar system (let them do this on their own; they should align so that the groups represent the eight—or nine—planets based on size and order). An option is to have the orbiting students form circles, approximating planetary spheres.

Assessment of Learning: Return to the classroom for discussion, having students review the idea that the solar system is made up of many diverse elements, an emergent property (not seen before), a new flow of energy, and a stable structure. Also discuss how the solar system has a life of its own—that it is ever-changing, has synchronicity, and has its own fragility (a change in the orbit of one planet, for example, can change the orbits of the others)—and that Earth just happens to be situated in the "habitable zone" where water can exist as a liquid.

Origin of Activity: This is an original activity.

Source Consulted: Christian, David, Cynthia Brown, and Craig Benjamin. *Big History: Between Nothing and Everything.* New York: McGraw-Hill, 2014. Print.

NOTES

1. David E. Fisher, *The Birth of the Earth: A Wanderlied through Space, Time, and the Human Imagination.* New York: Columbia UP, 1987. 22–23. Print.

2. David A. Weintraub, *How Old Is the Universe?* Princeton: Princeton UP, 2011. 198, 204. Print.

3. David Christian, Cynthia Brown, and Craig Benjamin, *Big History: Between Nothing and Everything.* New York: McGraw-Hill, 2014. 38–39. Print.

4. Ibid., 36.

5. Ibid., 39.

6. Ibid., 42.

7. Ibid., 44–45.

8. Ibid., 46–47.

9. Ibid., 48–49.

ELEVEN

· Teaching Threshold 5

The Evolution of Life on Earth

James Cunningham

> Through all this problem there runs a constant theme,
> and the theme is the flowing stream of time, unhurried,
> unmindful of man's restless and feverish pace. It is
> made up of geologic events, that have created mountains
> and worn them away, that have brought the seas out
> of their basins, to flood the continents and then retreat.
> But even more importantly it is made up of biologi-
> cal events, that represent that all-important adjustment
> of living protoplasm to the conditions of the external
> world.
>
> RACHEL CARSON, "Of Man and the Stream of Time"

I have been a biology teacher for over forty years, yet I continue to be amazed and awed by life, its complexity, and its variety. At the time of writing this I have taught Big History for four years. Teaching Big History has helped me tie the origin and evolution of life in with the rest of the universe's story. The origin and evolution of life is linked with the rest of the story by the framework of increased complexity. When teaching Big History I try to convey the wonder and awe of life to my students, many of whose experience with biology has not been particularly favorable. Often their only memory of biology is the dissection of dead worms or frogs.

So when teaching students about Threshold 5, the evolution of life on Earth, I begin by asking them to think about the characteristics of living things. What makes living things different from nonliving things? How is a frog or a mouse different from a rock? First, living things use energy to maintain themselves. They do this either by using energy from the sun or other sources to make their own food, or by eating other things that do. Second, living things make copies of themselves; they **reproduce.** Finally, living things change over many generations to **adapt** to their environment; they **evolve.** A discussion of what it means to be alive helps students understand how unusual life is. The **second law of thermodynamics** states that everything in the universe is moving toward greater **entropy** or randomness. But life is not like that at all. It is complex and growing more complex over time.

As mentioned above, all living things adapt to their environment; they evolve. At this point in the course I discuss the process of **evolution.** Charles Darwin and Alfred Wallace developed the idea that species evolve by a process that they called **natural selection.** Natural selection is the mechanism that causes species to change through time. Natural selection results from individuals in a **population** differing in the number of **offspring** they produce, specifically those offspring that are best suited to survive in the environment. Individuals that are best adapted to their environment leave the greatest number of offspring best suited to survive in their environment and are thus the most successful. This is often described as "survival of the fittest," where **fitness** is defined by the number of offspring produced. As an environment changes, species evolve to be better adapted to those changes.

In class, we go back and review the idea of increased complexity and what it means to say that something is complex. Complex things are made up of many parts, and these parts are arranged in a particular way. Complex things have new and unexpected properties. And complex things are maintained by flows of energy. The origin and evolution of life is a continuation of this theme of increased complexity. Simple atmospheric gases were transformed into more complex molecules, which in turn were the building blocks of still more complex **organic molecules**, such as **proteins** and **nucleic acids.** This transformation was driven by the flow of energy from the primordial Earth.

From these relatively simple organic molecules of proteins and nucleic acids, scientists believe that membrane-enclosed spheres may have formed that were capable of consuming other molecules from the environment in order to grow and reproduce. These spheres, called **heterotrophs**—that is, organisms that do not produce their own food but consume food already in the environment—were likely the first simple living organisms. This transformation from nonliving to living is an important example of the idea of emergent properties. From a collection of proteins and nucleic acids enclosed in a sphere of **lipids**, who would have guessed that a living organism would have emerged?

Competition among these early organisms may have been stiff, and so a further step in increased complexity may have occurred, resulting in the origin of organisms that can make their own food. We call these **autotrophs.** With this development, the diversity or complexity of life increased. At this stage, there were two kinds of organisms on Earth: those that made their own food and fed themselves, and those that fed on them.

The next major step in the evolution of life and increased complexity may have occurred when some simple organisms came to live inside the cells of others, in a process called **endosymbiosis,** which resulted in the creation of more complex **eukaryotic cells.** Eukaryotic cells contain parts (**organelles** derived from once-independent cells) that work together to function as a whole. Eukaryotic cells have a very important emergent property: they can clump together to form multicellular organisms. From this point life rapidly evolved into the vast diversity of species that we see today.

To conclude the teaching of Threshold 5 and to connect it with Threshold 6, the rise of *Homo sapiens,* I summarize the history of life on Earth using the eight stages described by Christian, Brown and Benjamin in *Big History: Between Nothing and Everything.*[1] Stage 1 consists of the first living organisms, which are thought to be similar to today's bacteria or **prokaryotes.** Since there was little free oxygen to efficiently break down food products, these organisms obtained energy through **fermentation.**

Stage 2 saw a new way of obtaining energy through the process of **photosynthesis.** These early photosynthesizers, called **cyanobacteria,** still exist today. An important development at this stage was the releasing of free oxygen, a by-product of photosynthesis. It was at this stage that oxygen began to accumulate in the Earth's atmosphere.

This abundance of oxygen led to the next stage, Stage 3, in which organisms began to more efficiently break down food through the process of **respiration.** In addition, 2.5 to 1.5 million years ago a second type of organism came along, the eukaryotes. Eukaryotic cells have a nucleus and organelles that are thought to have evolved from prokaryotic cells that came together through endosymbiosis to live in a **symbiotic relationship.**

During Stage 4 cells developed sexual reproduction. Prior to Stage 4, reproduction was accomplished by cells simply dividing into two, with each resulting cell being identical to the other. In **sexual reproduction,** special cells called **gametes** are produced. These cells have only half the genetic information of the original cell. During sex, two gametes from two different individuals come together to form a new individual whose genetic information is derived half from one parent and half from the other parent. Thus, the offspring is genetically different from the two parents. Sexual reproduction results in **variation,** the stuff of evolution.

These sexually reproducing eukaryotic cells were capable of clumping together to form **multicellular organisms.** This is Stage 5, which occurred 700 to 600 million years ago.

Because of the variation that resulted from sexual reproduction, multicellular organisms evolved rapidly during the Cambrian geological period. It was during this period, 500 to 400 million years ago, that the first vertebrates evolved during Stage 6.

Until this stage all life was aquatic. During Stage 7 we see the first organisms venturing out onto land, 475 to 360 million years ago.

Some 245 million years ago, many of the organisms that had moved out onto land, as well as those that had remained in the sea, went extinct. The cause of this extinction is unknown, but we do know that it resulted in the subsequent evolution of two important groups, **dinosaurs** and **mammals,** during Stage 8. A second large extinction occurred 65 million years ago, resulting in the elimination of the dinosaurs. This both freed ecological **niches** and allowed mammals to evolve to fill them. Some of these mammals eventually gave rise to our human ancestors.

KEY CONCEPTS IN THRESHOLD 5

CHEMICAL EVOLUTION

The experiments of Stanley Miller show that the building blocks of life (**amino acids** and **nucleotide bases**) could have formed spontaneously from the primitive Earth's atmosphere and the input of energy (in the form of lightning and heat). In his laboratory, Miller filled a glass container with the gases that were thought to be present in the early Earth's atmosphere: methane, water vapor, ammonia, and hydrogen gas. He then introduced a continuous electric spark into the gas mixture and allowed it to run for a week. After a week a brown sludge containing several amino acids and nucleotide bases had formed in the glass container. Amino acids are the building blocks of proteins, and nucleotide bases are the building blocks of nucleic acids such as **RNA** and **DNA.** It is important to emphasize here that there was no free oxygen in the early Earth's atmosphere. If there had been, these compounds would not have accumulated.

Several theories have been put forward to explain how these building blocks could have been assembled into more complex molecules. According to one theory, this process may have happened at the edges of the primordial sea, where clays may have acted as templates, bringing the building blocks together and allowing bonds to form between them. Another theory suggests that meteor collisions forced the building blocks together and, again, bonds were formed between them. Whatever the mechanism, gradually, the primordial sea accumulated more complex molecules, such as proteins and nucleic acids.

CELL MEMBRANE

An important condition of living organisms is that the environment inside a cell is very different from the environment outside that cell. A barrier that keeps the two environments separate but does allow some substances to move between them is necessary. Cells today do this with the **cell membrane,** which is made up of lipids or fats. Experiments have shown that in a solution of water and lipids, small, membrane-enclosed droplets will form spontaneously. These droplets will then reproduce, forming new droplets. In other experiments, droplets containing simple proteins that behave like **enzymes** (which cause chemical reactions to take place) will perform basic forms of **metabolism** or energy use. These small droplets may be similar to the earliest living organisms on Earth.

GENETIC MATERIAL

Another important aspect of life is that there needs to be some form of stored information that can be passed on from generation to generation. This information is used by cells to drive metabolism. In today's cells, information is stored in DNA, but most likely RNA preceded DNA in this function. Why is this? RNA is unique in that it can function as both an enzyme and a molecule that can store information. Because RNA is small and can store only a small amount of information, and because it is more fragile, eventually DNA evolved as the principle molecule of information storage. DNA is both more stable and larger, and thus capable of storing more genetic information.

ENDOSYMBIOSIS

The theory of endosymbiosis helps to explain the evolution of eukaryotic cells. According to this theory, organelles of eukaryotic cells were once small prokaryotic cells that began living within a larger cell, or host. The prokaryotic ancestors of these organelles probably gained entry to the host as undigested prey, and once inside, formed a mutually beneficial relationship with their host cells.

NATURAL SELECTION

Again, natural selection is the mechanism that causes species to change over time. Natural selection results from the difference in number of offspring that individuals in a population produce, specifically those offspring that are best suited to survive in the environment. Individuals that are best adapted to their environment leave the greatest number of offspring best suited to survive in their environment and are thus the most successful. This is often described as "survival of the fittest," where fitness is defined by the number of offspring produced. As an environment changes, species evolve to be better adapted to those changes.

THE FOUR FEATURES OF COMPLEXITY
IN THRESHOLD 5

DIVERSE COMPONENTS

A number of examples of increases in complexity during the early stages of the emergence of life stand out—for example, the transition from prokaryotic organisms to eukaryotic organisms by the process of endosymbiosis. Not only are eukaryotic cells made up of more parts, such as **mitochondria, chloroplasts,** and **membrane-enclosed nuclei;** those parts also work together in precise and coordinated ways. For example, the organism as a whole engulfs food particles, which are digested by packets of membrane-enclosed enzymes, and then the digested food is converted into useful energy (as **adenosine triphosphate,** or ATP) by the mitochondria so that the whole cell can move by the action of the flagella to chase down and engulf more food. All these parts and steps work together to enhance the likelihood of the organism's survival.

SPECIFIC ARRANGEMENTS

The specific arrangement of amino acids in cell proteins dictates the function of those proteins, whether they are structural or enzymatic. Also, the specific arrangement of nucleotide bases in DNA and RNA determines the genetic information that is encoded in these molecules.

FLOWS OF ENERGY

The flow of energy through living organisms is what allows life to maintain its complexity and resist entropy.

sun → cells of photosynthesizer (plant)
sun → cells of photosynthesizer (plant) → cells of consumer (animal)

EMERGENT PROPERTIES

First, the assembling of diverse molecules in the first cells allowed these cells to metabolize, reproduce, and adapt—in other words, to be alive, a truly amazing phenomenon. Second, eukaryotic cells were capable of clumping together to form multicellular organisms. There are no multicellular prokaryotic organisms.

HOW EMERGENT PROPERTIES GIVE RISE TO THRESHOLD 6

The new emergent property—the ability of eukaryotic cells to become multicellular—was crucial in leading to Threshold 6. As mentioned above, no prokaryotic cells are multicellular, so something about the arrangement of the diverse components of the eukaryotic cells, their ability to sexually reproduce, and their use of energy allowed them to clump together. All multicellular organisms, including humans, are a product of this emergent property of eukaryotic cells.

STUDENT LEARNING OUTCOMES AND POSSIBLE ASSESSMENTS

By the end of this unit, students will demonstrate the ability to

1. describe the differences between living and nonliving things (assessment: quiz or in-class writing assignment);

2. define what natural selection is and how it causes species to evolve (assessment: quiz or short essay);

3. describe the experiments of Stanley Miller and discuss how his findings fit into the story of the origin of life on Earth (assessment: reflective writing assignment or short essay);

4. explain the theory of endosymbiosis and how it fits into our concept of increased complexity and emergent properties (assessment: reflective writing assignment or short essay);

5. define and explain the differences between prokaryotic and eukaryotic cells (assessment: quiz); and

6. discuss and illustrate the advantages of being multicellular (assessment: reflective writing assignment or short essay).

CHALLENGES IN TEACHING THRESHOLD 5

From my perspective, there are two major challenges to teaching Big History in general and Threshold 5 in particular. The first is that many of our students are not science majors and many have had bad experiences taking science courses. Because much of the material covered in the first half of the course is science-based, this can be a problem. What I try to do when teaching science is to make it as simple and straightforward as possible without compromising accuracy. I keep the conversation light, avoid using complex jargon, and bring in humor as much as possible. This approach seems to work.

The second challenge is the sheer amount of material being covered. This is true not only of Threshold 5 but throughout the course. When I first began teaching Big

History I felt it was important to cover *everything*. Now, I know that what is most important is to effectively convey the story. Also, relying on the idea of increased complexity and the signposts of the eight thresholds helps students see the whole story and their place in the story at any point along the way.

CONCLUSION

One benefit that has come from teaching Big History is that the Big History perspective has crept into my other classes. Now, when I teach comparative anatomy to biology students, I talk about increases in complexity during the evolution of vertebrates. A Big History colleague and I teach a course in which we combine the human and natural history of California, helping students understand the influence of California's environment on the history of the state and how humans have influenced the environment of California. We take a very Big History perspective and begin with plate tectonics and how it shaped the geography of California, ultimately influencing one of the pivotal events of California's history, the Gold Rush.

I am now seeing the whole picture from the very beginning. I know where the story is going, and I can help students recognize the important points along the way that will help them, in the end, to see the entire story. The other day I was talking to another of my Big History colleagues about how comfortable I was feeling about teaching Big History this year. He felt the same way. I think that this feeling of comfort comes from the absorption and understanding of the whole story. We are now living Big History, as opposed to just teaching it.

Emphasizing increased complexity, specifically in Threshold 5, helps students see that the origin and evolution of life—and therefore their own lives—are a continuation of what came before. We truly are a product of supernovae explosions; we are made up of stardust. This focus on increasing complexity also emphasizes the connection we have with the rest of creation. We modern humans have become very detached from our environment. We live and work in buildings that insulate us from changes in sunlight and temperature. So anything that can help us see the connection between our own lives and the rest of the universe is good. This is what Big History does.

ACTIVITIES AND EXERCISES FOR TEACHING THRESHOLD 5

NATURAL SELECTION SIMULATION

James Cunningham

Threshold Related to the Activity: Threshold 5

Category of Activity: Learning activity

Objective: Students will comprehend and apply the principles of natural selection through a guided simulation.

Overview of Activity: Students divide into groups of five, each of which represents birds of the same species with different types of beaks. The different types of beaks are the knifebill, forkbill, and spoonbill, which are simulated by a knife, fork, and spoon, respectively. Each group of students will experiment with using these utensil "beaks" to pick up "seeds" from different surfaces. Each group's success rate in picking up food, surviving, and even producing offspring demonstrates how a group's physical suitability to a specific environment can, over the course of generations, lead to marked population increases of a specific group in that environment, that is, to natural selection.

Faculty Preparation for Activity:

1. Introduce students to the principles of natural selection. In all species of organisms, individuals vary—that is, not all individuals of the species look the same. Most of this variation falls into three categories: morphology (shape), physiology (chemistry), and behavior. Taking a specific case in point, in nature, both predators and prey exhibit variation. Some predators will be better at catching prey, and some prey will be better at escaping from predators. The predators that catch prey more successfully will, in turn, produce more offspring that look like them than those that do not. Thus, over time, the predator population will come to look, function, and behave like the most successful individuals. The same will also be true of the prey population.

2. Assemble your supplies. Your kit should contain the following: (a) floor mats of different textures (one per group) to simulate hunting in different environments; (b) knives, forks, and spoons; (c) "seeds" represented by dried beans, like garbanzos; (d) eight cups or small containers to hold the collected "seeds"; and (e) measuring cups (one per group).

Cost: About $100 for mats, beans, utensils, and measuring cups

Student Preparation for Activity: Read related chapter in course text.

In-Class Sequence of Activity:

Introduction (read to students): In winter, members of a small seed-eating bird species feed on seeds they find in their habitat. This species of bird has three different bill shapes and thus three different ways of feeding. Some individuals are knifebill feeders, others are forkbill feeders, and still others are spoonbill feeders. An individual *cannot* change its way of feeding—once a knifebill feeder, always a knifebill feeder. In this simulation the three types of feeding habitat will be represented by three different-textured floor mats. Students will represent the birds with different bill types, equipped with a plastic knife, fork, or spoon. The seeds that the birds are feeding on will be represented by one type of dried beans.

1. Break students into groups of five and have each group measure out the beans, representing its prey, with the provided measuring cup. Scatter the beans on the provided mat.

2. Equip three students in each group with a knife, spoon, or fork to represent the different bill types of the seed-eating birds. The fourth student will act as a timekeeper. The fifth student will observe until she or he can be added to the group as an offspring.

3. At the signal of "Go," given by the timekeeper, the birds proceed to pick up seeds. They must pick seeds off the carpet using their respective tool only and transfer the prey into their cup (they cannot put the cup on the carpet and push seeds into it). If the type of "bill" allows it, the bird may pick up more than one seed at a time.

4. At the end of sixty seconds, the timekeeper should signal "Stop." This is the end of the first generation. Each predator should count their number of seeds and record the results. The predator who has taken the greatest number of seeds is able to reproduce and adds one more of its bill type to the next generation, welcoming the fifth student into the group and giving him or her the appropriately shaped bill. The predator that picked up the second-greatest number of seeds survives but doesn't reproduce. Finally, the predator that picked up the fewest seeds dies. In a tie, either both birds survive and reproduce or both simply survive. In the latter case, the instructor could be added to the group if an additional member is needed.

5. Return all seeds and spread them out on the mat again.

6. With your new population of predators repeat steps 3 and 4 for two additional generations and record the results in the data grid in table 11.1.

TABLE 11.1   Data Table: Members of Bill Types in Each Generation

	Generation 1		Generation 2		Generation 3	
	No. of beans	No. of individuals	No. of beans	No. of individuals	No. of beans	No. of individuals
Knifebills						
Spoonbills						
Forkbills						

Suggested Discussion Questions:

1. The bird population started with an equal number of individuals of each variation (knifebill, forkbill, and spoonbill). Which variation became more common in the total population over time? Explain why.

2. Which variations were eliminated from the population? Explain why.

3. Compare the data between the different environments (types of floor mat). Were the results the same in the bird populations across all environments? Explain why.

Assessment of Learning: Students complete data charts and discuss.

Origin of Activity: This is an original activity.

DESIGN PRINCIPLES IN MICROORGANISMS (ART PROJECT)

Lynn Sondag

Threshold Related to the Activity: Threshold 5

Category of Activity: Studio art activity / project

Objective: Gazing through a microscopic lens, students can get a glimpse of the numerous ways the Earth is continually recycling and renewing itself. The natural process of *autopoeisis* creates unique visual patterns and functional design. Students make connections between nature's laws and forces and the influence, both visual and functional, of those laws and forces on artists' and designers' creations.

Overview of Activity: This project takes place over three or four class periods. Students select their favorite microorganisms, based on form or function or both. They find a visual depiction of the organism from which they create a design motif. With this motif, students organize and design an overall pattern that demonstrates the use of design principles. (See page 75 for an example of student work.)

Faculty Preparation for Activity:

1. Create a PowerPoint of images that illustrate the main principles of design: unity through variety, balance, repetition and rhythm, scale, contrast, and directional force.

2. Create a handout defining the principles of design.

3. Gather visual examples of pattern designs from various time periods and cultures.

4. Gather materials: (a) drawing paper (8″ × 11″), three or four sheets per student; (b) standard pencils, at least one per student; (c) several sets of colored pencils, which can be shared; and (d) drafting vellum or tracing paper (approximately 11″ × 14″), at least one sheet per student.

Cost: About $25 for drawing paper, colored pencils, and drafting vellum (a pad containing 100 sheets of 8″ × 11″ drawing paper is about $5.00; sets of 12 colored pencils are about $5.00 each; and a pad containing 25 sheets of 11″ × 14″ drafting vellum is about $15.00)

Student Preparation for Activity: Students should read about Threshold 5, the evolution of life on Earth, in the course text.

In-Class Sequence of Activity:

Class 1:

1. Discuss and review the emerging themes of the evolutionary story: geo-biological systems, microbial ancestry, DNA revolution, and the creativity of life expressed through autopoiesis.

2. Give visual examples of work by designers and architects who are known to design guided by and assimilating natural laws. For example, students could compare a cytoskeleton to Buckminster Fuller's geodesic dome through the concept of "tensegrity," in which structural balance is equal to tension and integrity. Another example is Joris Laarman's "Bone Chair."

3. Go over the principles of design and give students a PowerPoint presentation displaying examples of how these principles are used in designers' patterns.

4. Review the five earliest stages of life. Have students recognize the thousands of species of microbes, bacteria, protists, and multicellular organisms. They can list several to use as key words for an image and information search.

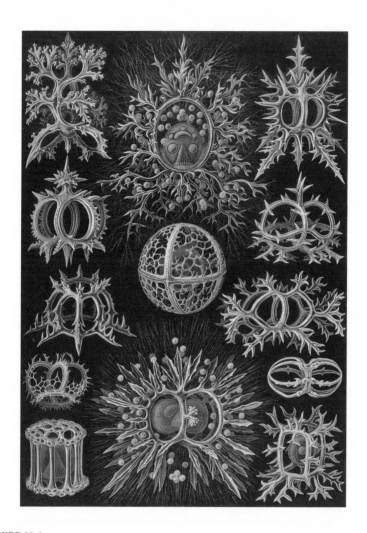

FIGURE 11.1

Sample student writeup of selected microorganism:

Image: Ernest Haeckel, *Radiolaria*, *Stephiodea*, *Illustration Kunstformen der Natur*. Leipzig: Verlag Des Bibliographischen Instituts, 1904. Kurt St BioLib.de. Web. 6 June 2014.

Form and Function: Radiolaria produce intricate mineral skeletons that are divided into inner and outer portions. These skeletons protect the unicellular body, but they also allow protruding bundles of microtubules, axopods, to extend outside and aid in buoyancy and movement. Photosynthetic algae called zooxanthellae provide much of the cell's energy.

Principles of Design: Symmetrical balance. Directional forces.

Homework:

Between class 1 and class 2, students should

1. read through the handout describing the principles of design;

2. gather at least five examples of patterns and designs created either by humans or nature, citing the source of each image;

3. research and select a microorganism for their design motif and find visual depictions of the organism (fossils, diagrams, photos, or drawings), citing the source of each image (author, title, medium, date, website, date visited); and

4. learn about their chosen microorganism's functions and describe its visual form as it relates to one or more of the principles of design.

Class 2 and 3:

1. Students should come to class prepared with their three sketches of the microorganism motif in three different sizes, one drawn on each sheet of 8″ × 11″ drawing paper.

2. Students consider and plan an organizational structure for their pattern, such as a grid, a spiral, a vertical line or horizontal line, or an *S* curve.

3. Using the larger sheet of drafting vellum, the students use pencil to trace their motifs multiple times to form a larger overall pattern. From here, they can color in the patterns using a simple color palette such as a trio of primary colors, complementary colors, or analogous colors.

Class 4:

Students engage in a critique conversation:

1. Working in pairs, students exchange drawings and describe the way their classmate employed at least two design principles in their work.

2. Each student shares her or his artwork in front of the whole class, naming her or his microorganism motif and describing its function and the intention behind the overall design. Their peers share out the design principles they notice when analyzing the piece.

Assessment of Learning: Critique conversation. Refer to table 7.1 on page 104 for the assessment rubric.

Origin of Activity: This is an original activity.

Sources Consulted:

Adam, Hans Christian. *Karl Blossfeldt: The Complete Published Work.* Cologne, Germany: Taschen, 2008. Print.

Christian, David, Cynthia Brown, and Craig Benjamin. *Big History: Between Nothing and Everything.* New York: McGraw-Hill, 2014. Print.

Gybter. "Proteus 2004" (video clip excerpt from Le Brun, David. *Proteus: A Nineteenth Century Vision.* Night Fire Films, 2004.) *YouTube.* You Tube, 9 July 2009. Web. 10 Feb. 2012.

Haeckel, Ernest. *Radiolaria, Stephiodea, Illustration Kunstformen der Natur.* Leipzig: Verlag Des Bibliographischen Instituts, 1904. BioLib.de. Web. 10 Feb. 2012.

Preble, Duane. *Artforms: An Introduction to the Visual Arts.* 7th ed. Upper Saddle River: Pearson Prentice Hall, 2004. Print.

Sachs, Angeli. *Nature Design: From Inspiration to Innovation.* Baden, Switzerland: Lars Muller, 2007. Print.

NOTES

1. David Christian, Cynthia Brown, and Craig Benjamin, *Big History: Between Nothing and Everything.* New York: McGraw-Hill, 2014. 69-76. Print.

TWELVE

Teaching Threshold 6

The Rise of Homo sapiens

Cynthia Taylor

"Curiouser and curiouser!"

LEWIS CARROLL, *Alice's Adventures in Wonderland*

When I was a graduate student, it was drilled into me that I must focus my study of American history on a particular time period, or on one historical event or individual, based on written historical documents found in archives. While history survey courses were for lower-division undergraduate students, graduate students searched to find areas of study in ancient, medieval, or modern history that had not been written about or were open to new methodological approaches and revisionist interpretations. Then, a few years ago, I was asked to participate in the Big History program at Dominican University of California. Faced with the challenge of teaching the ultimate history survey course, one that spans from the Big Bang to the distant future, I was forced to throw out all my past preconceptions of how to learn and teach history. Not only does Big History incorporate a world history perspective, it opens up ways to teach the "prehistory" of the **Paleolithic era**, previously accessible only to archaeologists, paleoanthropologists, and biologists. I felt like Alice in Wonderland, viewing history from a side of the mirror that I had never looked at before. Like Alice, I was finding history "curiouser and curiouser" and learning right along with the young college students I was teaching.

This chapter explores Threshold 6, the period when modern humans emerged as the most influential species Earth has ever known. I believe that Threshold 6 is the most critical of all thresholds because it follows the emergence of life on Earth in Threshold 5. Human history should not be separated from this earlier threshold, lest humans become disconnected from the larger story of their evolutionary past.

Another important aspect of this threshold is its emergent property, collective learning. In the following pages I will explore the basic components of Threshold 6 that led to a critical breakthrough and started human beings on the path that is the dominant feature of the Earth today: the emergence of *Homo sapiens* and their eventual domination of the Earth for the past ten thousand years. So dominant have humans been during this time that scientists now identify it as two geologic epochs: the **Holocene** and the **Anthropocene.** But without fully understanding the tremendous achievements of the earliest human beings, which yielded both harmful and beneficial effects, the last ten thousand years cannot fully be understood. This chapter will introduce the key concepts of this threshold, explore how complexity theory explains the course of its development, and then suggest some pedagogical strategies that may help teachers both learn and teach this critical threshold.

KEY CONCEPTS IN THRESHOLD 6

The basic historical content of Threshold 6 is most approachable by framing the whole period in a narrative with smaller turning points or mini-thresholds. Threshold 6 comprises the last seven or eight million years of life on Earth. In their analysis of this threshold in *Big History: Between Nothing and Everything*, David Christian, Cynthia Brown, and Craig Benjamin divide this period into three such mini-thresholds: "Hominine Evolution," "The Appearance of *Homo Sapiens*," and "The Paleolithic Era."

HOMININE EVOLUTION

The first mini-threshold is "Hominine Evolution," occurring from 8 million to approximately 200,000 to 250,000 years ago. Recent genetic research reveals that humans and chimpanzees, who share 98.5 percent of their genes, evolved from a **common ancestor** about 5 to 8 million years ago, the time necessary "to acquire the differences in genes between the species, based on an estimated rate of genetic change."[1] About 2.5 million years ago, early forms of the genus *Homo* began appearing, roughly coinciding with the **Pleistocene** epoch, an era of repeated glaciations or ice ages, as well as the period scholars identify as the **Lower Paleolithic.** Table 12.1 gives a more visual representation of the comparison of human history and evolution with the geologic time periods.

Many topics can be used to bring this first period in Threshold 6 alive. Introducing students to—or reminding them of—several of the recent findings of **australopithecine** fossils, especially Donald Johanson's 1974 discovery of "Lucy," the best-known and most-studied hominine of the twentieth century, is a good start.

TABLE 12.1   Threshold 6 at a Glance

Time span	Human emergence	Historical eras	Geological epochs
65 million years ago	Dinosaurs disappear		Paleocene (the Cenozoic era) begins
8 to 7 million years ago	Family of hominines emerges		Late Miocene epoch
			Pliocene epoch
2.6 million to 250,000 years ago	Genus *Homo* emerges	Lower Paleolithic	Pleistocene epoch (epoch of repeated glaciations or ice ages)
300,000 to 50,000 years ago	Species *Homo sapiens* emerges	Middle Paleolithic	
50,000 to 12,000 years ago	Hunter-gatherer cultures emerge	Upper Paleolithic	
10,000 b.c.e. to 1800 c.e.	Agrarian cultures emerge	Agrarian era	Holocene epoch
1800 c.e. to present	Global industrial culture emerges	Industrial era	Anthropocene epoch

Students can read Johanson's own version of how he found Lucy, whom he describes as the "woman who shook up the family tree."[2] Another interesting story is that of Raymond Dart, who in 1925 published his description of a fossilized cranium of an early human ancestor he found in South Africa. But it took years before Dart's findings were accepted by other paleoanthropologists as relevant to early human evolution and eventually identified as *Australopithecus africanus*.[3]

Besides the discussion of australopithecine fossils, it is important to stress that evidence shows that species of the genus *Homo* also lived as contemporaries of these early human ancestors. Students need to grasp the great variety of human species that existed during this period so that they can better appreciate that *Homo sapiens* is the last remaining species of the hominine family. ***Homo rudolfensis, Homo habilius, Homo erectus,*** and other species from the hominine family lived at the same time. *Becoming Human*, a recent documentary film from the NOVA series, does a fine job of introducing Turkana Boy, an almost complete specimen of a young *Homo erectus*. These were our first human ancestors to leave Africa and subsequently colonize the globe.[4] This film, as well as other useful sources on human evolution, provide both teacher and student with graphic details and evidence of early human ancestors in the era historians designate as the Lower Paleolithic. Historians usually cite the Lower Paleolithic as the beginning of human history, with the emergence of *Homo habilis* (or "handy man"), approximately 2.5 million years ago. *Homo habilis* created the **Oldowan cultures,** the oldest known **stone tools,** and a tremendous technological advance.[5]

THE APPEARANCE OF *HOMO SAPIENS*

The appearance of a new species, *Homo sapiens*—"wise" or "knowing" human—marks the second mini-threshold of Threshold 6. Our species emerged in the **Middle Paleolithic,** sometime between 250,000 years ago, the start of the Middle Paleolithic, and 50,000 years ago, the end of the Middle Paleolithic or start of the Upper Paleolithic. So "successful" was this new species at adapting to its environment that *Homo sapiens* became the only surviving hominine, and would go on to dramatically transform Earth's biosphere.

This mini-threshold explores the distinctiveness and uniqueness of humans in ways that can open up classroom discussions on such questions as: What made humans human? Was it their ability to walk on two feet? Was it their larger brain size? Was it their capacity for language and the ability to speak and communicate complex ideas? Table 12.2 compares characteristics of modern chimps with those of modern humans. This table might help students to think more deeply not only about the differences between themselves and their closest relative, but also about the amazing evolutionary structure of the human body.

In analyzing the characteristics of humans, students can more readily grasp four key developments in human evolution: **terrestriality** (coming to the ground from the trees), **bipedalism** (upright walking), **encephalization** (brain expansion in relation to body size), and **culture** (civilization and tool making/technology).[6] Ask students to ponder which of these developments came first. Because scientists and anthropologists differ on the order, it may be impossible for students to come to any definitive answer, but the real issue is to get the students to think more deeply about the distinctiveness of this species. *Homo sapiens* evolved because their blending of these traits into a highly efficient structure enabled them to evolve in a variety of environments, often in harsh and extreme conditions.

Another valuable and essential human trait is the ability to use **language.** Archaeologist Chris Scarre offers a helpful explanation on how *Homo sapiens* evolved sophisticated language by developing syntax, organized ways of stringing words together to give variable meanings, and tenses that allowed communication about things not necessarily present. Furthermore, this sophisticated language enabled humans to store

TABLE 12.2   What Changed Over Seven Million Years

Modern Chimpanzees	Modern Humans
Knuckle walking	Bipedalism
Smaller brain size (about 3 times smaller)	Larger brain size
Large teeth, jaws, and mouth	Smaller teeth, jaws, and mouth
Dark fur, light skin	Little hair, dark skin (now variable)
Sociable	More sociable
High larynx	Lower larynx
Unassisted solitary childbirth	Assisted social childbirth
Males 25 to 30 percent larger than females	Males 15 to 20 percent larger than females
Male and female hierarchies	Pair bonding
Solitary eating	Social eating
No use of fire (eats raw food)	Use of fire (eats cooked food)
Simple tools	Complex tools
Vocalization and gestures	Full speech with syntax
48 chromosomes	46 chromosomes

SOURCE: David Christian, Cynthia Brown, and Craig Benjamin, *Big History: Between Nothing and Everything*, 83.

considerable amounts of information on an unprecedented scale with great adaptive potential. With the recent discovery of FOXP2, a language-related gene in modern humans, scientists believe that language emerged rapidly among *Homo sapiens*, leading to an acceleration of **collective learning** about fifty thousand years ago, during what historians called the Upper Paleolithic and the emergence of **hunter-gatherer societies**. The recent discovery of this same gene in fossil **Neanderthals** reveals the great similarity between these two species of the same genus.[7] Because language is a characteristic shared by modern humans and Neanderthals, this can be a great way to encourage a classroom discussion comparing and contrasting these two species, and can also serve as a segue into the topic of how language was a critical step leading to collective learning, an emergent property of Threshold 6.[8]

THE PALEOLITHIC ERA

The third mini-threshold of Threshold 6 is the Paleolithic era, dating from approximately two hundred thousand years ago until ten thousand years ago, and the emergence of the Holocene, a new geologic epoch. One of the important contributions Big History makes to the expanding human story is the inclusion of this under-taught period of human history. The addition of the Holocene to the "world history narrative" is critical, because it was during this epoch that humans realized their full potential—physically, socially, technologically, and linguistically—and developed powerful hunter-gatherer cultures and societies on all the continents, except Antarctica. Not only is the Holocene the foundation of all subsequent world history, it constitutes 95 percent of all human history.[9]

Instructors have an opportunity to undo some of the stereotypes students have learned about the "Stone Age." They should emphasize the great intelligence and ingenuity that hunter-gatherer societies exhibited; their development of egalitarian communities; and the high standard of life Paleolithic people enjoyed as a result of reasonable population sizes, availability of resources, and their knowledge of their place in nature—in Earth's ecosystems. It should also be emphasized that from a Big History perspective, the Paleolithic era is relatively recent and the Paleolithic human experience is familiar. Young students need to understand how directly this history bears on their own lives today. This period of human development is pivotal because it is inextricably tied to the emergence of life on Earth. It provides teachers with a "teaching moment" and opportunity to show students how the study of only ancient and modern history can sever them from their larger history, as well as their place in nature, and most important, the critical impact their species is having on the Earth's fragile biosphere today.

DARING KINGS PLAY CHESS ON FINE GRAIN SAND

One way to link the three mini-thresholds of Threshold 6 together for students, while also connecting this threshold to the larger biological and environmental narrative of Big History, is to encourage students to learn scientific taxonomy, or the biological classification of human beings, as suggested by Christian, Brown, and Benjamin in *Big History: Between Nothing and Everything*.[10] Table 12.3 summarizes the categories of basic human taxonomy, from kingdom to species, and table 12.4 places this information into the historical framework of Thresholds 5 and 6. Although scientists still debate how these classifications should be arranged, these schemata do help students grasp, in a visual way, the evolution of human development on Earth, and they relate the classifications to the Big History thresholds. As table 12.4 suggests, teachers could use the well-known mnemonic device—Daring Kings Play Chess On Fine Grain Sand—to help students recall this complex information. From this, students begin to grasp that they are eukaryotic, or multicelled, mammals from the order of primates. Table 12.5 expands on this knowledge by tracing human evolution even further back, to the Cambrian explosion at the

TABLE 12.3   Basic Human Taxonomy Chart

Classification	Stage in human evolution	Description
Domain or super-kingdom	Eukaryota/eukaryote	Made of eukaryotic, (multi-)celled organisms
Kingdom	Animalia/animal	Not plants or fungi
Phylum	Chordata/backbone	Animals with backbones
Class	Mammalia/mammals	Chordates that suckle their young
Order	Primata/primates	Large, tree-dwelling mammals
Superfamily	Hominoidae/hominoid	Humans and all apes: chimps, bonobos, gorillas, gibbons, and orangutans
Family	Homininae/hominid	Including gorillas, chimps, bonobos, and humans
Subfamily	Homininae/hominine	Bipedal apes (upright, with two-footed posture): every species on the human side since the split from the chimpanzee line
Genus	*Homo*	Bipedal apes with brains larger than 800 cubic centimeters
Species	*Homo sapiens*	Anatomically modern humans: only remaining species of hominines

NOTE: This table was adapted from David Christian, Cynthia Brown, and Craig Benjamin, *Big History: Between Nothing and Everything*, 81.

outset of the **Paleozoic era.** Over the course of eight million years, the family of hominoids of this primate order set an evolutionary course for modern-day humans, so that today our species is the only species of hominine remaining. This is important information for students to understand. In order to comprehend the increasing complexities of the later agrarian and industrial societies that are the focus of Thresholds 7 and 8, they must begin to grasp that during the Paleolithic era an important course was set that directly impacts their lives today.

THE FOUR FEATURES OF COMPLEXITY IN THRESHOLD 6

Another pedagogical reason to teach all three mini-thresholds of Threshold 6 is that doing so provides students with an additional opportunity to analyze this eight-million-year history in relation to the Big History framework of complexity observed in the other major thresholds.

TABLE 12.4   Basic Human Taxonomy Chart within a Threshold Framework

Threshold 5					Threshold 6		
3.8 billion to 8 million years ago					8 million to 10,000 years ago		
Daring	*Kings*	*Play*	*Chess*	*On*	*Fine*	*Grain*	*Sand*
Domain	Kingdom	Phylum	Class	Order	Family	Genus	Species
Eukaryotic cells	Animals	Animals with chordata/ backbones	Mammals or chordates that suckle their young	Primates or tree-dwelling mammals	Bipedal apes (upright, with two-footed posture): every species on the human side since the split from the chimpanzee line	Bipedal apes with brains larger than 800 cubic centimeters	Anatomically modern humans: only remaining species of hominines

DIVERSE COMPONENTS

The basic building components or ingredients necessary for human life had emerged in the previous threshold, as explained earlier. Like other biological life-forms on Earth, humans possess the metabolic capacity to extract energy from their environment; they can reproduce exact replicas of themselves; and they have the capability to constantly adapt to their environment through slow change and the appearance of new forms through natural selection, but with a highly developed manipulative, perceptive, and neurological capacity.

SPECIFIC ARRANGEMENTS

The human body, a highly specific biological structure governed by human DNA, evolved unique characteristics, such as bipedalism and hand dexterity, a large brain, and full speech with complex syntax.

FLOWS OF ENERGY

With the evolution of a powerful body, humans transformed themselves into highly efficient energy-extracting machines—structures that are held together by the constant flows of energy available to them and extracted by their cells from the Earth's rich and bountiful resources. Humans experienced "ideal Goldilocks conditions,"

TABLE 12.5   A Brief History of Life on Earth

Eight Stages of Life of Threshold 5			
Pre-Cambrian explosion (4.5 billion to 600 million years ago)		*Post-Cambrian explosion* (600 million to 6 million years ago)	
Stage 1	First organisms emerge 3.8 billion years ago	Stage 5	First multicelled organisms emerge 600 to 550 million years ago
Stage 2	Photosynthesis emerges 2.5 billion years ago	Stage 6	First vertebrates emerge 400 million years ago
Stage 3	First eukaryotes emerge 2.5 to 1.5 billion years ago	Stage 7	Life moves from the sea to colonize land 475 to 360 million years ago
Stage 4	Sexual reproduction emerges 1 billion years ago	Stage 8	Dinosaurs emerge 245 million years ago, die off 65 million years ago
			Mammals emerge 200 million years ago, begin to flourish 65 million years ago

NOTE: This table was adapted from David Christian, Cynthia Brown, and Craig Benjamin, *Big History: Between Nothing and Everything*, 68–76.

in which they had the opportunity to develop over a long period of time (millions of years) the capacity to successfully adapt to environments conducive to their flourishing. This flourishing was rooted in their ability to both digest plants and subdue existing megafauna.

sun   →   cells of plants   →
sun → cells of plants → cells of animals →   cells of *Homo sapiens*

EMERGENT PROPERTIES

The three previous features of increasing complexity, held together by the powerful energetic resources of Earth, created and formed new, complex human cultures and societies made possible by collective learning. This capacity to share information precisely and rapidly allowed information to accumulate at the level of the community of the species, giving rise to long-term historical change.

HOW EMERGENT PROPERTIES GIVE RISE TO THRESHOLD 7

The new emergent property of collective learning is the key to explaining the continuing development and expansion of complex agrarian societies during Threshold 7.

Small human communities will expand from villages and towns to cities, states, and huge agrarian civilizations. These societies will fracture into complex communities that will develop hierarchies of power through the control of resources by powerful elites and divisions of labor through race, class, and gender. *Homo sapiens* will strengthen its position as the most powerful species on the planet during Threshold 7 through increasing collective learning networks, and humans will develop innovative ways to manipulate and extract resources from their environment. Threshold 6 remains the critical link between the beginning of life on Earth in Threshold 5 and the many types of agrarian civilizations emerging in Threshold 7, as well as later industrial civilizations that extract energy through the innovative use of fossil fuels, atomic reactors, wind- and water-driven electric turbines, and solar photovoltaic cells.

STUDENT LEARNING OUTCOMES AND POSSIBLE ASSESSMENTS

By the end of this unit, students will demonstrate the ability to

1. summarize the basic course of human evolution in Threshold 6 using the scientific taxonomy or biological classification of human beings (assessment: quiz or short writing assignment);

2. discuss aspects of the first mini-threshold, "Hominine Evolution," prior to the emergence of *Homo sapiens* (assessment: reflective writing);

3. analyze aspects of the Paleolithic era that led to the emergence of complex hunter-gatherer and foraging societies in Afro-Eurasia, the Americas, the Pacific, and Australasia (assessment: exam or essay); and

4. describe and evaluate the types of evidence scientists use to study human evolution, such as fossils, stone tools, the study of modern primates, genetic comparisons, and climate change (assessment: essay or exam).

CHALLENGES IN TEACHING THRESHOLD 6

The biggest challenge a teacher faces in teaching Threshold 6 is the considerable amount of information and material that constitutes this period. This threshold could easily be the topic for a full-semester course. But often, in a Big History course, one has only a few class days in which to present this eight-million-year history. That is why breaking it down into the three mini-thresholds has been helpful for me, as is connecting each of these mini-thresholds with one category of human development from the taxonomic table.

On the first day, I teach the students the mnemonic device "Daring Kings Play Chess On Fine Grain Sand." Then I introduce the students to the "family" of hominines, stressing their characteristics, their great variety, and the many famous and recent archaeological discoveries that are continually expanding our knowledge of these early humanlike ancestors. On the second day, I discuss the "genus" stage and tell students about the emergence of *Homo sapiens,* as well as other species of this same genus, such as *Homo erectus* and *Homo ergaster.* I emphasize that early humans lived among a great variety of other groups of their genus, giving special attention to comparing and contrasting *Homo sapiens* with Neanderthals. Again, there are many topics to choose from in order to expand the students' knowledge: the **Out of Africa hypothesis** (modern humans evolved in Africa) versus the **multiregional hypothesis** (modern humans evolved across Afro-Eurasia), or the various debates scientists engage in about where and when our species evolved. Another topic for this second day is a discussion of what makes humans human: their use of tools and ability to create technology, their upright stance and bipedalism, and their special skill of sophisticated speech. On the third day, I focus on the spectacular success of the human "species," stressing that *Homo sapiens* is the last of the hominine family still standing. I link this with humans' great success in the Paleolithic era, when they successfully navigated the globe and established powerful communities of hunter-gatherer or foraging communities on all the continents. With a clear focus, whether the teacher has three days or three weeks to present this material, Threshold 6 can be taught in a way that excites students to forward their own personal study of this otherwise broad topic.

Another challenge the teacher faces is the multidisciplinarity of this threshold. To teach it effectively, one must be familiar with many different academic disciplines: paleoanthropology, archaeology, history, physiology, anatomy, genetics, and evolutionary biology, to name just a few. One must also be able to interpret and analyze different forms of evidence, such as fossils, stone tools, modern primate studies, genetics, and climate change. Walking students through Threshold 6 adds considerable complexity to the story that is being told. This requires the teacher to be comfortable with a multidisciplinary approach to teaching. The best way to develop such an approach is to turn to the incredibly helpful scholarship that is being produced right now. Keep reading and keep teaching until you find your own way to explain this complex and critical threshold. Stay open, and let your own curiosity expand until you can teach this unit effectively and well.

CONCLUSION

As an American historian accustomed to focusing on just one aspect of human history, learning to teach Threshold 6 as part of our Big History program has opened me up to the great value of studying history on such grand scales. Now I can better grasp the greater context in which American history unfolds.

The most significant goal of teaching Threshold 6 is to get students to understand that the emergent property of this threshold is human culture; that because of humans' ability to learn collectively, human culture has now taken precedence over biological evolution; and that the next thresholds of agrarian and industrial civilizations have been built on the foundations of this pre-ancient threshold. I have come to see that leaving this threshold out of *any* history or humanities program is detrimental to the understanding of our future on this planet. Because Threshold 6 teaches the critical history of our family of hominines, our genus, and our species, if we don't recall this history and ponder its meaning, we may not long remain the "last ape standing."[11]

ACTIVITIES AND EXERCISES FOR TEACHING THRESHOLD 6

HOMINOID SKULL LAB

Adapted by J. Daniel May from "Hominid Cranial Comparison (The Skulls Lab)" by Martin Nickels, with images by Maia Kobabe

Threshold Related to the Activity: Threshold 6

Category of Activity: Brief lecture, learning activity

Objective: To equip students with the observational and deductive skills to understand how scientists have mapped the evolution of *Homo sapiens*. Also, to encourage the use of a hands-on, inquiry-based approach to learning and to "set the scene" for hunter-gatherers later in the course.

Overview of Activity: Students examine a set of anonymous model skulls and observe their skeletal features. With the instructor's guidance, the students take notes on the uniqueness of each skull, and then classify and identify each specimen. The entire activity should take approximately seventy minutes.

Faculty Preparation for Activity:

To familiarize yourself with the anatomy of the modern human skull, materials written for artists (rather than for medical students) are the clearest, quickest, and easiest to use, and are readily available both in print and online. One such source is *Classic Human Anatomy: The Artist's Guide to Form, Function, and Movement*, by Valerie L. Winslow.

To familiarize yourself with the hominoid fossil record, an excellent book is *The Last Human: A Guide to Twenty-Two Species of Extinct Humans*, by Esteban Sarmiento et al. *The Last Human* is divided into orderly, discrete chapters on each species, so it is easy to read up on only the ones you need to know about. (Note: the skull identified in the set as *Australopithecus boisei* is referred to in this book as *Paranthropus boisei*. It is the same fossil; the different genus name reflects a different interpretation of where this species belongs in our family tree.)

Gather the following materials for use in this activity:

- Hominoid skull set: seven resin skulls, half-scale. Skulls Unlimited (www.skullsunlimited.com/) is a great supplier. (Their "hominoid skull set" comes with a single page (double-sided) of "liner notes" that offers a succinct paragraph about each of the seven skulls.)
- In order to encourage students to examine the skulls for themselves, the species of each skull is not identified until the end of the lab. The skulls

are identified only by a number (use the last digit of the manufacture's serial number, which is imprinted on the base of each skull and inside each corresponding jawbone). It is best to label each skull and jawbone with the number on a small sticker so the pieces are not confused.

- Seven containers. Lidded, plastic boxes lined with bubble wrap are best because they encourage students to handle the skulls delicately and are great for storage. Label these boxes with the skull numbers, too, for easy reference.
- Handout packet (document titled "Hominoid Skull Comparison Lab: Student's Guide," included below)
- Diagram of human family tree. There are many excellent diagrams available from various sources, including National Geographic and the Smithsonian Institution(see http://humanorigins.si.edu/evidence/human-family-tree). You may choose to distribute copies of these to your students or project the diagram onto a screen.

Cost: About $550 for a skull set and $20–$80 for plastic boxes

Student Preparation for Activity: n/a

In-Class Sequence of Activity:

I. Instructor introduces and provides context for activity (approximately fifteen minutes)

   A. Prepare students to examine skulls (five minutes of introduction and one minute to explain each of the eight observable features).

      1. Remind students of earlier lessons on genetics, natural selection, and evolution.

      2. Direct students to note the uniqueness of every face in the room. Point: Although we all have essentially the same skull, minor variations in the shape of each individual skull gives us each a unique face. We take this variability so much for granted that we find it remarkable only when two faces are truly alike, as with identical twins. This is an example of normal, healthy variation within a population. Such variation is the raw material of natural selection.

      3. Draw students' attention to the essential sameness of all seven hominoid skulls; they share the same bone structures, same number of teeth, same holes for blood vessels and nerves, and so on. Point: One of the ways genes work is by acting like timers that start and stop the growth of physical features. Thus, the variability in

the hominoid skulls can be produced by small variations in rates of growth—some have a slightly more pronounced browridge, some have smaller canine teeth, and so on, and these variations can accumulate over generations.

    B. Direct students to the first page of the skull lab handout, which explains the skull features in more detail and includes a line drawing of Skull #4. Explain the skull features and demonstrate with Skull #4, highlighting the contrasting features on modern human skulls. Students follow along with instructor, circling the correct descriptive terms for Skull #4 on their handout.

II. Students examine and compare skulls (approximately thirty-five minutes total: five minutes for each group to examine six skulls, and five minutes for additional instructions at the halfway point)

    A. Students form six groups and examine skulls.

        1. Each group has one skull to examine at a time. Students record their observations on the corresponding line drawings for each skull.

        2. Instructor rotates among the groups to answer questions and direct observations as needed.

        3. Every five minutes students rotate skulls.

        4. Skull #4 (which was used for the introduction) should rotate in the opposite direction as the rest of the skulls. This insures that at some point each group will have two skulls to compare side by side.

    B. At the halfway point (i.e., after each group has examined three skulls), the instructor briefly halts the group activity to give further direction to the entire class.

        1. Students are directed to look at the "Grouping Your Observations" page, and are advised that they now have enough empirical observations to attempt some rudimentary interpretation of what they are looking at. Simply put, they should try to determine whether each skull is a modern ape, a member of genus *Australopithecus,* or a member of genus *Homo.*

    C. Students resume examination of the remaining three skulls.

III. Instructor identifies and discusses skulls (approximately twenty minutes: just under three minutes per skull)

    A. Instructor announces an end to the group activity and returns class to lecture mode

1. Instructor identifies each skull individually—the species, its distinctive features, and the significance of the observable features
2. Suggested order:
   Skull #4 Gorilla
   Skull #3 Chimpanzee
   Skull #2 *Australopithecus afarensis* ("Lucy")
   Skull #1 *Australopithecus boisei*
   Skull #5 *Homo erectus*
   Skull #6 *Homo neanderthalensis*
   Skull #7 *Homo sapiens*
3. Instructor sums up with remarks that connect this activity with what will be coming up in the following class (e.g., *Homo sapiens* as hunter-gatherers).

Assessment of Learning: Sheet at the end of the activity (identifying the skulls)
Origin of Activity: Adapted from the original activity by Martin Nickels, professor emeritus of anthropology at Illinois State University and adjunct professor of anthropology at Illinois Wesleyan University.
Source Consulted: Nickels, Martin. "Hominid Cranial Comparison (The 'Skulls' Lab)." Indiana University: Evolution & the Nature of Science Institutes, 1999. Web. 27 Dec. 2013.

Overview: You will examine seven hominoid skulls and compare each of the features described below. Your instructor will explain the terms as they refer to your own skull and the example pictured below (which we will refer to as Skull #4). Then, your instructor will divide you into small groups. Each group will examine all of the skulls in turn, and you will record your observations in this packet under the line drawings of each skull.

Terminology:

Canine diastema: the gap in the teeth that creates a "slot" for the opposing "fangs" when the jaw is shut.

Question to ask for each skull: Are any gaps visible, YES or NO?

Canine teeth, or "fangs": the four longest teeth at the outside "corners" of the mouth used for ripping and tearing

Questions to ask for each skull:

1. Are these teeth LONG or SHORT?

2. Are these teeth SHARP or DULL?

Chin: the bony protuberance below the mouth that forms the front of the jaw

Question to ask for each skull: Does the chin project FORWARD or slope BACKWARD?

Foramen magnum: the large opening in the underside of the skull through which the spinal cord passes

Question to ask for each skull: Is this hole closer to the REAR of the skull or the FRONT?

Forehead: the bony area above the browridge (Hint: Hold the skull so that you are looking at a side profile. Use the top of the browridge as a reference point.)

Question to ask for each skull: Is the forehead relatively HIGH, MEDIUM, or LOW?

Prognathism, or "snout": the protrusion of parts of the face below the eyes (as on a dog)

Question to ask for each skull: Is the "snout" LARGE, SMALL, or ABSENT?

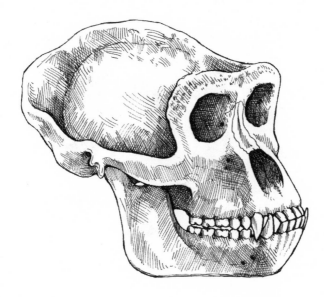

FIGURE 12.1
Skull #4 (example).

Sagittal crest: the bony ridge along the top of the skull

Question to ask for each skull: Is the *crest* LARGE, SMALL, or ABSENT?

Supraorbital browridge: the bony ridge protruding above the eyes

Questions to ask for each skull:

1. Is the ridge LARGE, MEDIUM or SMALL?
2. Are the ridges of each eye socket SEPARATE or CONTINUOUS?

SKULL #4 (Example)

Examine and compare skulls: For each skull, use the questions below to guide your observations and refer to the glossary of terms on the first two pages of this handout to help you answer the questions. Quickly but carefully examine each skull and discuss the questions with the other members of your group. Take notes in the spaces provided below.

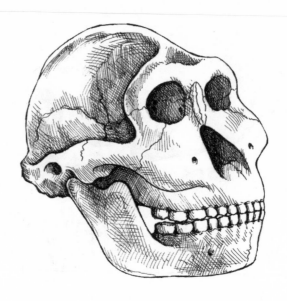

FIGURE 12.2
Skull #1

SKULL #1

1. *Canine diastema:* Are any gaps visible, YES or NO?

2. *Canine teeth,* or "fangs": Are they LONG or SHORT? SHARP or DULL?

3. *Chin:* Does it project FORWARD or slope BACKWARD?

4. *Foramen magnum:* Is it closer to the REAR of the skull or the FRONT?

5. *Forehead:* Is it HIGH, MEDIUM, or LOW?

6. *Prognathism,* or "snout": Is it LARGE, SMALL, or ABSENT?

7. *Sagittal crest:* Is it LARGE, SMALL, or ABSENT?

8. *Supraorbital browridge:* Is it LARGE, MEDIUM, or SMALL? SEPARATE or CONTINUOUS?

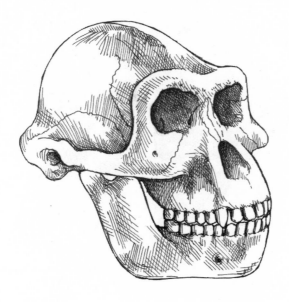

FIGURE 12.3
Skull #2

SKULL #2

1. *Canine diastema:* Are any gaps visible, YES or NO?

2. *Canine teeth,* or "fangs": Are they LONG or SHORT? SHARP or DULL?

3. *Chin:* Does it project FORWARD or slope BACKWARD?

4. *Foramen magnum:* Is it closer to the REAR of the skull or the FRONT?

5. *Forehead:* Is it HIGH, MEDIUM, or LOW?

6. *Prognathism,* or "snout": Is it LARGE, SMALL, or ABSENT?

7. *Sagittal crest:* Is it LARGE, SMALL, or ABSENT?

8. *Supraorbital browridge:* Is it LARGE, MEDIUM, or SMALL? SEPARATE or CONTINUOUS?

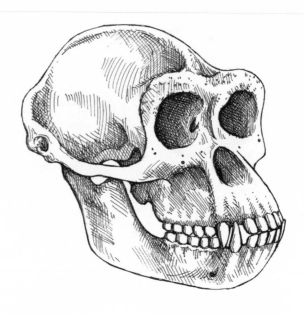

FIGURE 12.4
Skull #3

SKULL #3

1. *Canine diastema:* Are any gaps visible, YES or NO?

2. *Canine teeth,* or "fangs": Are they LONG or SHORT? SHARP or DULL?

3. *Chin:* Does it project FORWARD or slope BACKWARD?

4. *Foramen magnum:* Is it closer to the REAR of the skull or the FRONT?

5. *Forehead:* Is it HIGH, MEDIUM, or LOW?

6. *Prognathism,* or "snout": Is it LARGE, SMALL, or ABSENT?

7. *Sagittal crest:* Is it LARGE, SMALL, or ABSENT?

8. *Supraorbital browridge:* Is it LARGE, MEDIUM, or SMALL? SEPARATE or CONTINUOUS?

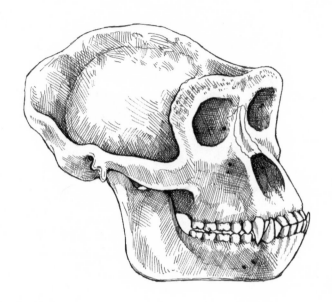

FIGURE 12.5
Skull #4

SKULL #4

1. *Canine diastema:* Are any gaps visible, YES or NO?

2. *Canine teeth,* or "fangs": Are they LONG or SHORT? SHARP or DULL?

3. *Chin:* Does it project FORWARD or slope BACKWARD?

4. *Foramen magnum:* Is it closer to the REAR of the skull or the FRONT?

5. *Forehead:* Is it HIGH, MEDIUM, or LOW?

6. *Prognathism,* or "snout": Is it LARGE, SMALL, or ABSENT?

7. *Sagittal crest:* Is it LARGE, SMALL, or ABSENT?

8. *Supraorbital browridge:* Is it LARGE, MEDIUM, or SMALL? SEPARATE or CONTINUOUS?

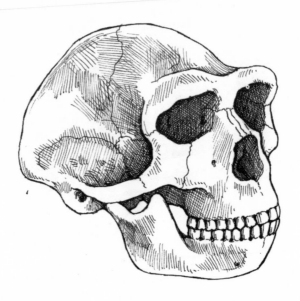

FIGURE 12.6
Skull #5

SKULL #5

1. *Canine diastema:* Are any gaps visible, YES or NO?

2. *Canine teeth,* or "fangs": Are they LONG or SHORT? SHARP or DULL?

3. *Chin:* Does it project FORWARD or slope BACKWARD?

4. *Foramen magnum:* Is it closer to the REAR of the skull or the FRONT?

5. *Forehead:* Is it HIGH, MEDIUM, or LOW?

6. *Prognathism,* or "snout": Is it LARGE, SMALL, or ABSENT?

7. *Sagittal crest:* Is it LARGE, SMALL, or ABSENT?

8. *Supraorbital browridge:* Is it LARGE, MEDIUM, or SMALL? SEPARATE or CONTINUOUS?

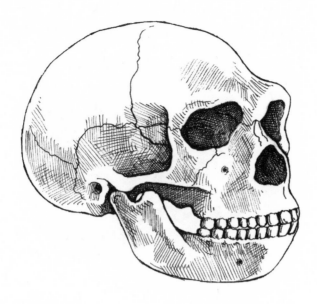

FIGURE 12.7
Skull #6

SKULL #6

1. *Canine diastema:* Are any gaps visible, YES or NO?

2. *Canine teeth,* or "fangs": Are they LONG or SHORT? SHARP or DULL?

3. *Chin:* Does it project FORWARD or slope BACKWARD?

4. *Foramen magnum:* Is it closer to the REAR of the skull or the FRONT?

5. *Forehead:* Is it HIGH, MEDIUM, or LOW?

6. *Prognathism,* or "snout": Is it LARGE, SMALL, or ABSENT?

7. *Sagittal crest:* Is it LARGE, SMALL, or ABSENT?

8. *Supraorbital browridge:* Is it LARGE, MEDIUM, or SMALL? SEPARATE or CONTINUOUS?

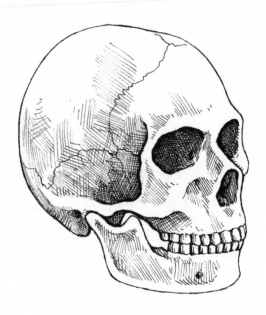

FIGURE 12.8
Skull #7

SKULL #7

1. *Canine diastema:* Are any gaps visible, YES or NO?

2. *Canine teeth,* or "fangs": Are they LONG or SHORT? SHARP or DULL?

3. *Chin:* Does it project FORWARD or slope BACKWARD?

4. *Foramen magnum:* Is it closer to the REAR of the skull or the FRONT?

5. *Forehead:* Is it HIGH, MEDIUM, or LOW?

6. *Prognathism,* or "snout": Is it LARGE, SMALL, or ABSENT?

7. *Sagittal crest:* Is it LARGE, SMALL, or ABSENT?

8. *Supraorbital browridge:* Is it LARGE, MEDIUM, or SMALL? SEPARATE or CONTINUOUS?

Grouping your observations: Now that you have examined all seven skulls, use your observations to place each skull within one of the three classifications below. Classify each skull as an APE, genus *HOMO*, or genus *AUSTRALOPITHECUS*. (Note: There are three LIVING SPECIES above the dotted line and four EXTINCT SPECIES below the dotted line.)

APES:

SKULL #_____

SKULL #_____

       HOMO:

       SKULL #_____

       living

----

       extinct

       SKULL #_____

       SKULL #_____

              AUSTRALOPITHECUS:

              SKULL #_____

              SKULL #_____

Neal Wolfe

Threshold Related to the Activity: Threshold 6

Category of Activity: Learning activity

Objective: While the Paleolithic era spanned many thousands of years, with a wide variety of manifestations of human communities, the following exercise may help students relate to what a hunting and gathering lifestyle may have been like. This exercise is not meant to suggest that all such communities would be organized accordingly. Rather, it is intended to prompt students to place themselves hypothetically within such a community, imagine what it would be like and what their role in it would be, and hopefully make a connection to how they might see their role in their own community today.

Overview of Activity: This activity begins with a brief introduction by the instructor, putting the hunting and gathering era in the context of Threshold 6. The instructor then guides the class through a brainstorming session about the roles and responsibilities of individuals in a hunting and gathering community, recording students' ideas on the board. Students decide which of these roles they would be most suited to, write a paragraph response to that decision, group themselves with other students in the same role, and reflect as a class on the benefits and drawbacks of a hunting and gathering lifestyle as well as what we still have in common with communities from that era. This activity takes approximately forty-five minutes.

Faculty Preparation for Activity: No preparation required beyond normal classroom preparation on hunting and gathering lifestyle

Cost: n/a

Student Preparation for Activity: No preclass preparation for students beyond assigned reading on hunting and gathering era

In-Class Sequence of Activity:

1. Stipulate that the class is a hunting and gathering community. Certainly the size of a Big History class would be a reasonable one. So we imagine that our class is such a community, while noting that in an actual Paleolithic community the age range would vary more.

2. Establish some context. Imagine that we live *here*, and that we may move from place to place as we follow sources of food, such as animals to hunt and fish and plants to harvest. We may also move according to seasonal changes in climate, and possibly in relation to the movements or actions of neighboring communities.

3. Ask what sorts of activities we need to accomplish in order to survive and, hopefully, prosper. Write on the board the various possible roles that are associated with those activities: hunters; gatherers; artisans (such as those who make nets for fishing or catching birds, spears for hunting, or pots for cooking); healers (those who know which plants provide medical benefits for particular illness or injuries, and how to prepare and apply them); scouts (those who may go ahead of the group in order to find a source of water or to see if a source of food is in season yet); possibly "seers" (those who have an ability to intuit when it is time to move, or which direction to go). Does the group need a leader? And so on. (Of course, in reality an individual may move from one role to another, as necessary or desirable.)

4. After all the likely roles are listed on the board, ask the students to think about and choose which role they feel they would be best suited for or most inclined to take on. Have them take a few minutes to write a paragraph or two on why they feel they would be best suited for that role. (This is important in order to make a connection to their understanding of how they see themselves fitting into their own communities today.)

5. Have the students who chose the same role join together in groups and discuss among themselves what they think a typical day or lifestyle would be like. Then have each group present its findings to the class. Discuss the ways in which we may coordinate our activities for the benefit of the entire community.

6. Emphasize that through our exercise, we have explored how a hunting and gathering community may have functioned, in a time when humans' relationship to nature was more direct and experiential. Suggest that even though the circumstances of our present-day lives are greatly changed, we still function as parts of a whole, with individuals serving in the various capacities required for our society to succeed.

Assessment of Learning: This activity culminates in a class discussion. Students are encouraged to reflect on how this exercise gives them a better sense of what the hunting and gathering lifestyle may have been like; how each person in a community contributed according to their strengths; how the community's survival and prosperity depended on this cooperative engagement; and finally, how we may see ourselves, even in a much more complex society, as necessary contributors to a healthy, prosperous community.

Origin of Activity: This is an original activity.

Source Consulted: Christian, David, Cynthia Brown, Craig Benjamin. *Big History: Between Nothing and Everything.* New York: McGraw-Hill, 2014. Print.

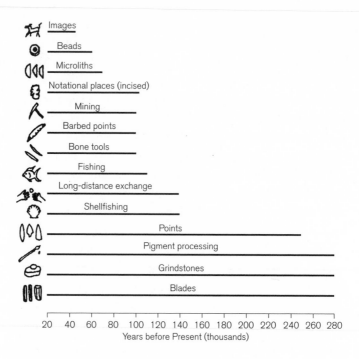

Images

Beads

Microliths

Notational places (incised)

Mining

Barbed points

Bone tools

Fishing

Long-distance exchange

Shellfishing

Points

Pigment processing

Grindstones

Blades

20 40 60 80 100 120 140 160 180 200 220 240 260 280
Years before Present (thousands)

FIGURE 12.9
Behavioral innovations of the Middle Stone Age in
Africa. (Image redrawn from Sally McBrearty and
Alison S. Brooks, "The Revolution that Wasn't: A
New Interpretation of the Origin of Modern
Human Behavior." *Journal of Human Evolution* 39
[2000]: 453–563.)

STONE AGE TOOLS (CREATIVE WRITING)

Judith Halebsky

Threshold Related to the Activity: Threshold 6

Category of Activity: Learning activity, writing activity

Objective: This activity asks students to reflect on how new technologies and behavioral innovations changed daily life and social relationships for early humans.

Overview of Activity: This activity joins skills in creative writing with content in Big History. Students connect aspects of writing monologue and internal monologue with developments in the Paleolithic era. The activity requires two class meetings: one that includes guided writing exercises and a short lecture, and a second in which students share their written work.

Faculty Preparation for Activity: Plan in-class writing prompts and prepare a brief lecture on Threshold 6.

Cost: The only cost is that of photocopying the handout (see figure 12.9).

Student Preparation for Activity: Assign students to read a short dramatic monologue before class. I use "French Fries," by Jane Martin, which is included in Janet Burroway's *Imaginative Writing: Elements of Craft*. Students should also read a section on character voice in the same textbook. This activity follows previous exercises that focus on character development.

In-Class Sequence of Activity:

1. Begin class with a character development writing activity. Have students close their eyes and imagine a character. Have students write a description of the character with the help of writing prompts such as:
    - What is your character's recurring dream?
    - What does your character do first thing in the morning? Describe where they sleep and what they do after they wake up.
    - Your character has a scar somewhere on their body. Write about how they got the scar and how they feel about the scar now.
2. Invite students to share their writing aloud with the class. (Allow students to "pass" if they are reluctant to share their work). Or, in a large class, invite students to share their writing with a partner.
3. Give a brief lecture on Threshold 6.
4. Have students write a monologue with their character as the speaker, responding to the following writing prompts:
    - Writing with your character speaking in the first person, take one point from the lecture and have your character describe it.
    - Imagine your character is living at the time of early humans. In the first person, have your character describe where they live.
    - Again, write with your character speaking in the first person. In what ways does your character adore their sibling and in what ways do they feel threatened by their sibling?
5. Again, invite students to share their writing aloud with the class. (Also, again allow students to "pass" if they are reluctant to share their work). Or, in a large class, invite students to share their writing with a partner.
6. Assign the following writing prompt as homework:

Two siblings are leaving their family to start a new life on a different island. They are each given one "treasure" from their parents. This treasure is one of the tools or skills listed in the *Behavior Innovations of the Middle Stone Age in Africa* chart. Write a monologue spoken by one sibling that discusses which treasure each sibling was given. Discuss what the treasure represents in terms of what their parents wanted to share with them. Also explain the life they will be able to live on the new island with this treasure.

Students should come to the next class with a 250-word monologue. They are required to bring photocopies of the monologue to share with the class.

7. At the beginning of the next class, have all students share the writing they did as homework. Students read their writing aloud and classmates share their responses to the writing.

Assessment of Learning: Students bring copies of their work to class. They share their work in a writing workshop format.

Sources Consulted:

Burroway, Janet. *Imaginative Writing: Elements of Craft.* 3rd ed. New York: Longman, 2011. Print.

McBrearty, Sally, and Alison S. Brooks. "The Revolution that Wasn't: A New Interpretation of the Origin of Modern Human Behavior." *Journal of Human Evolution* 39 (2000): 453–563. Print.

LANGUAGE CHANGE SIMULATION

J. Daniel May

Threshold Related to the Activity: Threshold 6

Category of Activity: Learning activity

Objective: To experience how a language both changes and divides over time, first into regional variations, or dialects, and eventually into a family of separate languages

Overview of Activity: Students are presented with an imaginary (but historically realistic) scenario: the entire class represents a Paleolithic society speaking the same language. Students are divided into groups that represent smaller regional populations. Using the language change worksheet (included at the end of this activity write-up), students progressively alter the pronunciation of a single sentence such that each regional group begins to sound increasingly different from every other group. Regional dialects emerge by the end of the simulation. In the process, everyone has a chance to sound silly and have some fun with words! This activity takes approximately one hour (varies with class size).

Faculty Preparation for Activity: The simulation can be conducted simply with the information below. For the instructor wishing for more thorough background knowledge, the best resource is a short essay by Stephen Jay Gould, "Grimm's Greatest Tale," included in his anthology *Bully for Brontosaurus* (1991).

Cost: Photocopying the language change worksheet (one page per student)

Student Preparation for the Activity: Textbook reading or a prior lecture should have brought students up to the time of Paleolithic *Homo sapiens* emerging in Africa, the spread of *Homo sapiens* out of Africa, and the subsequent peopling of the globe.

In-Class Sequence of Activity:

A. Introduce the activity in a lecture.

   1. Instructor establishes the relevance of the activity to the current threshold: Our ancestral population in Africa may have been as small as five thousand individuals, who may well have all spoken the same language (or at least a family of very similar languages). Yet today there are at least five thousand different languages spoken across the globe. Why don't we all speak the same language? How could one Paleolithic language turn into five thousand modern languages? Students will simulate the real-life language changes experienced by expanding and migrating human populations over many generations.

   2. Two critical facts about human language need to be made absolutely clear:

      a. All human babies are born with the same genetically endowed "starter kit" for learning a human language, sometimes referred to as "universal grammar."

      b. Language change across generations is a *normal* and *inevitable* result of the process of children learning to speak their native language. (In brief, this is because children do not produce perfect copies of the language of the adults around them; rather, children *re-create* the language for themselves, introducing tiny variations that accumulate over many generations.)

B. Explain the imaginary scenario.

   1. The class as a whole represents a Paleolithic human society whose members call themselves "the Homelanders" and all speak the same language. The modern English being spoken in the classroom represents the Homelander language.

   2. The Homelanders have an important legend about a hero named "Po-To-Key," and the legend is so well known that it is not usually necessary

to retell the entire story. For ceremonial purposes, the Homelanders simply recite the last line of the legend:

> "Po-to-key killed the great beast with one blow of that big rock!" (This sentence can be compared to stock phrases such as "Remember the Alamo!" and "Damn the Torpedoes!" which English speakers will invoke without context.)

3. Every year the Homelanders have a Great Gathering, bringing together representatives from all the regional tribes. The "head of ceremony" (the instructor) conducts a roll call of the attending tribes. For example, the head of ceremony will ask, "Is the north tribe here?" and a member of the north tribe will reply by saying, "Po-to-key killed the great beast with one blow of that big rock!" This provides a rationale for saying the same sentence over and over, and it allows the students to add a little play-acting to the activity.

C. Divide students into groups.

1. There should be at least four groups (north, west, south, and east), each representing a regional tribe of the Homelanders. The activity is designed for a class of twenty students, with five students to a group, but the numbers can easily be adjusted to fit class size. Groups can be larger or smaller, as the instructor sees fit. Ideally, at the final phase of the activity (see "The Great Gathering," below) each student should have at least one turn reciting the "Po-to-key" sentence out loud to the entire class.

2. For a larger class, two additional groups can easily be created: northwest and southeast. These groups simply use a mix of the sound changes under the cardinal headings on the worksheet. For example, the northwest group can use the first two sound changes from the north column and the last three sound changes from the west column.

3. Prior to handing out the language change worksheet, the instructor may wish to use a highlighter or colored pen to indicate which column of sound changes each student is using, especially if mixed columns are needed. This takes only a few minutes but pays off in clarity, making the activity flow faster and more smoothly.

D. Students apply the sound changes.

1. Now in their regional groups, the students rewrite the "Po-To-Key" sentence. For each "century" elapsed, one more sound change is added. Therefore, the sound changes accumulate. For example, this would be the final written result for the north group:

Today: "Po-to-key killed *th*e great beast wi*th* one blow of *th*at big rock*!*"

100 years [p] > [b]: Bo-to-keẏ killed *th*e great beast wi*th* one blow of *th*at big rock*!*

200 years [t] > [*th*]: Bo-*th*o-key killed *th*e grea*th* beas*th* wi*th* one blow of *th*a*th* big rock*!*

300 years [k] > [s]: Bo-*th*o-sey silled *th*e grea*th* beas*th* wi*th* one blow of *th*a*th* big ros*!*

400 years [d] > [*th*]: Bo-*th*o-sey sille*th* *th*e grea*th* beas*th* wi*th* one blow of *th*a*th* big ros*!*

500 years [r] > [l]: Bo-*th*o-sey sille*th* *th*e glea*th* beas*th* wi*th* one blow of *th*a*th* big los*!*

2. As students work in their regional groups, the instructor circulates to answer questions, observe, and keep the activity on track. Some tips:

  a. Remind students to pay attention to how a word *sounds,* not how it is *spelled.* For example, the sound change [w] > [v] means that the written phrase "with one" should be changed to sound like "vith vun."

  b. Written "c" (as in "club") and "ck" (as in "rock") are really just the sound [k], so all [k] sound changes apply.

  c. English has more spoken sounds than written letters, and compensates by using digraphs (two-letter combinations for one unit of sound). To draw attention to this, digraphs have been italicized: [*th*] (as in "*th*at") and [*ch*] (as in "*ch*urch").

4. Since students could do this activity by writing individually in silence, they may need to be encouraged to work together in their groups, to check one another's sound changes, and to take turns reading the new pronunciations out loud to one another.

D. Convene the Great Gathering.

  1. After all the groups have had time (about fifteen to twenty minutes) to write out their sound changes, it is time to act out the Great Gathering.

  2. The head of ceremony (instructor) calls all the tribes (groups) to gather as one (that is, to reassemble as a whole so that everybody is ready to listen to their classmates). The head of ceremony greets the gathered Homelanders, thanks them for attending the hundredth Great Gathering, and calls one representative from each tribe to stand. As the head of ceremony calls "north," "south," and so on, each representative recites the "Po-to-key" sentence with its changed pronunciation.

(Note: As head of ceremony, the instructor continues to speak in modern English. The rationale is that, as a keeper of tradition, the head of ceremony continues to speak "Old Homelander," while the tribes speak in their local vernacular. This is analogous to the continued use of Latin in church services, or singing the phrase "auld lang syne" at New Year's parties.)

3. Next, the instructor explains that we are moving ahead in time to the next century. A new set of representatives stands up, each now replying to the roll call with the two-hundred-year version of the "Po-to-key" sentence.

4. With the protocol established, the cycle can be quickly repeated through to the five-hundred-year versions. Moving quickly through the cycles makes the growing contrasts in pronunciation more obvious.

E. Conclude and process the activity.

   1. After the five-hundred-year cycle of recitations, the instructor declares the simulation concluded, and everyone can step out of character.

   2. In an immediate follow-up discussion, the instructor asks students for their observations about what just happened. (Points not made by students should be made by the instructor before the class ends.) The following observations should come out as "take-aways" by the end of class:

      a. Each group sounded progressively different as the sound changes accumulated. This simulates the normal and natural changes that any language exhibits over time.

      b. Each group sounded progressively different from every other group. This simulates the development of regional dialects of a language. (A relevant real-life example would be the different pronunciations of modern English as spoken in Ireland, the United States, and Australia.)

      c. Some groups became difficult to understand, in a breakdown of what linguists call "mutual intelligibility." This simulates the transition from a regional dialect to a separate language. (For English speakers, the relevant real-life example would be the development of prehistoric proto-Germanic into regional dialects, which eventually became the distinct modern languages of English, Swedish, and German, among others.)

      d. The elapsed time of the simulation (five hundred years) is about the same as the time from William Shakespeare to modern English, and the degree of simulated change is comparable.

Language Change Worksheet
The Legend of Po-to-key
A Simulation of Language Change Across Generations and Geography
J. Daniel May

	NORTH	WEST	SOUTH	EAST
100 years:	[p] > [b]	[p] > [f]	[t] > [d]	[t] > [*ch*]
200 years:	[t] > [*th*]	[t] > [s]	[k] > [g]	[k] > [*ch*]
300 years:	[k] > [s]	[k] > [h]	[l] > [r]	[d] > [t]
400 years:	[d] > [*th*]	[*th*] > [d]	[*th*] > [z]	[g] > [k]
500 years:	[r] > [l]	[b] > [p]	[w] > [v]	[s] > [z]

"Po-to-key killed *th*e great beast wi*th* one blow of *th*at big roc*k!*"
100 years: _____
200 years: _____
300 years: _____
400 years: _____
500 years: _____

---

    e.   Restating the relevance to Threshold 6, students should understand
that the simulation is an accurate portrayal of how human languages
must have divided and multiplied as our ancestors left Africa and
populated the globe.

Assessment of Learning: Immediate follow-up discussion in class. Additional
options could include reflective writing or short essay responses on a subsequent
quiz or exam.
Origin of Activity: This is an original activity.
Sources Consulted:
Cavalli-Sforza, Luigi Luca. *Genes, Peoples, and Languages.* Berkeley: U of California P, 2000. Print.
This is the primary source for the information in Gould's article. Chapter 5, "Genes
and Languages," provides the details most relevant to this simulation.
Curzan, Ann, and Michael Adams. *How English Works.* 3rd Ed. New York: Long-
man, 2011. Print.

This is an excellent introduction to linguistics for the non-expert. The technical details are clearly explained in chapters on English phonology, first-language acquisition in children, and the documented history of English.

Gould, Stephen Jay. "Grimm's Greatest Tale." *Bully for Brontosaurus.* New York: W.W. Norton and Sons, 1991. 32–41. Print.

Gould's short essay (ten pages in the paperback edition) is clear and elegant, and it illuminates the triple threads of parallel evidence in archaeology, genetics, and linguistics.

NOTES

1. David Christian, Cynthia Brown, and Craig Benjamin, *Big History: Between Nothing and Everything.* New York: McGraw-Hill, 2014. 81. Print.

2. Donald C. Johanson and Kate Wong, *Lucy's Legacy: The Quest for Human Origins.* New York: Harmony Books, 2009. 3–21. Print.

3. Roger Lewin, *Human Evolution: An Illustrated Introduction.* 5th ed. Malden, Mass.: Blackwell, 2005. 121–122. Print. Chris Scarre, ed., *The Human Past: World Prehistory and the Development of Human Societies.* 2nd ed. London: Thames and Hudson, 2009. 96–97. Print.

4. *Becoming Human: Unearthing Our Earliest Ancestors.* Dir. Graham Townsley WGBH Educational Foundation, 2009. DVD.

5. For excellent textbook sources with separate chapters chronicling human origins in Africa, later hominine dispersals in Eurasia, and the rise of modern humans, along with helpful maps, graphs, photos, and illustrations, see the texts by Roger Lewin and Chris Scarre mentioned above. For another useful illustrated textbook on human evolution, see Chris Stringer and Peter Andrews, *The Complete World of Human Evolution.* 2nd ed. London: Thames and Hudson, 2012. Print.

6. Lewin, *Human Evolution,* 7.

7. Scarre, *The Human Past,* 147.

8. Nicholas Wade, *Before the Dawn: Recovering the Lost History of Our Ancestors.* New York: Penguin Press, 2006. Print. Brian Fagan, *Cro-Magnon: How the Ice Age Gave Birth to the First Modern Humans.* New York: Bloomsbury Press, 2010. Print. Clive Finlayson, *The Humans Who Went Extinct: Why Neanderthals Died Out and We Survived.* Oxford: Oxford UP, 2009. Print. Steven Mithen, *The Singing Neanderthals: The Origins of Music, Language, Mind, and Body.* Cambridge, Mass.: Harvard UP, 2006. Print.

9. Christian, Brown, and Benjamin, *Big History,* 93–100.

10. Ibid., 81.

11. Phrase taken from Chip Walter's book *Last Ape Standing: The Seven-Million-Year Story of How and Why We Survived.* New York: Walker, 2013. Print.

THIRTEEN · Teaching Threshold 7

The Agrarian Revolution

Martin Anderson

And little Sir John and the nut brown bowl and his
brandy in the glass
And little Sir John and the nut brown bowl proved the
strongest man at last
The huntsman he can't hunt the fox nor so loudly blow
his horn
And the tinker he can't mend kettle or pots without a
little barleycorn

Traditional / STEVE WINWOOD[1]

One of the main difficulties of teaching the agricultural revolution in general, and agrarian civilizations in particular, is that most people in modernized societies, including students, are divorced from the nature of agricultural life. I had only a brief experience of it myself, as a child, having spent a few weeks one summer on my uncle's farm in Minnesota. I recall several milk cows, chickens, and a barn in addition to the house. I'm sure he was growing something, as I rode in a tractor. I saw the farm animals as pets and had a great time. I wasn't there long enough to experience the tedium of endless repetition of doing the same necessary work every day to keep the farm running.

In teaching about agriculture, one strategy is to ask students if any of them grew up on a farm. This can be a risky proposition. In the United States, part of the process of urbanization included images of rural people in popular culture as roughhewn farm folk, in contrast to sophisticated city people, particularly in movies such as the *Ma and Pa Kettle* series and television shows such as *The Beverly Hillbillies, The Farmer's Daughter,* and *Green Acres.* As urbanization and agribusiness wiped out family farms after World War II there was some nostalgia for farm life, too, provided in television shows such as *The Waltons.* Today, consumer capitalism in the United States promotes embrace of a mythical country life, mostly by people who have never experienced a rural, agricultural existence. The central theme of this lifestyle is listening to country-western music (which is neither country nor Western) and pride in being country, again allowing for comedic portrayal of such people as unsophisticated hicks. Consequently, the one or two students in a class who may have actually grown up on a farm may be reluctant to identify themselves as such. If these portrayals can be downplayed or used as a discussion point, it is useful to have a student describe farm life so other students can get some sense of what life may have been like for the vast majority of human beings who lived under an agrarian civilization. One can lead this into a discussion of what life must have been like for peasants, serfs, and agricultural day laborers.

Students also need to understand that they are still dependent on agriculture (far more than on oil), perhaps piquing their interest in the agricultural activity that goes on all around them, mostly without their knowledge. This could lead into discussion of the continued existence of day agricultural laborers and the reality of their lives. I have, usefully, gotten students to recognize that they are just modern-day hunter-gatherers, only they do their hunting and gathering in their refrigerators or in grocery stores.

When teaching the agrarian civilizations portion of Threshold 7, one must avoid teaching about the array of other civilizations that came and went from 2500 B.C.E. to 1800 C.E. In fact, looking at the achievements of human civilizations diverts one's attention from the agricultural activity that underlay and supported all the agrarian civilizations. At the bottom is the simple reality that humans need to eat food to survive. Food scarcity may have driven humans to invent agriculture.[2] From 2500 B.C.E. to 1800 C.E. (and even beyond), except for those humans who remained in cultures not under agrarian civilizations, the main economic activity of humans was agriculture. The period between 2500 B.C.E. and 1800 C.E. should be seen as one of "humans with command of agriculture." The focus of the agrarian civilization portion of Threshold 7 should be on the forms and necessary infrastructures of agriculture, as well as its impacts and consequences.

Focusing on agriculture is different from the typical approach to civilization prior to industrialization, which takes the excess wealth produced by agriculture for granted, talks little about actual agricultural processes, and emphasizes the achievements of humans freed from agricultural labor in the areas of intellectual thought (religion, philosophy, or science), art, or social engineering. The implication is that these achievements of civilization are what ultimately led to the Industrial Revolution. Whatever the merits of this assumption, it diverts students' attention from the actual ongoing impact of agricultural activity by humans and the problems created by the dependence of seven billion people on that agricultural activity. The approach taken here is that learning about agricultural activity and its implications is what is most valuable for students in Threshold 7, and that these points are more manageable pedagogically than a survey of the achievements of the large number of civilizations that came and went between 2500 B.C.E. and 1800 C.E. The useful achievements of the agrarian civilizations can be quickly summarized in the advancements in astronomy, engineering, weapons, shipbuilding, and architecture. In each of these areas there was increasing sophistication from 2500 B.C.E. to 1800 C.E. These can be pointed out by selecting a few examples from civilizations from the various regions of the world. Specific civilizations can also be highlighted by linking them to the grain crop that was the basis of their success—for example, southern China and rice, Egypt and wheat, Mexico and maize.

KEY CONCEPTS IN THRESHOLD 7

The key concepts of agriculture are (1) the nature of plants susceptible to cultivation; (2) the nature of animals susceptible to domestication; (3) the requirements of land, water, and nutrients; (4) the limits of human and animal labor; and (5) necessary technology. Points to examine are the processes of stripping the land to prepare for cultivation, developing irrigation to provide a water supply, and adding nutrients to the soil as needed for plants to grow (including understanding the plant cycle of life). Students should also be asked to consider the intense labor required, threats to the fragile process of agriculture (such as insects, plant diseases, and weather), and agriculture's actual results. Additional topics can be found in agricultural products used not for food but as raw material for clothing and drink (wine, beer, distilled alcohol). Overarching all this is that during this entire four-thousand-year period, humans did not command science, which means inventions were of a practical and local nature only.

PLANTS SUSCEPTIBLE TO CULTIVATION

Plants susceptible to cultivation include carbohydrates, which vary by locale, but in general were regionally focused on a single crop: wheat, rice, sorghum, millet, maize, and so on. Fruits, vegetables, and nuts are more difficult to grow and often require more water so were less widespread and were seen as luxuries. The quality of the human diet probably declined as a result of agriculture.[3] Through the eighteenth century many agricultural societies offered little variety in diet. Starting with Rome, for example, for centuries, virtually everyone was dependent on bread. Bread provides a high-calorie diet, supporting heavy labor, but is a poor diet overall.

ANIMALS SUSCEPTIBLE TO DOMESTICATION

Human labor is insufficient to sustain agriculture, so domestication of animals developed along with it. Some animals, such as oxen, provided plowing labor;

others, such as horses and llamas, provided transportation labor. Still others provided material for clothing (sheep), tended other animals (dogs), or provided food (cattle and chickens). In general, domesticated animals were so valuable for these uses that eating them for protein was limited.

REQUIREMENTS OF LAND, WATER, AND NUTRIENTS

The life cycle of plants dictates agriculture. Seeds must be planted at the correct time, and the food product harvested at the right time. Plant varieties require the right amount of water, nutrients, and sunlight; otherwise, they will not grow, or they will produce a poor crop. During this period of agrarian civilizations, crops were highly susceptible to climate; droughts, early freezes, or heavy rains could destroy them or limit their yield.

LIMITS OF HUMAN AND ANIMAL LABOR

Without science or machinery, increasing the food supply meant either trade or increasing the amount of land under cultivation (though toward the end of the period some focus was given to increasing crop yields). While the achievements of agrarian civilizations in organizing human and animal labor to build aqueducts, canals, grand roads, irrigation systems, and religious and public buildings is impressive, individual humans can do only so much. Consequently, the vast majority of humans in agrarian civilizations were part of the agricultural labor force, working as slaves, peasants, serfs, laborers, or farmers. Sharp population loss caused by disease outbreaks or famine would result in contraction of the economy due to lack of labor.

NECESSARY TECHNOLOGY

Though crude by the standards of humans in command of science and industrial machinery, agriculture did foster the invention of various tools, including plows, animal yokes, harnesses, and saddles; vases and barrels for storage; wheeled vehicles for transport; pitchforks, scythes, and threshing wheels; and eventually windmills and waterwheels. Forges and metalworking were necessary inventions to produce tools. Boats and ships were invented for transporting goods. Humans learned how to selectively breed plants and animals.

PREPARING LAND FOR CULTIVATION

This brutal process eliminates everything else living in the area of land, generally known as a field. There are limits on land that can be cultivated. It must be arable. The dictionary defines "arable" simply as "suitable for plowing"—apparently it comes from a Latin word meaning "to plow."[4] Human intervention can bring non-arable land under cultivation, but there are limits to this. A large percentage of the Earth's landmass is simply non-arable. Agrarian civilizations were reaching the limits of land that could be sustainably brought under cultivation.[5] Converting arable land to cultivation means driving some plant, animal, and insect species out (and possibly to extinction), as well as committing resources to sustaining artificial agricultural systems.

IRRIGATION

Plants require water. In natural cycles, plants adapt to rainfall, or grow along rivers, lakes, or streams, or tap groundwater. Consequently, agriculture began near major rivers in plains, or in areas with dependable annual rainfall. Nevertheless, even along the Nile, in Mesopotamia, and in the Indus Valley, irrigation systems were built to divert water for agriculture. Areas with smaller river systems, or those that were dependent on rainfall, were limited as to the type and amount of agriculture they could support. Despite water diversion systems, agrarian civilizations remained subject to the climate, droughts and floods again being the major difficulties. The collapse of the Indus Valley civilization may partially have been the result of an inability to maintain a complex irrigation system.[6]

FERTILIZATION

Plants require nitrogen. Though they lacked science, humans during the agrarian age learned that soil became unable to support agriculture after a few harvests. Recognition that soil needed to be renewed to continue growing cycles led to early fertilizing efforts, such as the use of manure, crop rotation, and the planting of crops like clover and beans. However, much land became non-arable due to nutrient loss, and a common solution was simply to plow new land.

THE FOUR FEATURES OF COMPLEXITY IN THRESHOLD 7

DIVERSE COMPONENTS

Humans, plants, animals. Land, water, nutrients, weather.

SPECIFIC ARRANGEMENTS

As agriculture takes hold, humans reorganize social structures to produce agricultural labor, domesticate animals for additional labor, reroute watersheds to irrigate fields, domesticate certain plants, engage in active destruction of the "wild" (undomesticated animals and plants, and humans not under agrarian civilization), and build up central locations (villages, then cities) that are surrounded by cultivated land and that redirect the excess wealth from agriculture into the activities of humans freed from agricultural labor (and organized into social hierarchies that justify their freedom from labor) and into trade of excess products. These trade routes, over land or water, became the early networks of exchange.

FLOWS OF ENERGY

Agricultural activity allowed for greater concentration of human population, and the increased food supply allowed for population increase. The new flows of energy harnessed by agriculture supported slow population growth across four thousand years. Humans seized control of the production of energy and concentrated it, generating new surpluses of energy—that is, wealth—beyond that possible by humans in societies without agriculture or practicing only limited agriculture.

sun $\rightarrow$ cells of plants $\rightarrow$ cells of *homo Sapiens* $\rightarrow$ work $\rightarrow$ surpluses/wealth
sun $\rightarrow$ cells of plants $\rightarrow$ cells of animals $\rightarrow$ cells of *homo Sapiens* $\rightarrow$ work $\rightarrow$ surpluses/wealth
sun $\rightarrow$ cells of plants $\rightarrow$ cells of animals $\rightarrow$ work $\rightarrow$ surpluses/wealth

EMERGENT PROPERTIES

Agricultural surpluses were converted into wealth within new social hierarchies. Networks of exchange organized around cities to distribute excess wealth. Armies were developed to defend cities and engage in conquest. Innovation led to improvements in metallurgy, agriculture, military arts, and architecture. Exchange contributed to the spread of collective learning. Smaller numbers of people could devote their time to philosophic, political, religious, and legal thought, along with mathematics and early science and contemplation of the stars through astronomy.

HOW EMERGENT PROPERTIES GIVE RISE TO THRESHOLD 8

It is unclear that agricultural activity alone led to the discoveries that underpin the Industrial Revolution. The best that might be said is that the relative success of agrarian civilizations freed at least some people to think about larger issues of the universe and the natural world, leading to a slow accumulation of knowledge, along with an increasing practical body of knowledge learned by solving immediate problems posed by agriculture, including the military arts.

A good many agrarian civilizations came and went during these more than four thousand years without any evidence that they were on the cusp of harnessing the power of steam engines to drive an Industrial Revolution. Some agrarian civilizations, especially China, appear to have accumulated enough collective learning to invent steam engines, yet they never employed them in any revolutionary, transformative way.[7] In addition, collective learning was spread unevenly across the globe during this time, as global connections between agrarian civilizations in different regions of the world were nonexistent or weak until literally the last three hundred years of the period (from 1500 C.E.), and really substantial only in the last hundred years (from 1700 C.E.).

Further, coal, a necessary raw material for steam engines, is unevenly distributed across the globe, and the need to produce heat, leading to knowledge of charcoal and coal, was in more temperate zones not a pressing problem that needed solving.

Finally, the relationship between the Scientific Revolution and its explanations for the universe and the natural world and practical inventions like iron smelting is unclear. In the end, the refinement of the steam engine seems to have been driven not by an agricultural problem, but either by a need to replace charcoal in iron smelting to meet an increased demand for iron, or by a trade opportunity in textiles.[8] By offering more profit, the latter seems to have been the main driver.

The British, who dominated the seas by the end of the 1700s, were in a position to suppress one source of textiles (India), grow the raw material (cotton) in another (the West Indies), and ship it throughout the world, if only they could produce enough finished product in their own country. The solutions needed—rotary piston, blast furnace, high iron production, and coal—drove the Industrial Revolution. Numerous additional applications were rapidly recognized, but not all agrarian civilizations were in a position to capitalize on them before they fell to those humans in command of industrial power in the latter part of the 1800s.

Within this configuration lay an additional problem: what led to Britain's domination of the oceans by 1800? The story of human command of the oceans seems separate from the story of agriculture, and, as the case of China demonstrates, command of agriculture by no means compelled agrarian civilizations to seek command of the seas. Human forging into the oceans has a long and varied history, including, for example, the Polynesian expansion across the Pacific. The main drivers of oceangoing seem to have been migration and trade.

How Britain came to dominate the seas at nearly the same time as Britons were inventing steam power is a long, complicated story in itself, and the relationship between the two is certainly unclear. What can be said is that in eighteenth-century Britain, all the diverse knowledge and materials needed to refine the steam engine existed, and these were ultimately arranged into a practical steam engine, with new properties that produced powerful new flows of energy, vastly aided by Britain's command of the oceans. Constructing a path to that point from the era of agrarian civilizations is by no means clear. In the end, perhaps all we can say is that the Goldilocks conditions needed to start the Industrial Revolution appeared in Britain sometime during the eighteenth century.

If the path to the Industrial Revolution across the four thousand–plus years of the era of agrarian civilizations is unclear, the emergent properties are not. In a very short time, steam power dwarfed the power of human and animal muscle and wind. The best example of this transformation is in Britain's relationship with China. When Britain sent the Macartney embassy to China in 1793, in hopes of improving trade, Emperor Qianlong dismissed both Britain and its embassy as insignificant. Less than fifty years later (during the 1839–1842 Opium War), Britain was able to impose her will on China—thanks to steam gunboats like the HMS *Nemesis*.

Less well understood is how the Industrial Revolution sparked interest in science, particularly in its practical applications, which led to a vast increase in humankind's collective knowledge. Understanding that science could greatly expand their power, rulers throughout the world sought to encourage it and to benefit from its inventions, especially in the military arts. While human knowledge in astronomy, physics, mathematics, and the natural world advanced during the Scientific Revolution and the Enlightenment, human advancement in engineering, medicine, and (unfortunately) weaponry took off only during the nineteenth century. Thanks to science, harnessing electricity and oil quickly brought on a Second Industrial Revolution around 1900. Just as humankind's use of energy from 1800 dwarfs its use in the prior four thousand years, humankind's increase in collective learning from 1800 also dwarfs that of the prior four thousand years.

STUDENT LEARNING OUTCOMES AND POSSIBLE ASSESSMENTS

By the end of this unit, students will demonstrate the ability to

1. describe the cycle of plant life (assessment: quiz or short essay);
2. discuss the impact on the biosphere of agriculture (assessment: quiz or short essay);
3. illustrate the dependence by an agrarian civilization on a particular grain or food type (assessment: short essay);
4. describe the labor needed for agriculture, both human and animal (assessment: quiz);
5. examine the relationship between agriculture and climate (assessment: short essay); and

6. select and connect agriculture to events and conditions in Thresholds 2 through 6; these events and conditions can include (but are not limited to) creation of elements (that make up life, plants), uneven distribution of elements, conditions for life (in the creation of the planet; as a result of plate tectonics; global climate), evolution (distribution of different plants and animals across the globe), *Homo sapiens* (who need to eat to survive) (assessment: essay).

CHALLENGES IN TEACHING THRESHOLD 7

The first challenge an instructor may face in teaching this period is that it is the same period traditionally taught in world history, which focuses on the facts and achievements of the many civilizations that existed during this period. The impulse to replicate that teaching in an extraordinarily condensed version may be overwhelming, and students may be expecting that. However, the main focus in teaching this period should be on agriculture and the critical idea of collective learning. A second challenge in teaching this period is that it is unlikely that the instructor herself or himself is familiar with agricultural life, the life cycle of plants, the tools of agriculture, the problems of agriculture, and the history of plants and agriculture. Secondary sources on these topics may not be readily available to many instructors. A third challenge in teaching this period is teaching the history of how humans came to command the oceans. This needs to be taught to some degree, as overseas trade became more and more a feature of agrarian civilizations and connected the human web. A fourth challenge is reminding students that not all humans lived in agrarian civilizations during this period. Some time will have to be spent on those humans, generally categorized today as indigenous peoples, who were outside agrarian civilizations. These people could have major impacts—for example, the Mongols. Finally, to help students understand the relevance of this period, they should be reminded that we are still in the period of agrarian civilizations because all human civilizations are still based on having enough food to eat.

Discussing agriculture should get students talking about their own eating habits. They can examine issues like organic farming; genetically modified foods; overuse of sugar in food; consequences of a diet high in beef consumption; consequences of overfishing or overharvesting; consequences of some societies becoming dependent on a single cash crop, like coffee; and what science today tells us is a balanced diet. The good news is that we do actually have the capacity to produce enough food for the world's population. Famines are generally the result of local conditions. The problems of agriculture have to do with water use, overuse of fertilizers and pesticides, and encroachment on the natural world with threatened extinctions of plant and animal life not subject to domestication.

TEACHING MODELS AND EXERCISES

After giving a broad overview, the best way to teach agrarian civilizations, in my view, is to go through some examples of staple crops, from their origin to their current form. For example, trace the origin of the potato in South America (the original potato bears little resemblance to what we consider a potato today), to its early cultivation in South America, to its transport to Europe in the sixteenth century, to its popularity as a food in Europe and the United States in the nineteenth century, to its form today. Another example would be sugar, which allows for discussion of the slave trade and our modern addiction to sugar and attendant health problems.[9] Another exercise is to have students categorize animals according to how they are perceived by farmers and ranchers. They can discuss why wild animals and "vermin" are considered "bad" animals and domesticated animals are "good" animals. Another exercise is to teach water systems, both natural watersheds and cycles and human-made irrigations systems, as agriculture is dependent on water. Finally, weather can be discussed as a part of agriculture, in terms of how floods and drought can impact crops. All of these relate easily to features of the prior thresholds.

At the end of this unit, a summary of the results of the agrarian age should be given: an increase in population from 2500 B.C.E. to 1800 C.E.; a large-scale impact on the biosphere, including extinctions and forest loss; growth of cities; lack of progress in real science; longer life expectancies; and continued emphasis on childbirth for women; but also valuable advances in collective learning and the closing of the human web.

ACTIVITIES AND EXERCISES FOR TEACHING THRESHOLD 7

COLLECTIVE LEARNING IN AGRARIAN CIVILIZATIONS

Richard B. Simon

Threshold Related to the Activity: Threshold 7
Category of Activity: In-class learning activity
Objectives:

- To experience how collective learning leads to various degrees of innovation in groups of different sizes—and how such groups interrelate;
- To analyze agrarian civilizations using the four features of complexity;
- To reinforce student understanding of the four features of complexity; and
- To consider how emergent properties in agrarian civilizations give rise to the features of modernity and the industrial age

Overview of Activity: This is a single-blind experiment—a meta-recreation of collective learning and urbanization in agrarian civilizations. Students are broken into groups of various sizes, in which they list the four features of complexity as they manifest in Threshold 7, agrarian civilizations. The students are unwitting subjects of an experiment in which they are re-creating the dynamics of collective learning and urbanization in the age of agrarian civilizations. The experiment is revealed at the end of the activity, and students process what they've learned by synthesizing their experiences in the re-creation with what they've read about agrarian civilizations. This activity will take about thirty to forty-five minutes.

Faculty Preparation for Activity: The instructor should be familiar with Threshold 7.

Cost: n/a

Student Preparation for Activity: Students should be familiar with agrarian civilizations and the many developments that characterize this period. In addition to reading the textbook on Threshold 7, this may include other Threshold 7 lectures, slideshows, discussions, and so on.

In-Class Sequence of Activity:

1. Students are divided, without explanation, into groups of various sizes, with a significant size difference between the largest group and the next largest, and a few that are quite small. In a class of twenty, this could mean

one large group of eight, a group of four, two groups of three, and one group of two students.

2. Each group is asked to brainstorm about the four features of complexity—diverse components, specific arrangements, energy flows, and emergent properties—for fifteen minutes, to come up with a list of each of these features as they manifest in Threshold 7.

   Diverse components might include humans, plants, animals, and seeds. Specific arrangements might include a central village—with family compounds, areas for specialized activities such as food preparation, a communal fire pit where religious ceremonies take place, and merchants' shops—surrounded by outlying agricultural fields, areas at the fringe where traders live, and roads that connect the village to other villages. Specific arrangements might also include social structures that arise in response to surpluses. Energy flows might include burning wood (from trees that capture solar energy through photosynthesis) for cooking and warmth; harnessing energy from plants (again, through photosynthesis) and from livestock for food; and the processing of energy by human bodies to fuel further agricultural production. Additional energy flows might include information, food, and other goods—including money, which represents energy (by giving those that have it the ability to purchase food). Emergent properties could include new forms of organized religion, social hierarchies based around defense of surpluses, and religious authority, as well as networks of exchange. Not to mention . . . heightened collective learning, and the exponential growth of innovation itself.

   The activity can get lively—and the largest group, with the largest population, is the liveliest. Members of the smaller groups (neighboring communities) may find the larger group either appealing or annoying. Some students may begin to communicate with other groups or want to join the larger group.

   Students are processing how agrarian civilizations work by directly analyzing what they've read and understood using the four features of complexity. They are also modeling the development of urban centers and the heightened innovation that happens when larger agricultural villages become nuclei in broader civilizations, and then nodes in larger networks of exchange. We might see specialization, especially within the larger group (a scribe, a leader, a group spokesperson). We also see that those other civilizations whose populations are too small run out of ideas, and they may be subsumed by the nearby growing urban centers. We might see the rise of envy and contempt for urban centers, as well as the allure of innovation and population centers for some of those who live in smaller nearby villages. We might also see some culture and cultural chauvinism develop within and among the groups. In one class, members of the largest group began, quite organically, to speak in their own "lingo" as they discussed and then explained their findings. A member of another group, eager to participate with the large group, began to insinuate himself as translator for the rest of the class!

Again, not setting any rules that prohibit (or encourage) communication between groups will add illustrative complexity to the experiment—should any of the groups innovate in this fashion (for example, by forming networks of exchange).

3. Once this phase of the activity has run its course, students remain in their groups and report their findings to the instructor, who records them on the board, group by group. The likely result is that the largest group, with the largest population engaging in collective learning, will have the most entries, and even the most innovative entries (and this group should probably report last). The members of the smaller groups may seem disheartened that their lists are not as large or as innovative. But this is the whole point of the exercise. They might need some reassurance. The big group's innovation is a function of its size—of the number of brains engaged in collective learning.

4. Engage the class in a discussion.

Assessment of Learning: Class discussion can focus on the disparity in the volume of entries (of collective learning) among the groups and tie the experiment back to the developments students read about in Threshold 7. After the activity, the instructor may wish to continue discussion of Threshold 7 or begin to prepare students for Threshold 8—as students think about how the emergent properties they see in Threshold 7 will give rise to the features of modern industrial civilization. In addition, students could be asked to write reflectively, in a way that synthesizes their experience in the re-created agrarian civilization with what they've read about the period. The students should have reinforced their understanding of how the four features of complexity manifest from threshold to threshold; the main features of agrarian civilizations, and how they function; and the processes that will lead to the coming emergence of modernity.

Origin of Activity: This is an original activity.

Source Consulted: Christian, David, Cynthia Brown, and Craig Benjamin. *Big History: Between Nothing and Everything.* New York: McGraw-Hill, 2014. Print.

NOTES

1. Traffic, "John Barleycorn." *John Barleycorn Must Die.* United Artists, 1970. Vinyl LP.

2. Cynthia Brown, *Big History: From the Big Bang to the Present.* New York: New Press, 2007. 75–76. Print.

3. Brown, *Big History,* 75.

4. "Arable." *Webster's New World Dictionary.* 2nd ed. 1970. Print.

5. David Christian, Cynthia Brown, and Craig Benjamin. *Big History: Between Nothing and Everything.* New York: McGraw-Hill, 2014. 240. Print.

6. John McKay et al., eds., *A History of World Societies.* 7th ed. Boston: Houghton Mifflin, 2007. 42. Print.

7. Christian, Brown, and Benjamin, *Big History,* 245.

8. Ibid., 248–249. Dennis Sherman and Joyce Salisbury, eds. *West: Experience Western Civilization.* New York: McGraw-Hill, 2012. 399–400. Print.

9. Sydney Mintz, *Sweetness and Power: The Place of Sugar in Modern History.* London: Penguin, 1985. Print.

FOURTEEN · Teaching Threshold 8

Modernity and Industrialization

Richard B. Simon

> This was a very innocent planet, except for those great
> big brains.
>
> KURT VONNEGUT, *Galapagos*

Inspired by our work at Dominican, I was teaching Cynthia Brown's book *Big History: From the Big Bang to the Present* in an advanced composition class at City College of San Francisco, a large community college with a particularly diverse student body. On the first day of this intensive summer course, "Big History and the Future," I introduced students to the concept of complexity and led them through the story, threshold by threshold.

Once we got to the Industrial Age, the bigger questions arose, without any coaxing on my part. On some level I felt like the class discussion was getting away from me—but that is because the material was so generative (as we say in creative writing) that the students were already thinking creatively.

We talked briefly about global warming, and about energy resources, and the move toward solar and wind.

A student raised his hand. "If energy is free," he said, "what does that do to our economic system?"

Running through this story, even at a rapid clip, raises questions about how larger systems work. And this is the point. Big History is a bit like one of those puzzles in which a series of geometric figures, each slightly different, suggest a logical progression, a pattern. The trick is a multiple-choice question: choose the next in the series.

When we get to our own time, we need to run through the transition from the age of agrarian civilizations to the very complex developments of the last 250 years or so. This is where the story is the most familiar to students, so they are more and more engaged, especially as they see how recent history fits into the larger scheme. (We spent one four-hour class period discussing the U.S. Civil War as a struggle between an agrarian civilization and an industrializing one whose values had changed—and connecting that understanding to a Supreme Court decision in that day's news.)

This is where, if we're doing it right, it will all be coming together for our students—they will finally see how the patterns we've observed throughout the 13.8-billion-year story have shaped the world in which they live right now.

We're doing the math, and the math is this: agriculture plus civilization plus industrialization yields a wealth of miraculous technologies and wondrous developments in human society and culture. It also leads to a host of existential environmental crises.

Students will be hungry to know what happens next.

THE TRANSITION

As with earlier thresholds, the boundary between Threshold 7 and Threshold 8 is not a bright line. Christian, Brown, and Benjamin discuss the transition to modernity and modernity itself both as part of the industrialization process and as key developments in the late agrarian civilizations.[1] Those developments built the structures through which the increased energy flows provided by fossil fuels began to animate global industrial civilization, thus yielding the Anthropocene era of geologic time.

The long Paleolithic dawn of human life was characterized by the migration of humans to nearly all continents. Then, the floods that marked the end of the last ice age, about ten thousand years ago, swamped the various land bridges that had connected the landmasses, separating Asia from North America, Spain from Africa, England from the European continent, Sri Lanka from India, and the Philippines and Taiwan from Korea.[2]

This meant that, in the process of **divergence,** those human tribes that had been loosely connected by land were now isolated. These isolated peoples spent the next few thousand years developing disparate cultures, lifeways, and—perhaps most fatefully—resistances (or a lack thereof) to various types of disease.

The development of trade routes across what could now be considered the Afro-Eurasian world zone meant an exchange of goods, ideas, genetic material, and diseases. Civilizations that were disconnected from these **networks of exchange** did not participate in the key biological exchanges—and their people were thus dangerously susceptible to diseases to which Afro-Eurasians had developed resistance.

Also during the age of agrarian civilizations, the lands that bound the Indian Ocean were linked in a massive trade network that spread Islam. When the Ottoman Turks prevent Europeans from accessing these networks, European explorers—Italians, many of them—headed west to try to find new routes to access the "East Indies."

Thus were the Americas "discovered," the globe circumnavigated, and the world connected for the first time in one massive, global network of exchange. The dominant trend became **convergence,** the coming back together of disparate human cultures in ways that spark innovation. This was the beginning of **globalization.**

Fueled by abundant new energy resources (timber, crops, fish, whales, and the silver and gold that bestowed the ability to purchase energy and thus *represent* energy), Europe—and then England—became the center of a new global trade network. Wealth accumulated and drove further innovation.

COLLECTIVE LEARNING AND INNOVATION

Christian, Brown, and Benjamin suggest that key drivers led to modernity and industrialization by yielding sharp spikes in the rate of innovation—new technologies and ideas that changed lifeways:

1. Bigger groups (cities, civilizations, empires, and expanding networks of exchange) increase collective learning.
2. Improvements in communication and transportation increase collective learning.
3. "The expansion of commercial activity," including competitive markets and capitalism, leads to innovation.[3]

In addition, **Malthusian cycles,** in which populations rise until they meet the limits of available food/energy resources (or disease), stifle innovation. But the unlocking of the energy in fossil fuels will break those cycles.

Christian, Brown, and Benjamin write that the "discovery" of the Americas, of new stars, and of new crops (not mentioned in the Bible!) undermined faith in the old ways of knowing and drove intellectual innovation, including the rise of **scientific empiricism.** Furthermore, these "discoveries" helped decouple the story that agrarian-age religious texts told from the new story that science and global exploration were telling.

In the empiricism of English scholar Francis Bacon . . . we find a sense that new knowledge gained through exploration and direct observation was the key to truth. Bacon saw the geographic discoveries of his time as a model of how science itself should proceed, not through the study of ancient texts, but rather by the exploration and careful study of the real world. In the philosophy of the French philosopher René Descartes . . . we find a sense of the importance of questioning established authorities so that knowledge could be reestablished on new and firmer foundations. Both the skepticism generated by new knowledge and the conviction that knowledge should be sought in exploration link the expanding intellectual horizons of Europe to what has traditionally been called the "scientific revolution" of the seventeenth century.[4]

Here, alas, are the roots of Big History itself.

With these new epistemologies, coupled with Gutenberg's movable type printing press, the stage was set for a major increase in collective learning—a process of moving new information (and therefore innovations) globally, which Christian, in *Maps of Time*, calls "diffusion" and has compared to an atomic chain reaction.

So while the period from 1000 to 1700 was not marked by extraordinary technological innovations, human civilization seems to have been suiting up for the next era: connecting, through complex networks of exchange, agrarian civilization–based world zones that had lain isolated for millennia. This convergence was aided by (and inspired) improvements, mostly in existing transportation and communication technologies. Collective learning increased rapidly.

What the world needed now was an increase in energy flows to match.

INDUSTRIALIZATION

Journalist Paul Roberts writes, in *The End of Oil*, that it was the invention of the Newcomen engine—a steam engine that drove a water pump—that sparked the modern energy economy. England had replaced wood with coal as fuel—but by 1712, England's remaining coal reserves lay largely beneath the water table, so the mines were underwater, and the coal was inaccessible. Thomas Newcomen invented a piston-driven pump that, while quite inefficient, was able to pump water out of a mine faster than groundwater flow rates could fill the mine back up again.

As Roberts explains it, "Newcomen's engine gave us our first real mastery over energy and set humanity on a course that would change the world forever. . . . For the first time, human beings had the potential to harness energy in quantities far greater than previously imagined, and the impact would be enormous."[5] The meta-linkage of machines and coal (in the form of a coal-driven machine, made of iron, which enables the mining of more coal, which provides more energy, which can be used to smelt more iron, which can be used to make more machines, which can be used to obtain more energy) generated a surplus of money—wealth—which could, of course, be used to build more machines, to mine more coal, to make more money, to build more machines, to buy more coal.

And here was the central action of industrialization: machines driven by **fossil fuels** produced wealth. So machines, fueled by energy from the sun concentrated over hundreds of millions of years as coal, oil, and natural gas, allowed humans to do lots of work.

Much of that work, at the time, had been done by slaves. Over the course of the Industrial Age, fossil fuels obviated the need for slaves, and generated freedom as well as wealth. (Again, viewed in this revealing context, the American Civil War was a conflict between industrializing Northern states that increasingly used coal steam to do work and agrarian Southern states, which relied on slavery and a feudal system of social control . . . and had no need to cross the threshold to modernity.)

Arbitrage, the practice of buying products or materials in one place and selling them elsewhere at a profit, gave rise to **capitalism.** Driven by machines and fossil fuels, capitalism—practiced by state-supported **corporations** that grew in size and power in the modern age—generated wealth and opened new markets, leading to further increases in collective learning.

Industrial market capitalism also yielded the transition of human labor from farm work to factory work; mechanical timepieces created mechanized, regimented time; humans working as cogs in what are essentially massive machines led to repetitive stress disorders and income inequality—as well as discontent and rebellion. **Socialism** and **communism** were born as responses.

During World War I, the industrialization of warfare gave us mustard gas and machine guns. The interwar period featured the first aerial bombardment—of the Basque village Guernica, by Nazi German airplanes, interceding in the Spanish Civil War on behalf of Francisco Franco's fascist national government. The war between capitalist liberal democracies, socialists, and communists on one side, and fascists on the other—World War II—yielded both the industrial horror of the Nazi death camps (slaughter of civilians by machine gun, gas chamber, and mass incineration) and the rise of a global middle class, whose members enjoyed comforts that had, for most of human history, been the purview of royalty.

Also during the twentieth century came the rise of the **internal combustion engine,** which burns gasoline produced from crude oil, another fossil fuel with a higher energy yield than coal—increasing energy flows through industrial civilization once more.

That's not to mention the atom bomb—which yielded nuclear power, and nuclear waste—and the forty-five-year Cold War, a competition between the capitalist United States of America and the communist Union of Soviet Socialist Republics (USSR). This contest fueled innovation in the form of a "space race" to perfect **rockets,** which carried humans into Earth's orbit, and then to the moon. Humans have, through unmanned spacecraft, ventured farther into space, so that in 2014, we have rovers on Mars, satellites photographing the outer planets, space telescopes that see back in time, and a *Voyager* probe entering interstellar space.

The contest between the United States and the Soviet Union led to one more key innovation: **the internet,** created by a U.S. military research agency. With the sub-innovation of the world wide web, as well as of personal computing devices that could access it (and satellites that allow global communications networks), the people of the world could suddenly exchange vast amounts of information in real time. The advent of widespread cellular networks, inexpensive phone technology, and smartphones that can access the world wide web has linked humans and the wealth of human knowledge like never before—person to person—through a massive increase in the amount of energy that our current global digital civilization uses to process information.

The exponential growth in computing power and related technologies (smartphones and other mobile electronic communication devices, bionic prosthetics) linked to an increasing ability to harness energy more directly from the sun, with **solar- and wind-powered electricity generation** and new electric automotive technologies that can decouple transportation from fossil fuels, certainly give the appearance of the verge of a new threshold. Or could it be that we will soon see such great changes in the nature of human civilization that we will rethink whether the Industrial Age was a threshold at all, or just a supercharged global agrarian civilization that led to some entirely new human arrangement?

THE ANTHROPOCENE

Coal, like oil and natural gas, is a fossil fuel. These energy sources derive from the bodies of microorganisms deposited in anoxic environments and buried in Earth's rocks for hundreds of millions of years.

Ultimately, the use of fossil fuels to generate wealth—by improving transportation, communication, and well-being—also generates waste, and with it a host of environmental problems, such as standard toxic **pollution** of water, air, and soil—which are likely contributors to cancers, which industrialized medicine seeks to cure.

The most urgent of these environmental problems is **global climate change.** Earth's atmosphere acts as an insulating blanket for the planet. The chemical composition of the atmosphere determines its insulating properties. Among the gases that increase the atmosphere's propensity to trap the sun's energy as heat are methane, chlorofluorocarbons, and even water vapor. The primary culprit is the carbon dioxide (CO_2) that is released into the atmosphere as waste when fossil fuels—also known as hydrocarbons—are burned to power human transportation, communication, and well-being.

When carbon dioxide is added to the atmosphere, the atmosphere retains more heat. That heat melts the ice caps and glaciers, raising sea levels—and leads to a host of **positive feedback loops** that make the problem worse. More energy in Earth's contained (by gravity) yet open climate system means bigger and more dangerous storms; more and more severe droughts; and expanding vectors for disease and insects that overwhelm forests and crops—which will have a difficult time dealing with more extreme temperatures.

Unfortunately, these things are happening at least a hundred years earlier than many scientists expected.[6] In spring 2013, scientists in Hawaii measured carbon dioxide levels at 400 parts per million for the first time—a level not seen for millions of years, since the Pliocene, long before humans appeared.[7] Global temperatures were at their highest in four thousand years.[8] And the first two decades of the twenty-first century were marked by many large, dangerous storms, heat waves and droughts just as climate models predicted.

In addition, oceans absorb carbon dioxide. This is a crucial **carbon sink,** a process that buries carbon, locking it into rocks. But the oceans are acidifying as absorbed atmospheric carbon dioxide becomes carbonic acid. Typically, microorganisms and mollusks pull large amounts of carbon out of the sea to form calcium carbonate shells. When they die, those shells rain to the sea floor and are buried to eventually become limestone—another crucial carbon sink. But such animals are having a difficult time building shells in acidifying seas. And they're not the only ones having a hard time. Scientists believe we are in the midst of a **mass extinction** of species that rivals those that doomed the dinosaurs.

What's more, large amounts of methane lie frozen beneath the arctic "permafrost," which is proving not so permanent—and beneath the sea floor. These frozen deposits amount to a bomb waiting to dump methane into the atmosphere as they thaw—and methane is a much more potent greenhouse gas than carbon dioxide.

Several decades ago, climate scientists believed that the Earth had begun its long slide into the next ice age. It now appears that we have staved off that ice age by dumping heat-trapping gases into the atmosphere. That could be good news.

The climate system is as complex as Earth's ecosystems, and it relies on a complex interplay among the atmosphere, the oceans, the rocks, and all living organisms. But humans have seized control of the atmosphere, at least. Our collective hand is now on the thermostat.

In addition, human activities such as the hydrofracturing (commonly known as "fracking") of deep rock formations that contain trapped natural gas (this, in an attempt to move away from coal for electricity generation) appear to be causing earthquakes. The rebounding of land forms currently weighed down by rapidly thinning ice sheets might also lead to earthquakes.

All this is why some scientists are now referring to the current period of geologic time as **the Anthropocene,** to note that we have entered a time in which the most significant geologic force on Earth is *Homo sapiens sapiens*—us. Christian, Brown, and Benjamin assert that, on the grandest scale, increasing human control of the biosphere is the most important outcome of the Industrial Age. To that, we might add humans' increasing effect on the atmosphere, the oceans, and the very rocks.

The big question now appears to be whether humans will act with urgency to decouple our ravenous energy appetites from the carbon atom.

We are linked together in real time, and digital technology has proved a potent democratizing force.

Will our ever-more empathetic, interconnected digital civilization act to stave off the worst of climate change, before the climate system itself slips farther across a dangerous new threshold? Will our late-industrial wealth restrain innovation and lead to further inequality and a new climate feudalism? Will we live in a digitally connected, high-tech world, bunkered against megastorms and adapting to the realities of life on a new Earth? Will some unforeseen, revolutionary energy technology—like fusion laser pulse technology—save the day? Or will the development of **geoengineering** technologies that remove carbon dioxide rapidly from the atmosphere be the key?

Are we even still in Threshold 8, or has the transition across some new threshold begun with the **Age of Information?** Will our species, interconnected digitally and ramping up collective learning exponentially, master new ways of producing exponentially greater flows of energy and organize itself into entirely new arrangements, thus changing everything?

THE FOUR FEATURES OF COMPLEXITY
IN THRESHOLD 8

DIVERSE COMPONENTS

Humans. Machines. Industrialized nation-states (and industrializing or developing states). Business corporations. Fossil fuels.

SPECIFIC ARRANGEMENTS

Humans as citizens of states, linked together in a global network of exchange—and person to person through wireless communications networks that link up via the internet.

FLOWS OF ENERGY

sun	→ cells of plants	→ fossil fuels →	machines → work →	surpluses/wealth
sun → cells of plants → cells of animals → fossil fuels → machines → work →					surpluses/wealth
sun	→ cells of plants →	cells of *Homo sapiens*	→ work →	surpluses/wealth
sun → cells of plants → cells of animals → cells of *Homo sapiens*			→ work →	surpluses/wealth

EMERGENT PROPERTIES

Convergence. Globally connected civilization. Global self-awareness. Global networks of exchange and communication. Wealth. Industrial food production. Industrialized medicine. Atomic energy. Domestication of microorganisms. Space exploration. Mapping of the human genome. Genetic engineering. Wealth inequality. Rebellion. Industrial pollution. Environmental degradation. Greenhouse effect. Climate-forced food shortages. Resultant social upheaval, aided by digital communications technology.

HOW EMERGENT PROPERTIES MAY
GIVE RISE TO A NEW THRESHOLD

Threshold 8 is often, for students, the most resonant unit of a Big History course. Big History students are most drawn to sections of the course and of the story that are the most relatable. That means that many who are alienated by the cosmological scale get drawn back in when we begin to learn about human beings. And when we talk about the Industrial Age, we are looking at the world as they know it.

In addition, students want to know about the future. Of course, that is how a Big History course works: we teach students the macro- and micro-scale patterns in the grand sweep of history that they might ponder the future, and their place in it—and how they might shape their individual futures and society's futures to their own advantage, and the advantage of their "children and grandchildren."

We have sometimes found that there is so much to cover in a single-semester Big History course, and so much to do at semester's end, that the biggest and most important questions—where do we go from here, and how?—have gotten short shrift. Be wary and leave extra time for the future.

Considering how the future—or how any of several possible futures—may arise from the emergent properties in Threshold 8 is the culmination of the course.

Instructors should ask students to consider the four features of complexity in Threshold 8, and how, in earlier thresholds, key emergent properties led to the big changes that mark the next threshold.

What are, for students, the most important emergent properties in Threshold 8? Which ones have the potential to define the future, and perhaps to give rise to a new threshold (as we'll discuss in the next chapter, these are not the same thing)? Given the patterns we see in earlier threshold transitions, what might a Threshold 9 look like (if we're not already in it)? And what might human civilization look like as we cross it (or *if* we cross it)?

SUGGESTED LEARNING OUTCOMES
AND POSSIBLE ASSESSMENTS

By the end of this unit, students will demonstrate the ability to

1. discuss how structures of Threshold 7 lead to Threshold 8 (assessment: exam or short essay);[9]

2. illustrate familiarity with fossil fuels and their links to earlier thresholds (assessment: exam or short essay);

3. list key developments of modernity and the Industrial Age (assessment: quiz or exam);

4. illustrate how fossil fuels drive industrialization and globalization (assessment: essay);

5. examine the causal link between industrialization and environmental issues (assessment: essay); and

6. explain the patterns that recur throughout the Big History story, using the four features of complexity, with an eye to the future (assessment: major essay, such as a Little Big History, research paper, or future-visioning project).

CHALLENGES IN TEACHING
THRESHOLD 8

Threshold 8 is *so* complex, there's so much going on—so many diverse components, linked together in so many arrangements—that the history trap, of trying to go over absolutely everything, lies in wait.

A good way to approach Threshold 8 is to use the four features of complexity as an analytic tool. Allow students to talk about something they know well—their campus (see chapter 6, "Teaching Complexity in a Big History Context"). Because the campus is itself a substructure of industrial civilization, it should serve well as synecdoche, the part standing in for the whole. Don't worry—students will instinctively draw connections to the larger global civilization; twenty-first-century students are connected to it, and aware of it, at all times!

As tends to be the case in Big History, Threshold 8, being the most recent threshold, is also the shortest—so let's say that complexity density is at its peak.

Think about what are the most important concepts that you want to be sure to get across. Don't try to do everything.

Students will want to talk about environmental degradation. Some who have learned that anthropogenic global warming is "a hoax" or impossible may resist this discussion. But the numbers of young people who believe this are dwindling. Students are now becoming angry at previous generations for having caused this problem and for failing to solve it. This *is* a civic-minded generation. They will want to know what they can do.

CONCLUSION

At the end of the aforementioned three-week reading of Cynthia Brown's *Big History*, my composition students led us into a discussion of the future. I broke them into groups of four and allowed them to ponder how the patterns we'd seen up to now might lead to whatever comes next.

At the end, we reconverged and discussed the matter as a whole class, with each group reporting out.

One group had imagined a high-tech future in which the economy was based on the actual value of the natural resources (soil, water, air, and sun) that underpin it. Food and energy would be free. "What would you do for competition?" a student from another group asked. "What would motivate people to work, and to innovate?"

"Creativity itself would be the currency" was the group's answer. Their needs largely met, humans would trade in their creative works.

So they had begun to answer the question from a few weeks before: what would happen to the economy if energy were free? Another group chimed in that they had been thinking along similar lines. Yet another group admitted that their efforts had led mostly to a discussion of their paper topics.

The students didn't all agree. One skeptic doubted that humans would be ready for the kind of existence that *everything is free* would imply, and cited our current individual isolation from our neighbors, pointing to the same smartphone that the earlier group had asserted would (and seemed to have begun to, in the still-nascent Arab Spring) democratize the globe.

"It's not like we still live in a world where your neighbor comes over to borrow a cup of sugar," she said.

"I do," said a young man. "My neighbor came over just yesterday to borrow a cup of rice. That happens at least once a month."

Allow it to be personal, and allow students to drive the discussion. Again, this may be the most important section of the Big History course for them (for empirical evidence of this, see chapter 4, "Assessing Big History Outcomes: Or, How to Make Assessment Inspiring"). Allow it to have meaning for them. Even though you may feel like there is *so much to cover,* leave some time and space for students to sort it all out, to find meaning and resonance, and to think of where to go from here.

Allow them agency, the sense that they can actually affect what happens next. That's the whole point.

Leave some room for the future.

ACTIVITIES AND EXERCISES FOR TEACHING THRESHOLD 8

TRADING ON THE STOCK MARKET

Martin Anderson

Threshold Related to the Activity: Threshold 8

Category of Activity: Learning activity

Objective: To give students a basic understanding of how the stock market works

Overview of Activity: Students will engage in three rounds of buying stock and assessing their performance. This activity will take an entire class period, about one hour and fifteen minutes.

Faculty Preparation for Activity: Bring a briefcase or something to pull slips of paper out of.

Cost: n/a

Student Preparation for Activity: A week before the activity, students divide into groups of four. Each group will represent a corporation, which they will name along with the product they make. Students design and create twenty stock certificates (shares) for their company. Students should be encouraged to be as creative as possible in making their certificates. Students make eight slips of paper, one saying "25 percent loss," one saying "10 percent loss," one saying "0 percent profit," one saying "$5 dividend," one saying "10 percent profit," one saying "25 percent profit," one saying "50 percent profit," and one saying "100 percent profit." Students must bring these to class on the day of the activity.

In-Class Sequence of Activity:

1. On the day of the activity, students divide into their groups. Each group will be given $500 with which to buy stock. This will simply be represented by writing $500 on the board next to the name of the group.

2. Each group then has five minutes to convince their fellow students to buy their stock, at either $25, $50, or $100 per share. Purchases are recorded on the board.

3. The instructor then places the slips of paper into the briefcase, mixes them up, and randomly draws one slip out per company and records the result— for example, "Company A had a 25 percent profit." The results are now recorded—for example, "Group 1 purchased 2 shares of Company A at $50.00 each, which now have a value of $62.50 each."

4. It is expected that students will groan when they lose money and cheer when they gain. Personalities will emerge, with students dividing into those reluctant to lose money despite the possibility of gain and those willing to risk money despite the possibility of loss. Ask them why some are more risk-averse than others.

5. Repeat the process of buying and selling stock three times.

6. Some students will recognize that some means of evaluating how good a company is should be provided. Explain that this is done by stock market analysts or by themselves in selecting stocks or mutual funds for a personal or retirement account.

7. After the exercise is over, explain that a method of analysis was available: of the eight slips of paper, only two were losses (25 percent), two (25 percent) were neutral, and four (50 percent) were gains; therefore, there was a 75 percent chance of no harm and a 50 percent chance of gain. Given this, everyone should have always risked all their money, even though it is likely that many did not buy stock for fear of losing money.

8. Ask them why they feared a 25 percent chance of losing money. (Caution: This may invoke some personal difficult financial experiences students have had in their families.) Explain that because of inflation, the problem they face is that saving for retirement means they must invest in the stock market and gain money. However, this will generally be done by the managers of their company retirement accounts.

Assessment of Learning: There is no formal assessment of this activity other than a discussion of students' experience and thoughts about the stock market at the end of the activity.

Origin of Activity: This is an original activity.

MARX AND ENGELS ON THE ANTHROPOCENE

Cynthia Taylor

Threshold Related to the Activity: Threshold 8
Category of Activity: Primary source analysis and class discussion / debate
Objectives:

- To familiarize students with the Big History concept of the Anthropocene—and to think more critically about the "new geological epoch . . . in which [*Homo sapiens* have] come to dominate the biosphere";[10] and

- To enhance reading and critical thinking skills and openness to multiple perspectives

Overview of Activity: This activity may take about two weeks or four class sessions (in a class that can afford to have two weeks to discuss the modern era). Students do some background reading on Threshold 8 from the class text, define the term "Anthropocene" as used in the class text by analyzing several geologic time charts, read and study the document (Karl Marx and Friedrich Engels's *The Communist Manifesto*) in small groups, take some reading comprehension quizzes on the document, have a class discussion or debate, and write up a response paper.

Faculty Preparation for Activity:

1. Prepare reading assignments from class text and document.
2. Prepare reading comprehension quizzes.
3. Prepare or adapt PowerPoint lectures for Threshold 8.
4. Set up debate guidelines so that all students verbally participate in final class discussion.

Cost: Cost of texts

Student Preparation for Activity:

1. Background reading from the course text
2. Reading of *The Communist Manifesto* in groups
3. Participation in class debate

In-Class Sequence of Activity:

1. Class activity that familiarizes students with geologic time periods and the 4.5-billion-year history of the Earth. Introduce and identify the Holocene era with Threshold 7 and the Anthropocene era with Threshold 8. Class discussion should be based on the discussion of the term "Anthropocene" from the class text.
2. In a lecture, discuss the differences between the capitalist world and the communist world. The discussion should also cover other issues pertinent to Threshold 8, such as globalization, critical twentieth-century technologies, population growth, Malthusian cycles, the biosphere, and sustainability.

3. Read and analyze the document (*The Communist Manifesto*) in groups; conduct an interactive session in which students take comprehension quizzes in order to make sure they understand terms used in the document, as well as the document's organizational structure.

4. Make up statements for students to agree or disagree with, then facilitate a class debate based on the question: Is it possible that thinkers such as Marx and Engels were partly right? Could modern capitalism's destructive sides, including reliance on exploitation of the biosphere, eventually bring the entire system down? Students take sides and should sustain at least a thirty-minute debate in class.

Assessment of Learning: Students will report out on the class discussion debate, then each student will write a personal response paper supporting her or his own position or thesis regarding the following topic: In light of the issues raised by Marx and Engels in *The Communist Manifesto*, what is the relationship between twentieth-century capitalist economic and political systems and the emergence of a new geologic epoch, the Anthropocene?

Origin of Activity: This is an original activity.

Sources Consulted:

Christian, David, Cynthia Stokes Brown, and Craig Benjamin. *Big History: Between Nothing and Everything*. New York: McGraw-Hill, 2014. Print.

Marx, Karl, and Friedrich Engels. *The Communist Manifesto*. 1848. New York: Penguin Adult, 2002. Print.

NOTES

1. David Christian, Cynthia Brown, and Craig Benjamin. *Big History: Between Nothing and Everything*. New York: McGraw-Hill, 2013. Print.

2. Cynthia Brown, *Big History: From the Big Bang to the Present*. New York: New Press, 2008. 66. Print.

3. Christian, Brown, and Benjamin, *Big History*, 214–241.

4. Ibid., 239.

5. Paul Roberts, *The End of Oil: On the Edge of a Perilous New World*. New York: Houghton Mifflin, 2004. 23. Print.

6. Mark Hertsgaard, *Hot: Living through the Next Fifty Years on Earth*. Boston: Houghton Mifflin Harcourt, 2011. Print.

7. Justin Gillis, "Carbon Dioxide Level Passes Long-Feared Milestone." *The New York Times*, 11 May 2013. Web. 23 June 2013.

8. Justin Gillis, "Global Temperatures Highest in 4,000 Years." *The New York Times,* 7 March 2013. Web. 23 June 2013.

9. Martin Anderson, "Teaching Transition: Thresholds 7 to 8," Dominican University of California Big History Summer Institute, 21 June 2013. Lecture notes.

10. Christian, Brown, and Benjamin, *Big History,* 285.

FIFTEEN · Threshold 9?

Teaching Possible Futures

Earth's thirty-seven space elevators all had their cars full
all the time, both up and down. There were still many
spacecraft landings and ascents, of course, and landings
of gliders that then reascended on the elevators; but, all
in all, the elevators handled by far the bulk of the earth-
space traffic. . . . It was almost an ice-free planet now,
with only Antarctica and Greenland holding on to much,
and Greenland going fast. Sea level was therefore eleven
meters higher than it had been before the changes. This
inundation of the coastline was one of the main drivers
of the human disaster on earth. . . . Earth was a mess, a
sad place. And yet still the center of the story. It had to
be dealt with, as Alex had always said, or nothing done
in space was real.

KIM STANLEY ROBINSON, *2312*

INTRODUCTION

Richard B. Simon

When the group tasked with envisioning a new general education program for our
university first sat around the long oaken table and began to sketch out what a core
curriculum built around Big History would look like, we imagined an elegant course
sequence that would introduce students to the wealth of human knowledge, use that
knowledge to focus within a discipline of interest, and then follow up with a choice
of more advanced courses designed to give students agency in "shaping the future."
Three and a half years later, as we studied the assessment data we had gathered at
the end of the third year of our two-semester Big History program (see chapter 4,
"Assessing Big History Outcomes: Or, How to Make Assessment Inspiring"), to
our great astonishment, we found that, indeed, our students felt the greatest ownership of the content when it addressed matters of meaning or belief, when it contex-
tualized their studies by stressing the interconnectedness of the disciplines, and when
it foregrounded the future.

When we talk about the future, we are really talking about the present, and how the reality of the present might change over time. Because there is no set of events or facts that can be ascribed with certainty to the future—no fixed content—we think in terms of key concepts and approaches that allow us to address the problems of considering what the future could look like and give students agency in thinking about the futures they'd like to see. Because there are so many possibilities when it comes to teaching possible futures, we thought we might speak to this task in some of the many voices of our program. That's why, although this chapter follows a similar structure to the previous chapters, the discussion feels like a seminar—it's a bit like the discourse at our summer institutes.

First, Martin Anderson lays out some of the key concepts in teaching the future. Next, J. Daniel May clarifies the crucial distinction between "the future" and "Threshold 9." Richard Simon addresses the four features of complexity in considering a Threshold 9. Then, Neal Wolfe discusses important challenges in teaching the future. We include two brief essays that use a Big History context to ponder a next threshold—Kiowa Bower projects a technological revolution, perhaps the "singularity," in which biology and technology merge, and Philip Novak proposes a Threshold 9 that is a revolution in human consciousness. Finally, Debbie Daunt presents her approach in "Big History through the Lens of Health and Healing," a course popular with students in our nursing program, in which she connects *each* threshold to the future, and in particular, to students' own futures as health-care professionals.

More advanced students—those with an interest in becoming Big Historians, and those with specialized knowledge in the disciplines that inform Big History—may be more drawn to other areas of interest within the story: to complexity theory, to planetary geology, or to social hierarchies in agrarian civilizations. But in a general or liberal education context, at least, students are drawn to the future. As we stressed in the previous chapter, it is important to leave room for the future—for many students, this is the most significant part of the Big History story. Certainly, it is the most personal and the most relevant to them as individuals. It's also, potentially, the most fun. If the past is history and the present is politics, then the future is the realm of science fiction. (For some ideas on using sci-fi in the classroom, see J. Daniel May's "Post-Apocalyptic Film Festival" activity, at the end of this chapter.)

Of course, even advanced students (not to mention professors) enjoy using their new knowledge and understanding to attempt to project a possible future. Consider backloading your syllabus to foreground the future, and leave ample time (or perhaps a major writing assignment) in which students might apply their new Big

History understandings to try to envision what might come next—and to think about how to act accordingly. This is where they get to reflect on the sweep of the entire narrative and draw important conclusions that really tie the course together.

KEY CONCEPTS IN TEACHING POSSIBLE FUTURES

Martin Anderson

Teaching the future is an integral part of teaching Big History, largely because we now understand that processes not influenced by humans are at work and we need to understand how those processes impact human life. We also now understand more fully how human activity influences otherwise natural processes. It is the latter that gave rise to the idea of the Anthropocene, a geologic era in which humans are the dominant agent of change. Irrespective of their education, students are likely to be aware of the growing attention in public media to general questions of **sustainability** and **the environment**. While traditional history courses generally do not engage in speculation about the future, Big History more naturally addresses the future as part of closing the loop for students on why learning about Big History is relevant to them.

One way to approach teaching the future is to divide processes into **those not influenced by humans** (e.g., increasing expansion of the universe) and **those influenced by humans** (e.g., deforestation). You can use this distinction to foster debate among students as to what processes are or are not influenced by humans. Discussion of the two categories can be divided into three time frames: (1) **the near future,** about one hundred years out; (2) **the middle future,** one thousand to ten thousand years out; and (3) **the remote future,** one million to several billion years out.

THE NEAR FUTURE

Students are likely to be most interested in this period, as it is the most relevant to them. Being that they are likely around twenty years old, you can discuss their **life expectancy** (probably near ninety years, so they will live another seventy years into the hundred-year period, and any children they might have will probably live until near the end of the hundred-year period). In this period of time, students will be most interested in the processes that humans influence. However, basic ongoing processes like weather and the water cycle should also be reviewed.

One difficulty with this period is reconciling troubling developments with what is truthfully relatively good news in the short term. For example, we do produce enough food for seven billion people and will be able to do so for nine to ten billion people. **Famines** are generally local and often artificially created by human conflict, not by a global **Malthusian crisis;** our dependence on **artificial fertilizer** and increasing reliance on **genetically modified food** are the problems. It is also clear that in the short term there are enough reserves of oil, coal, and natural gas that we could continue to rely on them, particularly in **modernized societies,** to maintain a comfortable **standard of living.** This is part of the problem: many of those who are relatively well-off, and who live in modernized societies, have not experienced any dramatic impact on their lives from depleting fossil fuels, pollution, degradation of the natural world, or water shortages. Nor does any dramatic impact seem to them to be in the immediate offing. The problems seem like mere management issues. This makes it easier for those who do not wish to address questions raised by sustainability or the environment to claim that important environmental problems are exaggerated. Nevertheless, we do know that, absent some dramatic technological advance, all the problems we have now with **population,** energy, pollution, and destruction of the natural world will only get worse over the next hundred years, and it is questionable whether we can successfully manage them using only current technology, lifestyle changes, and management approaches.

One approach to demonstrating to students that the problems are real is to go over the various ways conservation is imposed upon them and to ask why, if there is no problem, they are being asked to conserve. This can be done by looking at the cost of goods and services and efforts made to restrict usage. For example, highly efficient cars and higher gas prices are really nothing more than a reflection of the need to reduce the demand for oil and to mitigate **air pollution,** including the release of **greenhouse gases** that cause **global warming** and the ensuing changes to the **climate system** that follow a warming world. Energy-efficient appliances and water inhibitors on showers are other examples.

Another approach is to discuss the standard of living achieved in modernized societies—especially in Western Europe, the United States, and Japan after World War II—in terms of **consumption.** The link between pollution and post–World War II consumption levels, particularly in Western Europe, Japan, and what economist Angus Madisson has called the "Western offshoots" (the United States, Canada, Australia, and New Zealand) is clear. People in those regions consume the most resources and produce the most pollution relative to their percentage of the overall **world population.** The problem with the increasing world population is not feeding people or delivering health care to them; it is delivering an acceptable standard of living, with adequate housing, access to clean air and water, personal safety (security from crime, war, and political oppression), and a meaningful life through employment or education. Students can research the percentage of the world population that has access to these things and those who do not. We know that the laudable effort across the globe to bring more people a standard of living approximating that in the West will only put more pressure on the environment, on energy and water resources, and will result in more pollution. On the other hand, while people left out of that standard of living have a far smaller **carbon footprint,** they engage in serious ecological damage through deforestation, destruction of wildlife through **habitat loss** and direct killing, **overfishing, soil depletion** through overgrazing and overplanting, and pollution through fouling water sources with human and animal waste in their desperate daily struggle to survive. Their burgeoning population numbers have destroyed any prior balance with nature that the remaining preindustrial societies may have had.

It is critical to introduce students to the idea that managing consumption levels, energy resources, and the water supply, and arresting pollution and ecological destruction while bringing an acceptable standard of living to as many people as possible will probably be the dominant challenge human civilization faces during their lives. The problem is not made easier by the fact that an additional two billion people will inevitably be added to the world population in the next forty years.

The question of global warming should be addressed in this section and in the next section when discussing the middle future. In this section, introduce students to the idea of greenhouse gases and the effect of trapping heat within a planetary atmosphere. One key measure of greenhouse gas concentrations is the volume of carbon dioxide, or CO_2, in the air. The instructor should explain advancements in the means of **measuring atmospheric CO_2** concentrations today as well as in the geologic past. How the accelerated burning of **hydrocarbons** or **fossil fuels** because of industrialization contributes to CO_2 in the atmosphere should also be discussed. Ask students to contemplate the consequences of continuing to burn fossil fuels without implementing systems to remove CO_2 from industrial discharge. In the near future, the consequences are likely to be uneven, with some areas of the world already experiencing devastating weather events, drought, and rising sea levels, while other areas may feel relatively little discomfort as a result of the rise in global temperatures. The **impact on other species** should also be discussed. The speed of change in the physical environment in some areas as a result of warming will be too rapid for evolutionary adaptation. As a result, many species will be doomed by global warming. Given the uneven impacts of global warming in the near future, students may be asked to discuss the likelihood of a concerted global effort to reduce human-induced global warming. Because scientists believe that failure to reduce global warming in the present will have impacts that last a thousand years or longer, such issues can also be raised in the next section.

In this period the focus may shift more to processes not influenced by humans, as consideration of the human side becomes speculative. Again, **global climate** is one issue that can be discussed here. Considering the longer-range effects of human-induced global warming, students can discuss the impact of rising or falling temperatures, rising seas, changing ocean currents, or the possibility of encroaching glaciers caused by shifting oceanic currents.[1] We know that throughout Earth's history, there have been ice ages, mini ice ages, and interglacials, or periods of warming. Students might ponder how natural climate cycles might begin to reassert themselves in a warmed world. They can also discuss the probable impact of volcanic eruptions (they will happen), earthquakes and tsunamis, continental drift, and a possible magnetic pole reversal.[2]

Relative to human activity, students may be encouraged to indulge their imaginations: space exploration, futuristic cities, world peace, collapse of civilization; it is all the stuff of science fiction.

THE REMOTE FUTURE

One of my favorite cartoons is one in which the first panel depicts a typical science class, with the professor droning on and on in front of a class of students fast asleep. The second panel shows the professor stating, "and two billion years from now the Earth will fall into the sun and be destroyed in a ball of fire." The third panel shows one student awakening with a frightened start, saying, "Wwwhat did you say?" The fourth panel shows the professor saying, "I said in two billion years . . . " The final panel shows the student saying, "Oh, I thought you said *one* billion years," and promptly going back to sleep.

This section should focus on nonhuman processes. The main ones are **evolution, continental drift,** the **life cycle of stars**—particularly our sun—and the expanding universe. It is clear that fairly substantial evolutionary changes can occur over the course of one to four million years. Students can speculate as to what might happen with bacteria, viruses, insects, plants, and animal life. Then, they can also speculate about changes in humans: larger brains, larger bodies, and longer life expectancies are all possibilities, as are human-caused or technology-driven evolutionary changes.

Sixty-five million years from now, the current continents will have shifted by approximately one thousand miles. Students can look at **plate tectonics** and imagine continents crushed together or spread apart, with the oceans reconfigured. They can also speculate about any human, animal, or plant life around by that time.

In this section, you should also review what we know is the likely process of the **death of our sun** several billion years from now, and that, long before that happens, Earth will be rendered uninhabitable and then destroyed by changes in the sun.[3]

Ultimately, because our universe continues to expand—and its expansion appears to be accelerating—everything in the universe will move farther and farther apart, until the forces that led to the forms of complexity that we find on Earth will no longer do so. Complexity will break down, as if in reverse, until the universe is again a somewhat homogeneous cloud of matter and energy.

Alternatively, the near, middle, and remote futures may be taught in reverse order, so the longer-term processes can be more briefly summarized before focusing on the more immediate and relevant near future. Some faculty find that ending on the largest scale leaves students feeling insignificant and without a sense of agency. Keep such affective learning outcomes in mind as you think about how to structure your discussion of the future.

THE FUTURE IS NOT (NECESSARILY) A THRESHOLD

J. Daniel May

In a new academic field such as Big History it is inevitable that terminology will vary from one expert to another. Often, that reflects a difference in vocabulary more than a difference in thinking. But that is not the case with the varying and inconsistent labeling of the future as a "threshold"—Threshold 9. Varying uses of this key term express different meanings. Here are a few:

1. The reflexive use of "threshold" to mean "chapter" or "unit"
2. The subjective use of "threshold" to mean "my own individual future"
3. The deliberate use of "Threshold 9" to signal an assertion that the next new thing is about to happen

First, let's consider the use of "threshold" as nothing more than a synonym for "chapter" or "educational unit." I'm particularly struck when I hear my own students talking about the future as "Threshold 9," as this terminology is not in the textbook we use, nor do I ever use the term "Threshold 9" myself. My students seem to spontaneously coin this usage. It can also be seen in a variety of print and online resources—but this use can reflect a basic misunderstanding of the specialized meaning of "threshold" in Big History: a transition in which something fundamentally new appeared in the universe.

When we use the thresholds as a pedagogical organization scheme, it is easy to conflate "threshold" with "chapter" or "unit." But treating "threshold" as nothing more than a chapter heading or a line in a syllabus obscures the significance of such a major change.

The subjective use of "threshold" is another reason why I hear my own students talking about the future as "Threshold 9." Of course first-year college students think of *their own* future as a critical threshold. On the level of subjective personal experience, they are absolutely right. They are just starting their adult lives, and naturally think a great deal about the education they are engaged in, the possible career that may follow graduation, and concerns about falling in love, getting married, having children, and caring for aging parents. They also worry about the state of the world: climate, ecology, politics, economics, war, and so much more. Indeed, from their point of view, the future naturally appears to be just one big threshold after another. But to conflate that sort of personal life stage with the story being told in Big History misses the point. In the Big History classroom, we need to be careful to draw the distinction.

Finally, referring to the future as "Threshold 9" is sometimes used to express a genuine conviction that something new really is about to happen. But we are often overconfident in our ability to make accurate predictions about the future (where's my jet pack? the moon base? the nuclear apocalypse?) While the future is always on its way, we can't be certain that a new *threshold* is upon us.

For one thing, we overestimate the long-term importance of trends we experience within our own lifetime. We are hyperaware of the details of change from decade to decade, often even from year to year. But Big History is about a large-scale context. One thousand years from now, the distinction between dirty nineteenth-century factories and sleek Information Age office buildings may not be the big deal that we think it is. One thousand years from now, historians might well view themselves as still in Threshold 8, seeing steam engines, motor cars, and computers as all part of the "machine threshold," or they may see the period from 1800 on in a new context that we can't even imagine.

What's more, many of our students already feel a crushing sense of despair about current events that they have no control over—yet. Labeling the future as "Threshold 9" promotes thinking of the future as a tangible, fixed reality—a reality as unchangeable as the evidence-based narrative of our fourteen-billion-year past. That's not the message we should give our students.

Regardless of what the instructor thinks, isn't it *pedagogically preferable* to not label the future with certainty as a "threshold"? Shouldn't the distinction between "the future" and "Threshold 9" be an open question that we invite our students to reflect on and discuss? Questions, not answers, should dominate this conversation. "Now that you've spent all semester studying past thresholds, what do you think about our future? Are we truly on the edge of a new threshold? Is the new threshold already here? What do *you* think?" and "What do you want to make of *your* future?"

THE FOUR FEATURES OF COMPLEXITY AND THRESHOLD 9

Richard B. Simon

Drawing the distinction between the future and a threshold doesn't mean we shouldn't think about what a Threshold 9 might entail. In fact, it's a good critical thinking and problem-solving exercise for students to attempt to extrapolate just what a possible Threshold 9 might look like.

You might ask students to consider the eight thresholds of the Big History story and try to make sense of the patterns that progress through the thresholds, as well as those that recur. The first three thresholds have to do with what Fred Spier calls **complex nonadaptive systems**[4]—stars, galaxies, and the universe—systems that don't maintain their own homeostatic states in response to perturbations. Thresholds 4 and 5 come to Earth, where, ultimately, the biosphere captures the sun's energy, and living things use that energy to actively maintain their own complex structures (life, and, as James Lovelock argues with the Gaia hypothesis, the Earth itself, are **complex adaptive systems**). Thresholds 6, 7, and 8 all have to do with human beings and our increasing hold on the biosphere as our **society** becomes more complex and as **cultural evolution** appears to outpace **biological evolution**. Would a Threshold 9 be another revolution in human civilization? Or would it involve something as yet inconceivable?

In any case, any Threshold 9—if it follows the pattern established thus far— would be marked by changes in the four features of complexity.

DIVERSE COMPONENTS

Across the story, we have moved from mere energy and matter to subatomic particles *made of* energy and matter, and new forces; to clouds of hydrogen and helium atoms *made of* subatomic particles; to galaxies filled with stars *made of* hydrogen and helium that churn out complex elements; to a star with planets *made of* complex elements and chemical compounds—including one planet whose chemical compounds give rise to living cells; to animals, vertebrates, mammals, primates, hominids, hominines, and *Homo sapiens made of* living cells; to agrarian civilizations *made of Homo sapiens* interacting in complex ways to trade energy and DNA and bacteria and viruses; to a globalized industrial civilization *made of*—initially, at least—interconnected networks of agrarian city-states. What comes next in the pattern?

SPECIFIC ARRANGEMENTS

Here, students might consider the patterns that repeat from level to level, from threshold to threshold—a dense, nuclear core, surrounded by energy flows, connected through networks to other, similar structures—be they atoms or cells or cities or empires or world zones or solar systems or . . . ?

FLOWS OF ENERGY

What we see from threshold to threshold, throughout the story, is an increase in energy flows. In *Cosmic Evolution,* Eric Chaisson writes about "free energy rate density," or the amount of energy that is used per unit of time, per unit of mass, within a particular form of complexity.[5] As Christian, Brown, and Benjamin frame the story, what marks a new threshold is essentially a revolution: a new form of complexity appears that can process more and more energy per unit of time, per unit of mass. Chaisson calculates that the most complex entity in the known universe is modern industrial civilization—us, right now. Whatever comes next, if it represents a new threshold, it will incorporate immensely higher flows of energy, which are used to maintain its unique structure, and which lead to new and heretofore unfathomable . . .

EMERGENT PROPERTIES

Thinking about what the emergent properties could be in some proposed, possible Threshold 9 is a great paper topic. It's also ripe for dynamic class discussion. At the end of this chapter, you can see a few approaches to leading this discussion—as exemplified in Kiowa Bower's projection of a technology-based Threshold 9 and Philip Novak's proposal that Threshold 9 may be a revolution in human moral consciousness.

Let students work to comprehend the patterns and to extrapolate from them to make predictions as to what might come next—or to think about what they might like to see come next. Predicting what might happen based on past experience and observation is a key life skill—and just the kind of complex, adaptive critical thinking and problem-solving skill we want students to have. It's what science fiction writers do, certainly. It's also what leaders do. And people we call "visionaries." Engaging in this type of thought experiment is how we prepare the next generation to solve the problems they will certainly face in their lives, in their careers, and as members of a complex, adaptive human society.

CHALLENGES IN TEACHING POSSIBLE FUTURES

Neal Wolfe

Teaching about the future is, for obvious reasons, a more difficult task than teaching about the past (despite the challenges inherent in that enterprise, as well). Certainly, we can say much about the challenges we face today, which will need solving as we go forward. It is reasonable to project into the next several decades, although even in the relatively short term, unforeseen developments are likely to skew our projections—possibly even dramatically so.

Clearly, as a global community, we face daunting challenges. There is nothing new about humanity fouling our own nest, nor are the repercussions of the inequities inherent in a surplus-driven economic system of recent origin, but now we have reached a point where we are effecting change on a much larger scale and with unprecedented speed, and there is ample reason to be alarmed. While students are, generally at least, aware of this, it is important to make sure they fully appreciate the challenges we face—and their contributions to them. (In a recent course, I found that while students knew that we live in a consumer society that generates a lot of waste, when they had to keep everything they normally throw away for several days, they were astonished at how wasteful their own habits were. For the first time, many of them realized their own complicity in a larger societal, even global, problem.) Through Big History we also learn how the situation we face today developed to this point, providing context for making the necessary choices going forward.

While, for good reason, the near future is of greatest interest to most of our students, it is fascinating as well to project into the middle and even remote futures. What will the fruits of genetic engineering of food sources and human beings bring about? What about bioengineered body parts? Will we be able to prolong life indefinitely? (I like to ask my students if they would choose to live forever, if that became possible. An interesting discussion follows, with virtually no one ultimately choosing eternal life—especially when I stipulate that everyone else will also live forever. When we reflect upon why we wouldn't choose to live forever, we find that our very mortality gives our lives meaning.) Will we reach the "singularity," where technology will render human life as we know it a thing of the past? And while it may not matter to us in any practical way that the sun will die and incinerate Earth in another five billion years or so, does knowing that will happen alter our perspective on how we see ourselves now and in relation to the future? Will we have accomplished the extraordinary feat of finding our way to a new home outside the solar system by then, or perhaps live in self-sustaining "arks," drifting through space? And what of the ultimate end game of the universe? Will it go on forever more or less like it is? Will it collapse upon itself? Or will it slowly but surely, over countless billions of years, go dim, as star-making material is used up? All this is fascinating to speculate about, although many students are made uncomfortable even considering the possibility of a "dead" universe. Still, let's not ignore the fact that changes are inevitable on a universal scale. It is also interesting, and perhaps useful, to consider that the universe is still "young."

Students are particularly interested in the unit on the future because it pertains most directly to their own lives. They are anxious about their personal futures, especially given the circumstances of the world they find themselves in. But it is counterproductive to educate them about the mess the world is in and then send them off with a hearty "Now go fix it!" (Try the "Opinion Snake" activity, on page 255, to gauge your students' optimism or pessimism.) The future is the part of the Big History narrative where students hope for an answer. "Okay, I've learned all about the history of the universe from the Big Bang until now; I understand that it is all one connected, continuous development. Now what? Where do I go from here? Give me some reason to be hopeful!"

First of all, make sure to emphasize the many positive actions and approaches already being taken, which unfortunately don't often make it into our popular media. Especially given our unprecedented interconnectivity through the internet and cellular phones, we now share information and work (and learn) collectively like never before. We are able to personally help disadvantaged people in developing countries, and nongovernmental organizations work without waiting for governments to act. Challenges create possibilities! Do your homework and find positive examples of solutions that are already being advanced. Encourage your students to consider the possibility of becoming part of those solutions.

Beyond doing what we can to make students aware of the challenges they face, and the positive, often exciting efforts at solutions under way today and in the years to come, now is the time to make sure the lessons of the Big History narrative are fully appreciated. Don't assume that by covering the entire narrative, students "get" the full implications of what they have learned. If we have done our job and effectively facilitated an understanding in our students of the seamless continuity of development from the Big Bang forward, they (and we!) must see ourselves in a profoundly new light. Whatever we are, we are part of the entirety of the universe, for all time. Obviously, it is easier to say this than to fully realize and sustain it. But doing so should be our goal. And while this realization is not easily measurable or assessable, instilling such perspective in our students, at least partially with the hope of further maturation, is a very meaningful kind of preparation for the future. We can do only so much, in a Big History course, to prepare students to take specific steps to contribute to solutions as they go forward. But if we facilitate a reshaping of their perspective on who they are and their relationship to the world around them (and this includes our modeling of our own perspective for them, of course), we will have done much to prepare them for their futures.

So, this is the time to clarify our current situation and how it developed to this point, our challenges now and going forward, and innovative, exciting actions toward solutions that are already happening— as well as speculating about the near, middle, and remote future. This is also the time to make sure that everything that has come before in the semester is fully integrated into a holistic understanding that includes the student from the very beginning, and projects ever after into the future. How to do this? Options include reflective writing and class discussions. Do what feels right to you with your particular class—the important thing is to make students reflect on what they have learned as a whole and what the implications are.

As visionary poet Walt Whitman observed in verse, "Immense have been the preparations for me" and "I am an acme of things accomplished, and I am an encloser of things to be."[6] Isn't this the message of Big History? We find ourselves (and everything else) as the culmination of everything that has come before, and the vanguard of all that follows. If we truly embrace the Big History narrative, how could it be otherwise? If our students understand this, they will be better prepared for whatever the future holds.

PROJECTING THRESHOLD 9: TECHNOLOGICAL REVOLUTION

Kiowa Bower

Our species faces the technical and political challenge of continuing to expand our population and increase our standard of living despite dwindling access to vital resources. Some of our problems seem intractable within the context of current technology and infrastructure. However, technological changes in the next fifty to one hundred years may be significant enough that humanity has means to overcome many of the present challenges and thrive in ways it never has before. There is a potential for such profound changes to society that this period in history could easily represent the passage through the next threshold of complexity.

To help students understand how such dramatic changes could be possible in the next fifty years, it is important to communicate the exponential nature of technological progress. The most familiar illustration of this trend is **Moore's law,** the observation that computational capacity doubles about every two years. This exponential trend has resulted in a million-fold increase in computational power in the last forty years.[7] Maintaining this doubling rate beyond this decade will, however, require implementation of new technologies, such as molecular computation or quantum computation. The overall technological progress (i.e., general human tool use) is difficult to quantify, but the long-term trend is clearly exponential. One explanation for this is that previous advances are used to facilitate the development of the inventions that follow. For example, the availability of a welding torch facilitates the creation of things like a steel crane, which in turn can help construct a skyscraper. You can reinforce the concept of accelerating change by pointing out its presence in past thresholds. The developments associated with agriculture led to a threshold-level transformation of human society within a few thousand years, and those of the Industrial Revolution transformed society within just a few hundred years.

Turning the discussion back to the future, you could ask students how long they think it could be until the next threshold is crossed. What are some elements that would suggest that such a transformation is taking place? Increased connections? Every human has an immediate connection to every other human and to every piece of information. The **Information Age** has certainly already significantly changed our society in just a few decades, but is this only the beginning? Collective learning itself is accelerating exponentially. Soon we may be connected to one another and to the entirety of all human knowledge through a **head-up display** on our retinas.

Biotechnology, nanotechnology, robotics, and **artificial intelligence** are the areas of technological development that are likely to have the most significant impacts on the near future. While the following ideas are speculative, they deserve consideration because they pose such serious implications for society. Advances in biotechnology are already showing a capacity to extend our life-span, by suggesting that we will eventually be capable of eliminating disease and significantly slowing or reversing aging. We will soon have the capability to genetically engineer our offspring and thus take direct control of our biological evolution. Nanotechnology will likely be a part of this biomedical revolution, as molecular-sized machines could target diseased cells and regenerate damaged tissue. Nanotechnology could also make possible the production of inexpensive solar panels that could provide society with sufficient supplies of energy. Advances in robotics will mean the automation of almost everything.

One exercise to consider is asking students to weigh the risks and benefits of future technological developments. Some important risks worth mentioning include access to advanced technology being limited to the wealthy, the social implications of **genetic engineering,** the risks of **bioterrorism,** the **nanobot "grey goo"** scenario, and the robotic replacement of the workforce, leading to unemployment. The more utopian visions incorporate an era in which human and machine intelligence merge, disease and aging are eliminated, and environmental issues are solved. Proponents of the technological **"singularity"** concept, such as Vernor Vinge and Ray Kurzweil, point to the emergence of greater-than-human general intelligence as being by far the most powerful force in such a transformation.

It is not unreasonable to think that artificial systems will one day surpass **biological intelligence,** and there are already functional **brain / machine interfaces:** the close relationships that many of us have with our mobile devices. To a certain extent, we have already begun the process of merging with our **information technology.** Artificial intelligence (AI) is a fascinating topic because it has the greatest potential to transform the world as we know it. However, this discussion can get into more speculative areas depending on how you define AI. We already have narrowly **limited AI** systems running all around us, performing Google searches, facial and voice recognition, and even the landing of jet airliners. This is not the same as **general AI,** which many believe will come at some point. General AI is the creation of a system that is more intelligent than humans in all ways, even in terms of **emotional intelligence.** These artificial systems would have the potential to accomplish things we cannot yet imagine, but if we are not careful, we could end up competing with them for resources.

In class, you may wish to consider the relevance of major technological advances to political, social, and environmental issues and future projection in the context of past accelerating trends.

Not only has technological progress played a dominant role in shaping human history but the rate of development of new technologies has been accelerating ever since hominids began shaping the first stone tools. It is pretty clear that the accelerating trend of technological development is continuing, and this is an immensely important factor when considering what our future might look like.

The **accelerating nature of technological change** is a critical concept to communicate to your students. The human brain has evolved to make linear projections, and the significance of change that results from an **exponential trend** is not intuitive. It may be helpful to illustrate to students the difference between a linear and an exponential curve. Again, a common example of a technological process following an exponential trend is Moore's law.

Collective learning can also be discussed in the context of technology. The development of better transportation, writing, the printing press, the telegraph, the telephone, and now the internet represents ever-increasing interconnectedness, and, thus, ever-increasing rates of collective learning. As we look into the future it appears that we are moving toward even greater interconnectedness. With access to all human knowledge in our pockets, and the ability to communicate with anyone in the world, we are in one sense moving toward becoming one large network.

Consider examining with your class whether human society is on the verge of (or in the midst of) the next threshold of complexity. It is, of course, difficult to see the shape of this transformation or to make accurate predictions about the outcome if such a transformation is taking place. Students will have strong and varied opinions on the evils or promises of future technologies—resulting in energetic class discussion. It is also a useful exercise to have students try to distinguish between those ideas that are realistic projections of current trends and those that are speculative. Understanding the potential implications of continuing technological progress on our future is vital and can help students prepare for a future that may be vastly different from the present.

THRESHOLD 9: AN ADVANCE IN CONSCIOUSNESS

Philip Novak

Big History courses end with a consideration of possible human futures. I have been proposing to my students that the next great evolutionary emergence of complexity ("Threshold 9") could very well be a species-wide advance in emotional and moral maturity, that such an advance would first require broad conscious intentionality toward it, and that the latter itself would require as a precondition an array of collective symbolic evocations of it. I ask my students to imagine that (1) such a Threshold 9 would arise from the same four general conditions as all prior thresholds, that is, plural elements or diverse components in a novel structure or specific arrangement capturing sufficient flows of energy under Goldilocks conditions (about which, more below); and that (2) such a Threshold 9 is no more implausible than any of Big History's other thresholds (about which, more right now).

Currently, Big History texts imagine the future in a time-honored manner, by extrapolating existing trends. Yet as the Epic itself shows, there could be no worse method for envisioning the emergence of a *new threshold*. Big History reserves the term "threshold" for something *radically* new—in principle unforeseeable—coming into being. Current Big History is nothing less than an account of eight major, near-miraculous emergences of new, complexly ordered somethings out of their own virtual absences, the whole amazing sequence of which runs inexplicably *against* the universe's entropic grain. It is the stunning strangeness of an entropic universe displaying a multibillion-year trend toward intelligent order that moves a complexity scientist like Stuart Kaufmann to exclaim: "We live in an emergent universe in which ceaseless unforeseeable creativity arises and surrounds us."[8] From the Big Bang itself; to the emergence of matter from pure radiation; to the de novo appearance of oligonucleotides, without which life would not have been possible;[9] to the emergence of life out of its own absence; to the apparently sudden emergences of protein folds, major groups of viruses, principal lineages of prokaryotic archaea and bacteria, eukaryotic supergroups, and animal phyla;[10] to the emergence of **human consciousness**; and finally to an entirely new dimension of evolution—cultural evolution—Big History teaches us that the universe is not a static cosmos but a continuous and irreversible cosmogenesis—a recurrent and partially lawless birthing of the radically new.[11] As biologist J. B. S. Haldane famously remarked, the universe is not only queerer than we suppose; it is probably queerer than we *can* suppose.

I therefore urge my students to experiment with the following thoughts: what if the next major evolutionary emergence is less about the human mind finding ways to unlock the power of matter (the essential theme of both Thresholds 7 and 8) and more about the human mind finding ways to unlock its own power—a collective advance in **moral intelligence?** This would be appropriately radical and in principle unforeseeable, would it not? Yet, in order to argue persuasively for its sheer possibility, we would have to be able to imagine it as arising from realistically conceived *plural elements*, in a *novel structure*, with access to sufficient *energy flows*, all within *Goldilocks conditions*.

The plural elements may be hypothesized as the vast array of *ethoi* (ethical worldviews) of the world's cultures. This data has been available to the world's scholars in all its depth and richness for less than a century; this collective experiment in moral self-consciousness likely has a long way to run. The specific arrangement, or novel structure, may be hypothesized as forming from the countless threads of interfaith conversations (deep and shallow, personal and institutional, idea-based and cooperative action–based, arising from intolerable conflict or from present harmony seeking new intensity) made possible by our shrinking global village and the internet's annihilation of distance. This new structure would be expected to throw off (as sparks) new universalistic symbolic constructs, quite new in human history, and (perhaps) corresponding new universalistic practices. The necessary flows of energy might be hypothesized simply as the possibly enormous aggregation of efforts by individuals who, seeking a more fulfilling relationship with reality, turn consciously and en masse toward ideals of moral flourishing with clear interspecies benefits. The Goldilocks conditions (which can be seen as a friendly expression of the same thing other cultures have called "grace" or "luck") would, of course, have to include the decent satisfaction for all the world's people of their material needs (as per the lower levels of Maslow's hierarchy of needs) and a near-miraculous enantiodromia of current, seemingly maximal, religious divisiveness in certain locations.

I confess to wanting to be part of a Big History staff that teaches its students not only that they are the astonishing material products of a fourteen-billion-year evolutionary sequence but also that, for all we know, the cosmos might actually be *predisposed* to favor the arising of conscious, rational creatures capable of moral development, like ourselves. If we can share such hopes with our students without violating any scientific canons of rationality, shouldn't we?

But an intellectually adequate proposal for the possibility of a Threshold 9 as described above logically requires nothing less than telling the Story somewhat differently from the very beginning, registering from the outset a frank skepticism about the rational adequacy of a materialist worldview. As Thomas Nagel has written in an important book addressed to this very topic, "My guiding conviction is that mind is not just an afterthought or an accident or an add-on, but a basic aspect of nature [itself]."[12]

TEACHING TO THE FUTURE: FOREGROUNDING STUDENTS' FUTURES IN EVERY THRESHOLD

Debbie Daunt

My course "Health and Healing through the Lens of Big History" appeals to many students, but the students most attracted to it are pre-professionals looking toward careers in health care. The course explores the concepts of health and healing through the thresholds of Big History, with topics including death and life; the changing health and healing of the Earth, organisms, and communities; the role of health and healing in evolution; disease in humankind; and cultural diversity issues as related to health and healing through time.

The academic demands on future health-care professionals are high. These students are experts at digesting information and correctly choosing the right answers on multiple-choice tests. But while they are required to read hundreds of pages a week, they may have difficulty intelligently discussing what they have read; the introspection and personal growth that one is expected to acquire in college is rare in this student population. Between the heavy coursework, labs, and the required hours they spend in clinical rotation, there is little time to sleep, let alone think. It is not surprising that these students often lack the ability to think critically, and to form and express opinions.

"Health and Healing through the Lens of Big History" fills that gap by presenting each threshold in terms of the future—or, more specifically, the students' own futures. When health-care students can see how the material connects to their future practice, they are engaged, and ready to discuss, reason, and think. The methodology is simple. The challenge for the teacher is to help students discover a personal link to the future in each threshold. As our assessment data shows, students feel the greatest ownership of the Big History material when they perceive the interconnectedness of the various disciplines of study (in this case, between health care and cosmology, or evolutionary biology, or human history); when they are able to ask larger questions about meaning; and when they have a chance to consider the future. That's why teaching *to* the future increases engagement in our future-focused students.

Here are a few examples.

THRESHOLD 1: THE BIG BANG

Near the beginning of the course, I encourage students to take a broader look at the concepts and establish an environment in which they will feel comfortable expressing themselves and their ideas. Again, this is something they rarely have the opportunity to do in their science classes. The class is asked to agree on operational definitions for the terms "health," "healing," and "disease." They are given the parameters that each definition must be broad enough to fit each threshold. One class defined the terms thus:

Health: the general condition of the entity that is moving toward a state of being without injury or illness

Healing: restoration to optimum integrity given the current condition; making something good out of something bad

Disease: a condition that impairs functioning due to the homeostatic imbalance of a system

Using these definitions, the class came to the consensus that the Big Bang caused a "disease" state that progressed to "healing," and eventually "health" was restored to the universe. We then equated the Big Bang with the course of a disease. Because these future healers will play a part in restoring health by combating disease that threatens a body, thinking about Big Bang cosmology in these terms allowed them to understand the course material.

Not only that, thinking about the Big Bang in terms of health, healing, and disease created a link between Threshold 1 and the future. The analogy to their work allows students to understand how the universe changes over time, just as a human body does.

The Big Bang is now an event that is *conceptually linked* to students' future practice in health care. By their definition, the Big Bang was a "disease," because the universe's homeostasis was disrupted. As order was restored over time, the "disease" was eradicated. *Healing* happened—and in this analogy, as we travel in time away from the Big Bang, the universe is in a state of general good "health."

THRESHOLDS 5 AND 6: THE EVOLUTION OF LIFE
ON EARTH AND THE RISE OF *HOMO SAPIENS*

Thresholds 5 and 6 offer great opportunities to shake things up and promote thinking. Our students have all studied basic biology, and most have been taught that the pyramid of life has at its base bacteria, viruses, and the like. Humans are the "most evolved" life-form, and are therefore at the top. Or are we? In class, we turn the pyramid upside down, and by the end of a couple of days of discussion, students gain a true appreciation not only for the emergence of life and the evolution of humans, but also for the power of the simplest organisms.

For health-care majors, this is of immense importance, in that the causative agents of most disease states are bacteria or viruses. Microbes kill the majority of us humans—and many of these students will spend their entire professional careers in battle with them. Students who are not in health-care programs relate to the power of the microorganism as they remember their required immunizations and how they all got sick when one friend came to the party with the flu. The power of the microbe is also illustrated in popular films—for example, *Contagion* (2011), which validates the idea of the potential elimination of humans by a virus.[13]

The survival techniques of those traditionally at the base of the pyramid are superior to those of humans. They are extremophiles. They will survive . . . anything. They will not go away and we will not eradicate them. In the fight against bacterial and viral diseases, we humans are losing—and it appears that we will continue to lose well into the future. Yet there is an essential symbiosis. Humans are made up of hundreds of types of bacteria. One could say that our bodies are not our own but belong to the single-celled organisms that inhabit them.

Health-care majors, with their backgrounds in microbiology, can easily see the connections between the long-ago emergence of life from microorganisms and their own futures, engaging with those microorganisms in their own work. That direct connection allows them to understand the origins of life *and* the role that microorganisms might play in the future.

An added benefit is that students become much more aware of their own health and transmission of disease. The *immediate future* benefits from a healthy respect for microbial transmission. The smallest organisms are the simplest yet most powerful part of life. We learn in Big History that they were here first. When we think about life and the future, we realize that they'll probably be here last, too.

THRESHOLDS 7 AND 8: THE AGRARIAN
REVOLUTION, MODERNITY, AND INDUSTRIALIZATION

Health-care providers are often confronted with situations and decisions that test the very core of their principles. A large part of my presentation of Threshold 8 revolves around the "Plague Simulation,"[14] an exercise in which we ponder the decisions humans had to make during the period of the Black Death in fourteenth-century Europe. The student learning outcomes focus on self-exploration and thought; students are provided the opportunity to question values and beliefs that they may have taken for granted.

If some of that questioning takes place in this course, the student will be better prepared to face the hard challenges that will come in the life of a caregiver. But the benefit of questioning one's core values is not limited to those who will be caring for others; knowledge of self is the foundation for all future growth.

In the plague simulation, I guide an exploration of people's thinking during the time of the plague and equate decisions made then with decisions that students might face in the future. We begin with a little background reading and video on the plague so that students can build an overall knowledge base. Next, we imagine ourselves in that situation. Discussion begins, driven by difficult, values-challenging questions:

What would you do if you knew death was imminent?

Would you party till you dropped—as many did?

Would you give your last days to the sick and dying and care of others?

Would you pray?

What would you do with limited resources?

Would you withhold food from the infants and old people because they have the least resistance and will most certainly die?

Would you hoard food to be sure you and your family have enough?

Would you give your food to the sick in the hope that they will survive?

Would you labor to grow, gather, and prepare food for others at the risk of depleting your own immunities?

Where would you go?

Would you stay in your town and face devastation with your friends and family?

Would you go off alone and fend for yourself, knowing that human contact could cause your death?

Would you stay and provide for your community but send your children to a less-populated area where they might be safer?

Would you take your immediate family into hiding and care for them alone?

People living through the plague faced all these questions, and many of them are applicable to the future. Health-care providers will certainly be faced with making such values-laden decisions in the course of their work, but all students will have to make decisions that test their values at some point in their future.

These discussions are the most engaging and rewarding experiences in the course for both my students and me. It is said that the best predictor of future action is to look at the past. Students who have at least considered such a challenging situation, even in an exercise, will be better prepared to face difficult decisions in the future.

The real-life question I use to link these newly considered values to the future is a scenario seen almost daily in most hospitals: A geriatric patient in the hospital is refusing all food. This patient is likely somewhat demented and therefore the decision-making responsibility falls to the family. It is possible to put a tube down the person's nose and flow liquid nutrition into the gastrointestinal tract. A more permanent solution is to put a tube into the stomach and insert the liquid food directly through this tube. With intervention, the patient will not die of starvation and could live for years, provided nothing else (such as a microorganism) causes death.

I ask the students to discuss what they would do if the patient were their own grandparent. Generally, there is no consensus, and the class is divided between giving treatment and allowing nature to take its course. I then ask if they could deliver the information about the proposed treatment to a patient's family without bias. I introduce the concept of "bracketing": could they set aside their own beliefs to honor the decisions of others, or would they give a biased explanation of the situation and the tube procedure to sway the decision?

In the future, when faced with a difficult decision or a conflict of opinion, the student will have had the experience of "practicing" a similar thought process in class; they will know themselves and recognize their own biases, and thus will understand when they need to "bracket" because of those biases.

Frankly, when the course draws to a close and the last topic of study is the future, there is little left to discuss. Because *each* threshold has related to students' future lives, the class has moved through time without ever losing sight of the future. This is exactly what our students are doing during their education. They move through our courses in the hope that what they learn will prepare them for whatever they'll encounter down the road. When we teach to the future by drawing the connection from each threshold to students' own individual futures, and thus to our collective future, we engage students more deeply in the material, connect it to their own lives, and prepare them to solve problems and face the challenges that the future will surely pose. Yes, the students learn about the Big Bang in Threshold 1, about the evolution of life in Threshold 5, about the Black Death in Threshold 8—but they also learn about themselves, and they acquire tools to carry with them into the future.

ACTIVITIES AND EXERCISES FOR TEACHING POSSIBLE FUTURES

OPINION SNAKE

Cynthia Stokes Brown

Threshold Related to the Activity: The future

Category of Activity: Learning activity

Objective: To encourage all students to express their opinions orally in class and to present concretely a continuum of possible opinions

Overview of Activity: Students can engage with any controversial topic through this activity. On the topic "What expectations do you have for the future?" proceed with the steps below, encouraging students to line up on a spectrum, from "most pessimistic" to "most optimistic," and then share their thoughts with the class.

Faculty Preparation for Activity: n/a

Cost: n/a

Student Preparation for Activity: n/a

In-Class Sequence of Activity:

1. After presenting on and having students discuss the challenges that face humanity today, ask students to think about how optimistic or pessimistic they are about whether or not humans will be able to solve the key challenges facing them over the next twenty-five years (or however many years you wish).

2. Give these instructions: "When I give the signal, each student will find a place in a single line, with one end representing the most extremely optimistic opinion and the other end the most extremely pessimistic one. The optimistic end will begin at————(the door, the desk, etc.) and stretch around to the————."

3. Students will take their approximate places, then talk with neighbors about what their opinions are, exchanging places as necessary to form a continuum of opinions.

4. After five to ten minutes of talking with neighbors, students will stay in place, and each will give a short statement of her or his opinion, starting with the pessimistic end so that the activity ends with the most optimistic statement.

Assessment of Learning: n/a

Origin of Activity: This is an original activity.

J. Daniel May

Threshold Related to the Activity: The future

Category of Activity: Film clip series, with supporting brief lectures and discussion

Objective: To demonstrate how fictional narratives of the future are rooted in the present reality of the storytellers, and thus become a modern mythology reinvented for relevance to each new audience

Overview of Activity: Following the use of the atomic bomb in World War II, a new subgenre of futuristic fiction mushroomed (pun intended!), featuring stories set in a post-apocalyptic world caused by nuclear warfare. This series of films highlights how the basic premise varies with the prevailing anxieties and concerns of the time in which each film was produced. The total class time required for this activity is thirty-five to forty-five minutes, with five to ten minutes of introductory remarks, a twenty-minute film clip, and a follow-up discussion lasting ten to fifteen minutes.

Faculty Preparation for Activity: Watch each of the films, select a desirable scene of about twenty minutes in length (suggestions made below), prepare introductory remarks to draw attention to key features and connect with course content, and prepare questions to prompt discussion following the screening.

Cost: All of these films are readily available on DVD, generally for under $15 each, or they can be borrowed for free from various libraries.

Student Preparation for Activity: n/a

In-Class Sequence of Activity: This film series is intended to be used over several class meetings, typically one film per meeting, but two or three films per meeting is possible if the instructor wishes.

On the Beach (1959). Based on the 1957 novel by Nevil Shute. The novel and film established the convention of starting the story in the aftermath of nuclear war. Australia has been left untouched by direct warfare, but a cloud of radioactive fallout is gradually enveloping the entire planet. The survivors know they are doomed, but they maintain social order and face their end with honor, military discipline, and dignified acceptance.

Suggested Scene: The crew of an American submarine visit the ghostly remains of San Francisco and San Diego, searching in vain for other survivors. Dialogue exemplifies Cold War nuclear fears.

Original Audience: The novel and film were received by a post–World War II audience that could identify with the sacrifice, discipline, and basic decency of the characters.

A Boy and His Dog (1974). Based on the 1969 novella by Harlan Ellison. A short prologue of atomic explosions and title cards establish that politicians "have finally solved the problem of urban blight," followed by the first shot of a desolate desert landscape. This is the film that established the look and feel of post-apocalyptic films to follow, right up to the present day. The protagonist is a young man concerned solely with scavenging food and finding women. His remarkable companion is a telepathic dog who is considerably more educated, erudite, and civilized than any of the humans. Humanity has indeed sunk lower than the animals; human savagery and anarchy abound.

Suggested Scene: First twenty minutes of film

Original Audience: The film was aimed at people who had experienced the social upheavals of the 1960s, for whom a descent into anarchy seemed highly plausible.

The Road Warrior (1981). Written and directed by George Miller, this sequel to *Mad Max* opens with a prologue establishing nuclear warfare and social decay, but with a special emphasis on the scarcity of remaining oil and gasoline reserves in the desolate, anarchic, post-apocalyptic world. The influence of *A Boy and His Dog* is evident throughout. Writer-director George Miller deliberately employed mythological themes based on the work of Joseph Campbell. When the reluctant hero (played by Mel Gibson) allies himself with a small group holding out in an oil refinery, his struggle is elevated to a fight to save the last remnants of civilization.

Suggested Scene: First twenty minutes of film

Original Audience: The premise of the film is superficially silly: people who ought to be fighting over food and water are depicted fighting over gasoline (and wasting a lot of gasoline in unnecessary roadway combat). But the film touched the fears of Americans who had seen gasoline prices skyrocket, lined up for blocks to fill up their gas tanks, and heard a series of dire predictions about the world's oil supply being rapidly depleted.

Mad Max Beyond Thunderdome (1985). The third *Mad Max* film, in which Mel Gibson returns as the title character and the themes in the story have matured. No longer fighting over gasoline, people have begun to rebuild, and knowledge of renewable resources has become more important. This is neatly personified by an engineer who is producing energy from pig feces (methane), and the film's climax is a struggle between two factions for possession of the engineer.

Suggested Scene: There are several possibilities, but the richest is the revelation of the Tribe of Lost Children and the new mythology they have created to explain their life after the "Pox-Eclipse." Watch with subtitles on to catch all the details of the dialogue, delivered in thick Australian accents.

The Road (2009). Based on the 2006 novel by Cormac McCarthy, which won the 2007 Pulitzer Prize. McCarthy takes a genre that has been done to death and infuses it with something that hasn't been done so well since *On the Beach* —realism. The story follows a father and his young son trudging through a world that appears to be in the worst throes of nuclear winter. Savagery, anarchy, and even cannibalism are commonplace. All that is left is for the pair to hold onto the last shred of human decency by not resorting to cannibalism.

Suggested Scene: First twenty minutes of film

Audience: The modern audience knows so many ways the world could end: nuclear war, bioterrorism, global warming, and more. Therefore, the exact cause of the environmental collapse is never specified—the audience needs no explanation.

The Book of Eli (2010). This film reuses many elements from previous films: the desolate, junkyard desert landscape, the lone wanderer who is also a skilled warrior, the ravaging gangs of murderers, rapists, and cannibals. What distinguishes this film is that the lone wanderer, Eli (played by Denzel Washington), is carrying what may be the last copy of the Bible, and the hints that he is under divine protection. Gary Oldman plays a brutal gang boss who is, ironically, the only person creating any semblance of social order, and who wants to gain possession of the Bible so that he can use its words to reinforce his control over the illiterate populace.

Suggested Scenes: Eli is a "guest" of the gang boss, who sends a young woman to "entertain" Eli. But instead, Eli teaches the woman how to say grace before a meal. She asks Eli about the world before "The Flash" (presumably nuclear war), and Eli replies that people threw away things that they now kill for. This is immediately followed by Eli's attempt to escape, and his dialogue with the gang boss about possession of "The Book."

Audience: The overtly religious components of the story obscure the subtler fact that this is also a story for the Information Age. The struggle between the holy man and the gang boss is very much a struggle over who controls the flow of critical information.

Assessment of Learning: Immediate follow-up discussion in class. Additional options could include reflective writing or short essay responses on a subsequent quiz or test.

Origin of Activity: This is an original activity.

Sources Consulted:

The Book of Eli. Dir. The Hughes Brothers. Perf. Denzel Washington, Gary Oldman. Warner Brothers, 2010. Film.

A Boy and His Dog. Dir. L. Q. Jones. Perf. Don Johnson, Jason Robards. Sling Shot Entertainment, 1974. Film.

Mad Max Beyond Thunderdome. Dir. George Miller. Perf. Mel Gibson, Tina Turner. Warner Brothers, 1985. DVD.

On the Beach. Dir. Stanley Kramer. Perf. Gregory Peck, Ava Gardner, Fred Astaire, Anthony Perkins. United Artists, 1959. Film.

The Road. Dir. John Hillcoat. Perf. Viggo Mortenson, Charlize Theron. Dimension Films, 2009. DVD.

The Road Warrior. Dir. George Miller. Perf. Mel Gibson. Warner Brothers, 1981. DVD.

NOTES

1. David Christian, Cynthia Brown, and Craig Benjamin, *Big History: Between Nothing and Everything.* New York: McGraw-Hill, 2014. 293–295. Print.

2. Cynthia Brown, *Big History: From the Big Bang to the Present.* New York: New Press, 2007. Print.

3. Christian, Brown, and Benjamin, *Big History,* 302–303.

4. Fred Spier, *Big History and the Future of Humanity.* Chichester, West Sussex, U.K.: Wiley-Blackwell, 2011. 20–21. Print.

5. Eric Chaisson, *Cosmic Evolution.* Cambridge, Mass.: Harvard UP, 2001. 136. Print.

6. Emory Holloway, ed., *Walt Whitman: Complete Poetry and Selected Letters.* London: Nonesuch, 1938. 77. Print.

7. "Moore's Law." *Columbia Electronic Encyclopedia.* MAS Ultra—School Edition. 6th ed. 2013. Web. 22 September 2013.

8. Stuart Kaufmann, *Reinventing the Sacred.* New York: Basic Books, 2008. 130. Print.

9. Gerald F. Joyce and Leslie Orgel, "Prospects for Understanding the Origin of the RNA World." In *The RNA World: The Nature of Modern RNA Suggests a Prebiotic RNA World.* Ed. Raymond F. Gesteland, Thomas R. Cech, and John Fuller Atkins. Woodbury: Cold Spring Harbor Laboratory Press, 1999. 49–78. Print.

10. Eugene Koonin, "The Biological Big Bang Model for the Major Transitions in Evolution." *Biology Direct* 2.21 (2007): n.p. Web. 14 January 2014.

11. Brian Swimme and Thomas Berry, *The Universe Story.* New York: HarperCollins, 1992. 223. Print.

12. Thomas Nagel, *Mind and Cosmos: Why the Materialist Neo-Darwinian Conception of Nature is Almost Certainly False*. New York: Oxford UP, 2012. 16. Print.

13. *Contagion*. Dir. Steven Soderbergh. Perf. Matt Damon, Jude Law, Kate Winslet, Marion Cotillard, Gwyneth Paltrow, Laurence Fishburne, Josie Ho, Demetri Martin, Jennifer Ehle, Bryan Cranston, and Elliott Gould. Warner Bros., 2011. Film.

14. The plague/cholera simulation I use is adapted from: Cory M. Wisnia. "The Black Plague: A Hands-on Epidemic Simulation." n.d. PDF file. 3 March 2013.

SIXTEEN

· # Reflective Writing in the Big History Classroom

Jaime Castner

> Reflection is an attitude which makes the difference
> between twenty years of experience or only one year of
> experience repeated twenty times.
>
> GILLIE BOLTON[1]

In the spring of 2012 I had just graduated from Dominican University of California with a bachelor's degree in literature, and I was in love with classrooms. To me, there is nothing like the intellectually saturating, paradigm-shifting, out-of-place-and-time kind of classroom colloquy that one finds in college. That is, I was in love with the kind of classroom colloquy one finds in college until I was tasked with facilitating one. Then, as I entered my first semester as a teaching assistant in Mojgan Behmand's "Myth and Metaphor through the Lens of Big History" course at Dominican, it seemed my only friends were the crickets. The times at which I was given rein to lead the class on my own were at first disastrous and riddled with uncomfortable silences. In my hour of need I turned to a former professor whose sage advice has since proven invaluable: when in a pickle, get students talking about themselves. Indeed, she was right. My passion for writing led me to the idea of sparking robust conversation with reflective writing exercises in the classroom—not a novel concept, I know, but a revelation to me. I put my theory to practice later that semester.

At the moment in the curriculum when Dr. Behmand was presenting at a conference and I was again to fly solo, the students were reading *Nisa: The Life and Words of a !Kung Woman*, by the anthropologist Marjorie Shostak. The stories of Nisa, a member of a modern hunter-gatherer community, were intended to help the class comprehend Paleolithic lifeways. I started the class by asking the students to write quietly for ten minutes in response to this prompt: "Describe one of your earliest memories, one that has stuck with you for any reason and remains relatively clear in your mind." I asked them to note any distinctive images, smells, sounds, tastes, or feelings associated with the memory. After writing in silence, I asked the students if they cared to share and they were eager. Students spoke of vivid memories—everything from birthday parties with scary clowns, to the birth of a younger sibling, to a favorite aunt's wedding, to the first day of school.

As a class, we detected a trend. Our reflective writing exercise had generated that college colloquy I was missing. There were common themes running through these stories: fear, coming-of-age milestones, and rituals. Another common thread was food and the sense of taste; the students often recalled what home-cooked meal they were eating or what flavor their birthday cake was. When we turned to Nisa's childhood stories we detected the same themes. Food was especially central. Fear, aging milestones, and rituals were also prominent in Nisa's recollections. So, we asked ourselves, as a class, what is it about these concepts and practices that appear so central to the human experience—even within human communities that, at first glance, appear very different from one another? What followed was an enlivened conversation driven primarily by the students and fueled by their own ideas. They emerged prideful from class that day and, I believe, surprised by the relevance of Big History content to their own sense of humanness. I, in turn, restored my own affection for the classroom and launched a research project on best practices regarding in-class reflective writing.

To the teacher of Big History, the lasting benefits of student engagement with a universal narrative are numerous and self-evident. Such a foundation offers a comprehensive understanding of our collective past, an interdisciplinary approach to undergraduate, graduate, and professional life, and a broader canvas on which to overlay future learning. Big History fosters critical thinking and inquiry, nuance, and open-mindedness. Still, despite educators' confidence that students' experience with Big History will prove its worth in ongoing and increasing ways as they mature, the earlier young scholars understand its merit, the earlier they may be able to process and apply its content. The thoughtful integration of reflective writing in the Big History classroom has the potential to support this process and application through

1. improved content retention and facilitated achievement of student learning outcomes;
2. application of "why factor" and "big picture" frameworks to content;
3. enrichment of active and experiential learning components and connection between content and the "real world"; and
4. stimulation of discussion and connection between content and students' individual human experiences.

In what follows, you will find pedagogical models for effective inquiry and examples for the application of these models in the Big History classroom.

The effectiveness of reflective writing in the classroom is widely recognized and the topic warrants little time here. Still, it is worth noting that the "Principles of Excellence" established by the Association of American Colleges and Universities (AAC&U) encourage the practice of consistent, reflective inquiry. Principle 2, "Teach the Arts of Inquiry and Innovation," encourages instructors to "Teach through the Curriculum to Far-Reaching Issues—Contemporary and Enduring—in Science and Society, Cultures and Values, Global Interdependence, the Changing Economy, and Human Dignity and Freedom," and Principle 3, "Engage with Big Questions," urges teachers to "Immerse All Students in Analysis, Discovery, Problem Solving, and Communication, Beginning in School and Advancing in College."[2]

First, note how exquisitely the Big History narrative is aligned with these ideals of higher education in America, addressing "Far-Reaching Issues [of] Science and Society . . . Global Interdependence . . . and Human Dignity," among others. Second, note that the points at which Big History is most aligned with these principles are in its pursuit of "Inquiry" and the tackling of "Big Questions." Then it comes as no surprise that the effectiveness of reflective, inquiry-based writing has been documented extensively across the disciplines in higher education. The practice of reflective writing in the classroom also has collateral benefits, such as internalization of the writing process and enhanced ability to organize thoughts on paper. And because writing is a solitary practice, it enables reserved students to express their thoughts safely and at their own pace. In graduate programs of law and medicine it has proven particularly effective.[3]

The first three inquiry-based models presented below are drawn from established higher-education programs with documented success to help educators derive the most value from reflective writing in the Big History classroom. Each model speaks to one of the four central benefits of reflective writing in Big History introduced above. It is recommended that instructors of Big History mindfully determine the objective of their reflective writing practice for the day's lesson and then implement the appropriate model.

MODEL 1: BLOOM'S TAXONOMY[4]

Benefit: Improved content retention and facilitated achievement of student learning outcomes

The classification of learning objectives developed by educational psychologist Benjamin Bloom in the 1950s provides a useful framework for the development of

TABLE 16.1   Bloom's Taxonomy as Applied to Big History

	Category	Definition	Question words	*For Big History*
Level of complexity	Evaluation	Judging, making value decisions about issues	Judge, appraise evaluate, assess	*Is the growth of our human community no longer sustainable?*
	Synthesis	Combining ideas, creating an original product	Compose, construct, design, predict	*Make a prediction about our distant future.*
	Analysis	Subdividing into component parts, determining motives	Compare, contrast, examine, analyze	*Compare forager lifeways with agrarian lifeways.*
	Application	Problem solving, applying for information	Interpret, apply, use, demonstrate	*Apply the theory of natural selection to the presence of forward-facing eyes in predators like us.*
	Comprehension	Interpreting, paraphrasing	Restate, discuss, describe, explain	*Describe how our moon was created and explain its role in sustaining life on Earth.*
	Knowledge	Memorizing, recalling information	Define, recall, list	*Define the Doppler effect.*

reflective inquiry in the Big History classroom. (And when there is an axis devoted to increasing levels of complexity, can there be any question?) Table 16.1 shows how we might apply Bloom's taxonomy directly to Big History. This blueprint assists instructors in crafting meaningful questions to retain learned content and achieve student learning outcomes as stated in the course syllabus. Suggestions for implementation will conclude this chapter.

MODEL 2: COLUMBIA COLLEGE'S LEARNING PORTFOLIO PROGRAM

Benefit: Application of "why factor" and "big picture" frameworks to content

This model offers a way to explore broader considerations of general Big History themes and their role in achieving students' personal and educational goals. In this model lies the potential for students to appreciate the value of Big History in their formal education and personal development before their studies conclude. Alternatively, if a student responds that Big History does not enrich or inform study in his or her field of interest, the reflective writing practice provides an opportunity to discuss this shortcoming or identify evidence to the contrary. Instructors may customize their inquiry based on the provided model questions listed below. Suggestions are italicized.

1. What difference has the learning made in my intellectual, personal, and ethical development? Ex. *How does my study of Big History enrich or inform study in my field of interest?*

2. In what ways is what I have learned valuable to learn at all? Ex. *What have I learned from the Big History story that might help me plan for my future?*

3. How does what I have learned fit into a comprehensive, continual plan for learning? Ex. *Has my introduction in this class to a vast array of disciplines affirmed my prior interests or informed what other types of classes I might like to try in college?*[5]

MODEL 3: BRIDGEWATER COLLEGE'S FOUR-YEAR PERSONAL DEVELOPMENT PROGRAM

Benefit: Enrichment of active and experiential learning components and connection of content to the "real world"

While this model will likely not apply to the lecture-style Big History classroom, it remains a valuable approach, particularly as active and experiential learning practices gain popularity in higher education. For the purposes of this chapter, one of Dominican's signature in-class Big History activities, "The Hominoid Skull Lab," developed by Martin Nickels and adapted by J. Daniel May, is used as an example. Again, a suggestion is italicized below each model question.

1. Exploration: Discuss your thinking prior to the experience. Ex. *Describe your understanding of hominoid evolution before this activity.*

2. Analysis: Consider why the experience was meaningful. Ex. *How did the activity change or enhance your understanding of where humans come from?*

3. Synthesis: Discuss the significance of the experience for your life. Ex. *How does understanding the lineage of* Homo sapiens *inform your sense of belonging*

in the Big History story? How can you apply this new knowledge to study in your field of interest?[6]

MODEL 4: DOMINICAN UNIVERSITY'S BIG HISTORY PROGRAM

Benefit: Stimulate discussion and connect Big History narrative to students' individual human experience

This final model draws on aspects of the three models above, with the distinct objective of calling on students' personal experience to illuminate or inform the Big History narrative. In surveys administered after the launch of our program in 2010, students revealed that they found the Big History content too abstract and impersonal. In the following year, I developed and refined this reflective writing model to help connect the program to students' lived experience. In response to students' discomfort with a nonreligious story of creation, a suggestion to "process conflict" has been added.

The following model questions and examples offer students the opportunity to share their personal thoughts on a range of topics. All may be used to prepare for a class discussion if students are comfortable sharing their work. (An expanded list of model questions, arranged by threshold, can be found in the activities section at the end of this chapter.)

1. Stimulate discussion. Ex. *What do you believe to be the qualities of a successful ruler?*

2. Connect to individual human experience. Ex. *Consider what steps you take at home or in your dormitory to minimize waste and preserve resources. Based on your knowledge of the Earth's history, do you feel that these measures are insufficient, sufficient, or unnecessary?*

3. Process conflict. Ex. *Do you believe the Big Bang theory and faith-based origin stories are necessarily incompatible? How do you plan to address conflicts of personal faith, should they arise?*

Modes of implementation for the models above include a range of possibilities. Of course, any of these questions may be used as a simple class or small-group discussion topic; they may also form the basis of take-home reflective assignments or even be used rhetorically in a lecture setting. However, when used as writing prompts, consider the following suggestions:

1. Embed the reflective writing practice, if not the specific questions, in the course syllabus.

2. Identify the learning objective or benefit and reflective question(s) to be asked in advance of class.

3. Initiate reflective writing practice before a discussion or after an active learning activity.

4. Allow students five to ten minutes to "free write" silently in response to a clearly articulated prompt.

5. Offer students two additional minutes to reread their work and underline or highlight their best ideas.

6. Request that students keep all their written reflections together in a dedicated folder for future reference. Free writing can produce wonderful ideas for upcoming research papers or presentations.

Big History content is immense and its implications immeasurable. While instructors in this burgeoning field are advancing a significant frontier in education, let us continue to encourage students to play a central role in their own learning. The practice of reflective writing has the unique potential to address a range of objectives in the Big History classroom, including content retention and the achievement of specific learning outcomes, understanding of "big picture" frameworks, enrichment of active and experiential learning practices and "real world" applications, and the stimulation of productive discussion about relevance to individual human experience. With the added benefits of supplementing the writing process and facilitating the organization of thoughts on paper, reflective writing practices are an intelligent component of any Big History syllabus.

So, I repeat: when in a pickle, get students talking about themselves. The practice of reflection need not be peripheral nor unproductive; reflection is a powerful tool that when mindfully embedded in a Big History curriculum can be the difference between crickets and colloquy.

ACTIVITIES AND EXERCISES FOR TEACHING REFLECTIVE WRITING

REFLECTION QUESTIONS

Jaime Castner

Threshold Related to the Activity: Thresholds 1–8 and the future

Category of Activity: Reflection prompt

Objectives:

- To instill in students the habit of daily writing and use of the writing process to generate and organize ideas;

- To enhance classroom discussion and information retention by encouraging students to reflect on course content; and

- To facilitate the personal connection students might develop with course content, enabling them to locate themselves in the universe story and envision their own futures

Overview of Activity: The following questions and prompts can be used (and customized) to suit your needs and those of your students. They are best presented at the beginning or end of class to generate ideas or further inquiry. It may be beneficial for students to keep these exercises in a folder so they can reference them when writing their papers. (Hint: When students' papers are missing the larger context "why factor," encourage them to look back at these reflections for inspiration.) Using these questions and prompts frequently should assimilate your students to a habit of writing that will enrich their academic careers (as well as their future professions) and foster meaningful and inquisitive reading and research. Questions and prompts are organized by threshold for your convenience.

Faculty Preparation for Activity: No prop or preparation is necessary for this activity. If you are using a PowerPoint presentation in class on the day of the activity, you may find it helpful to project the prompt on a slide for students to reference while they are writing.

Cost: n/a

Student Preparation for Activity: n/a

In-Class Sequence of Activity: Offer students five to ten minutes to free write a response to the prompt and one to two minutes to reread and underline their best ideas. Do not grade these exercises or require students to share their work; however, if you use this prompt at the beginning of class, you may find it useful to begin a conversation on what the students produced during this exercise.

Assessment of Learning: The success of this exercise is difficult to evaluate, though you may see signs of its effectiveness through stronger graded writing assignments and richer class discussion.

Origin of Activity: This is an original activity, created with guidance from Dr. Mojgan Behmand.

Source Consulted: Christian, David, Cynthia Stokes Brown, and Craig Benjamin. *Big History: Between Nothing and Everything.* New York: McGraw-Hill, 2014. Print.

Reflection Questions:

Introduction: What Is Big History?

1. Based on the introduction to your Big History textbook or your first day of class, what do you expect to take away from this course? What knowledge or skills do you hope to acquire? What questions would you like to answer?

2. What does the word "complexity" mean to you? Is this generally a positive or negative concept? How would you explain it the context of Big History?

Crossing Thresholds 1–3

1. How do you believe the universe began? Do you foresee any conflicts between the content of this course and any religious or spiritual beliefs you hold? Explain how you plan to handle these conflicts, if they arise. If you do not foresee any conflicts, briefly describe your familiarity with the Big Bang theory and evolution.

2. Imagine your average day . . . without gravity! Start at the moment you wake up and describe your routine. How would you eat your breakfast, get to school, and go to bed? Imagine what other basic laws of nature you might take for granted on a daily basis.

3. Which of the revolutionary minds studied so far in this course are you most intrigued or inspired by? Einstein? Galileo? Copernicus? Newton? Why?

Crossing Threshold 4

1. Think for a moment about *your* place in the time and space of our universe. How does this perspective make you feel? Big? Small? Powerful? Trivial? Energized? Nervous?

2. Meditate for a moment on the following excerpt from Christian, Brown, and Benjamin's *Big History: Between Nothing and Everything*: "Many scientists now view Earth as an interconnected system in which all components, organic and inorganic, work in harmony to sustain the planet and biosphere."[7] What does this statement mean to you? What can it offer us by way of advice for how to coexist on this planet?

3. Do you believe that the Big Bang theory and faith-based origin stories are necessarily incompatible? Where, if at all, do you see room for God or another higher power in this scientific story of the universe? If you believe the Big Bang theory cannot coexist with religious origin stories, explain why.

4. In the 1960s humans saw the first images of Earth from space and observed, among other things, that "Earth is extraordinarily beautiful."[8] Consider for a moment a place in nature or a natural object that you find to be especially beautiful. First, describe it in as much detail as possible. Then, explain why you think this place or object is so extraordinary. Is it peaceful? Complex? Does it appeal to your sense of sight, touch, hearing?

Crossing Threshold 5

1. How do you define "life"? What does being *alive* mean to you?

2. Consider some of the processes in our natural world that you have been studying in this class—photosynthesis, natural selection, genetic variation, the colonization of eukaryotic cells—and discuss what you think humanity could learn from observing them. (Hint: How about the recycling of dead matter into fertilizer by some fungi . . . doesn't this sound a lot like recycling used paper and plastic?)

3. Picture your favorite animal in your head. It can be wild or domesticated. If it is wild, describe which of the animal's physical qualities have helped it to thrive in its environment. If it is domesticated, which qualities have humans bred into the species by artificial selection, and why?

Crossing Threshold 6

1. What do you think sets humans apart from other living things? What makes us *exceptional?*

2. How has your understanding of history changed since the beginning of this course, if at all? How did you define "history" before?

3. Why do you think the ability to communicate is such a central factor in distinguishing humans from other animals? Think of and explain several ways in which advanced communication has helped humans to improve the planet. (Hint: How about the fight to end hunger, illness, homelessness, cruelty, pollution, and human rights offences?)

Crossing Threshold 7

1. Consider the reciprocal relationship between farmers, animals, and plants. What do you think non-farming humans can learn from these relationships? Can you think of any ways in which this relationship has been exploited or the practice of agriculture misused?

2. Why do you think humans desire power?

3. Do you think some humans are natural-born leaders or that leadership is an acquired skill?

4. Do you think centralized power works? If so, why? What are the benefits and drawbacks of centralized power?

5. What role does language play in your life? How do you think humans can benefit from being multilingual? If you are multilingual, what other language(s) do you speak, and why? If not, which language(s) would you like to learn, and why?

6. Where is your family originally from? Explain what role your familial heritage plays in your life and what it means to you. Do you speak other languages, share customs, or travel to visit relatives in other places?

7. The people of the Indus River Valley are known for being basically peaceful; archaeologists have found no evidence of warfare in their surrounding environment. Can you imagine a world without war? What does it look like? Does it work?

8. You have now learned about a number of different social structures from human history, including pastoral, hunter-gatherer, and agrarian models. Which of these models most appeals to you, and why?

9. Based on what you have learned so far in this class, how do you think the study of history can be used to prepare humankind for the future?

10. Are there any social, intellectual, cultural, spiritual, or artistic practices of communities outside your own that are particularly attractive to you? Perhaps there is an aspect of Native American spirituality, a quality of African art, or a concept in Greek philosophy that speaks to you. Explain this practice and its potential value for you or your community.

11. What do you think are the qualities of a successful leader or ruler?

12. How do you think the civilizations you've studied thus far might help modern cultures to understand or influence gender politics? Consider female pharaohs in ancient Egypt and musical courtesans in the courts of the Tang dynasty in East Asia, among other instances of relative equality between men and women throughout history.

13. In the absence of decipherable written language or records, what role do you think *art* plays in our understanding of ancient civilizations? What do you think is the value of art in interpreting history? Why is art still important today?

14. Even in just fifty years, historians' perception of the Western Hemisphere before 1492 has changed dramatically. The discovery of key artifacts and reinterpretation of existing historical evidence has revolutionized our understanding of history. What does this teach us about the study of history? Is our work of uncovering and understanding the past ever over? Why or why not?

Crossing Threshold 8

1. Spend thirty seconds free-associating a list of words you think of when you hear the term "modern." Now, spend five to ten minutes explaining what "modern" means to you. Is it good? Bad? Progressive? Overrated?

2. Britain evolved from an agrarian society to an industrial one quite quickly, with the effects of the Industrial Revolution fully felt by the mid-nineteenth century. Compare the work experience (as you imagine it) of a land-tilling laborer to that of a factory worker. Which appeals to you more, and why?

3. What do you think will be the next big innovation by humans? Time travel? Immortality? Vacations on Mars?

4. The environmental impact of industrialization was illustrated by the condition of Britain in the mid-nineteenth century, when the Thames

River was contaminated by sewage, the air rank with smog, and the foliage covered in the residue of burnt coal. How important do you think it is to consider the environment as our global community continues to grow and innovate? What measures do you observe being used today to protect the environment? Do you think they are enough? Or do you think damage to the environment is an inevitable by-product of human progress?

5. Consider the dramatic innovations in medicine made by scientists in the 1920s through 1940s. Compare these strides to some of the progress being made today in the health-care field. What do you think will be the most important medical breakthrough of the twenty-first century? Stem cell therapy? A cure for cancer? Alzheimer's? Diabetes?

6. Consider the innovations in technology and communication made by great minds of the twentieth century (telegraph, radio, telephone, computers). What do you think will be the next great breakthrough in this field? Domestic robots? Computers that read your mind? Flying cars?

7. Today, we have to ask ourselves this difficult question: can humans continue to consume at the current rate? Explain your answer.

The Future

1. Describe some of the things you do in your daily routine that reduce your "carbon footprint." Why do you think this is important? How, if at all, have you been affected by the message that our Earth is in danger?

2. Do you predict a Malthusian crisis in our near future, middle future, remote future, or ever? Explain your answer.

3. Reflect on how this course has altered or enhanced your perception of world history. Explain your answer and how this change could affect the way you look at your other coursework (in biology, art history, musicology, and religion courses, for example).

NOTES

1. Gillie Bolton, "Write to Learn: Reflective Practice Writing." *InnovAiT* 2.12 (2009): 752–754. Web. 30 March 2012.

2. Association of American Colleges and Universities, "Principles of Excellence." *Association of American Colleges & Universities.* AAC&U, 2012. Web. 20 December 2012.

3. Carolyn Grose, "Storytelling across the Curriculum: From Margin to Center, from

Clinic to Classroom." *Journal of the Association of Legal Writing Directors* 7 (2010): 37–61. Web. 30 March 2012. Johanna Shapiro, Deborah Kasman, and Audrey Shafer, "Words and Wards: A Model of Reflective Writing and Its Uses in Medical Education." *Journal of Medical Humanities* 27.4 (2006): 231–244. Web. 30 March 2012. Hedy S. Wald and Shmuel P. Reis, "Beyond the Margins: Reflective Writing and Development of Reflective Capacity in Medical Education," *Journal of General Internal Medicine* 25.7 (2010): 746–749. Web. 30 March 2012.

4. The Teaching Center, "Asking Questions Based on Bloom's Taxonomy." *The Teaching Center.* Washington University in St. Louis, 2009. Web. 3 June 2012.

5. John Zubizarreta, "The Learning Portfolio: Reflective Practice for Improving Student Learning." *Columbia College,* n.d. Web. 2 June 2012.

6. Edward W. Huffstetler, Nan R. Covert, Catherine L. Elick, and Harriett E. Hayes, "Achieving Better Horizontal and Vertical Integration in the Liberal Arts Curriculum: Bridgewater College's Four-Year Personal Development Portfolio Program." *Association of American Colleges & Universities.* AAC&U, n.d. Bridgewater College. Web. 2 June 2012.

7. David Christian, Cynthia Brown, and Craig Benjamin, *Big History: Between Nothing and Everything.* New York: McGraw-Hill, 2014. 42. Print.

8. Ibid.

SEVENTEEN

· Activities for Multiple
Thresholds

The activities that follow are meant for use with more than one threshold, at any point throughout the Big History course, or at the end of the course as a culminating project.

Neal Wolfe's "Walking the Big History Timeline" exercise may be most useful at the beginning of the course, in conjunction with Threshold 1. It would also be effective after teaching Thresholds 1 through 4, or it could be visited or revisited at semester's end.

Our "Amateur Astronomers Star Party" is an annual event held in the fall, as the days grow shorter, in which a group of local enthusiasts bring their telescopes to our campus, set them up in a dark field, and give our students a tour of the night sky.

Bill Philips' "Eight Thresholds Video Project" is intended as a culminating project that might resonate particularly with students savvy with multimedia and social media.

The "Annotated Bibliography" assignment is something we use in our "Through the Lens" courses to teach information literacy, to tie students' understanding of information literacy to the course content in Big History, and to prepare them for researching and writing research papers. (We meta-model the form at the end of this book, beginning on page 361, as an extensive resource for Big History teachers.)

Finally, the "Little Big History Essay" is just such a research paper, based on a concept developed by Esther Quaedackers at the University of Amsterdam. This essay assignment, which we have used as a culminating project in the Big History survey course, is a paper focused on a subject—often an object or system that exists in the present or in students' lives (beer, sweet potatoes, a smartphone, one's mom). Students use the structure of the Big History narrative to lay out an argument of cause: here is why this object exists, beginning with the Big Bang, and moving through each threshold until the paper's subject comes into being. (Cynthia Stokes Brown meta-models the form in chapter 19, "A Little Big History of Big History," on page 296.)

As with all our activities, feel free to adapt these or use them as inspiration for your own pedagogical innovations.

ACTIVITIES AND EXERCISES FOR
TEACHING MULTIPLE THRESHOLDS

WALKING THE BIG HISTORY TIMELINE

Neal Wolfe

Threshold Related to the Activity: Thresholds 1–8

Category of Activity: Learning activity

Objective: To help students gain a more realistic sense of the Big History timeline through kinesthetic activity. When we read or hear about such long spans as millions and billions of years of history, we have difficulty grasping their true significance in relative terms—they strike us as largely abstract concepts. When students experience the timeline through the movement of their own bodies, the learning is more easily integrated.

Overview of Activity: This activity can take place on nearly any outdoor walkway, such as a block of the school's campus. The students and instructor walk the length of the campus, for example, with the placement of the eight Big History thresholds (plus any other significant events in universal history that the instructor may choose to include) predetermined by the instructor. The relative distances between the various events in sequence are noted from one end of the block to the other. Students will see—in fact, experience—the extended distance between the first three thresholds and the last five, especially the final three.

Faculty Preparation for Activity: The instructor should determine the distance of the walking route in advance. The simplest way to do this is for the instructor to first walk the route and count the steps. Then the position of the thresholds (and any other historical events) along the way can be calculated.

For example: If there are 600 steps overall, those 600 steps represent the 13.8 billion years of history. Threshold 4, the formation of the solar system, took place 9.1 billion years after the Big Bang; so, using an algebraic proportion, the number of steps (x) from the beginning is found by constructing the proportion x divided by 600 equals 9.1 billion divided by 13.8 billion, then cross-multiplying and solving for x. (In other words, x is to 600 as 9.1 billion is to 13.8 billion.) Or, in this case, simply divide 9.1 by 13.8 and multiply by 600. In this example, Threshold 4 is found 399 steps from the start of the walk. And so on for the other events along the route. (Of course, having some math skills comes in handy!)

No supplies are necessary, although one option is to place markers (such as surveyor's flags) at the appropriate positions along the walking route, rather than having students count steps while walking.

Cost: No cost, unless event markers are used, which would be a minimal expense

Student Preparation for Activity: n/a

In-Class Sequence of Activity:

1. Tell the students that the very start of the walk, before any steps are taken, represents the Big Bang, and the beginning of the universe as we know it. Tell them that the entire walk will represent the history of the universe to the present.

2. Then, begin walking at a casual pace, counting the steps as you go. It is suggested to have the students count the steps along with the instructor, for the following reasons: (1) it engages them more in the process of marking off the timeline, especially when they notice the gaps between thresholds; (2) it makes them more aware of the movement of their own bodies, as they focus on their steps; and (3) it is more interesting for them!

3. At each of the predetermined threshold points, depending on the number of steps from the beginning, stop and ask the students which threshold it is, and how the previous one led to this one. Have them take note of the distance since the last one, and since the beginning of the walk.

4. After proceeding for the entire length of the walk, take the final step into the present. Note that (depending on the chosen distance you have walked), the final three thresholds, from the emergence of *Homo sapiens* forward, are likely all to be found within the final step!

Assessment of Learning: After the activity, engage the students in a discussion. Have them reflect on how their experience may have made the timeline more graspable or altered their view of the timeline and how they see themselves within it. Remind them that the timeline is in continual motion, extending not just all the way from the beginning to now, but forward indefinitely as well. We are as integral to its movement as any of the events marked along our walk.

Origin of Activity: This is an original activity.

Source Consulted:

David Christian, Cynthia Brown, and Craig Benjamin. Big *History: Between Nothing and Everything*. Boston: McGraw Hill, 2014. Print.

*Neal Wolfe, Jaime Castner, and Dominican University
of California Big History Faculty*

Threshold Related to the Activity: Thresholds 1–4

Category of Activity: Learning activity

Objective: To help students connect their classroom and textbook learning about the cosmos with the real thing, through active observation of the night sky

Overview of Activity: It is necessary only to have an outdoor space with relatively low ambient light from streetlights, buildings, and so on, and enough room to set up several telescopes. It is good to begin the activity at dusk, just as the evening sky begins to reveal the first stars and planets that are visible. Choose a date when the moon will not be very full, in order to maximize the visibility of other celestial objects. A crescent moon that is setting just after sunset can provide a terrific opportunity to view it telescopically, and then have it disappear as the stars and planets emerge. As described, the overall activity takes about two hours.

Faculty Preparation for Activity: Faculty will need to arrange for a group of astronomers (amateurs are fine!) to bring their telescopes and have them set up in time to train on the first objects revealed in the evening sky. Check to see if there is an amateur astronomer association in your area. They love to share their passion.

Cost: The main cost of this activity is whatever compensation is offered to the astronomers.

Student Preparation for Activity: Schedule the event to coincide, if possible, with coverage of the first few thresholds.

In-Class Sequence of Activity:

1. Students, faculty, and astronomers congregate on the field where the stargazing activity will take place, preferably a half hour or so after sunset.

2. While astronomers set up and train their telescopes on a variety of the first seen celestial objects, an astronomer offers a talk about the nature and scale of the universe, the variety of things to be seen—stars, planets, galaxies, globular clusters, even satellites. This educates the students and provides context for what they will be viewing as the evening progresses.

3. At the conclusion of the astronomer's presentation, and when the telescopes have been focused on objects, students will take turns visiting the various telescopes in order to view what each is trained on and to learn further from the astronomers.

4. After students have had a chance to visit the various telescopes, an effective activity is to have the them congregate in one place while an astronomer uses an astronomy laser pointer to identify constellations, galaxies, twin stars, stars that no longer exist (but whose light is still reaching us!), clusters, and so on. In this way, rather than look at just one thing in isolation through a telescope, viewers can see the object of interest in the larger context of the sky.

5. While this larger group activity is taking place, the other astronomers are retraining their telescopes on different objects.

6. Students return to visiting the various telescopes to view new objects.

7. The activity ends when the last conversation with an astronomer does!

Assessment of Learning: Since the stargazing activity takes place at night, class discussion occurs at the next class meeting. Students may be asked to come to class with a written reflection. In particular, it is important to have students reflect on how actually looking at the objects they have been learning about in class has influenced their sense of connection with the cosmos through the night sky.

Origin of Activity: This activity was developed by the program faculty as a whole.

EIGHT THRESHOLDS VIDEO PROJECT

William Phillips

Threshold Related to the Activity: Thresholds 1–8

Category of Activity: Learning activity and assignment

Objective: To foster creativity in describing and reviewing one of the thresholds in order to share it with the class. Each class should cover all eight thresholds.

Overview of Activity: Students are to choose one of the eight thresholds and create an original four- to six-minute video describing the key elements of that threshold. Students are instructed as follows: "You may be as creative as you wish (incorporating skits, songs, and so on). As a suggestion, you could simply create a PowerPoint presentation of four or five slides and record the presentation with your voice explaining the slides. You may use a video device of your choice."

The project may be done in pairs or groups of three. The thresholds should be distributed on a first-come, first-serve basis so that all are taken. (This will accommodate classes of sixteen to twenty-four.)

Faculty Preparation for Activity: Make sure a video recording device is available to be checked out of the library. (Most students will use their mobile devices.)

Cost: n/a

Student Preparation for Activity: n/a

In-Class Sequence of Activity:

Make sure that students understand that each threshold brings with it a new complexity comprised of diverse elements, a stable structure, new energy flows, and emergent properties,.

Give the students a minimum of two to three weeks to complete the project. It is helpful to assign the project four to five weeks in advance. All selections of thresholds must be made two weeks prior to the day selected to view the videos.

Assessment of Learning: After viewing the videos, discuss the features of complexity that exist at each threshold, and how the thresholds build on one another. Since it is difficult to include all the features, have each group expand on what features they left out.

Origin of Activity: This is an original activity.

Source Consulted: David Christian, Cynthia Brown, and Craig Benjamin. *Big History: Between Nothing and Everything*. Boston: McGraw-Hill, 2014. Print.

ANNOTATED BIBLIOGRAPHY

*Dominican University of California Big History
Faculty*

The Assignment   Create an annotated bibliography in accordance with MLA guidelines. List five sources, none of which may be readings from class. The mandate for this assessment is: *Categorize. Summarize. Connect.* That means the entry for each source should have three parts: (1) category: identify the type of source and describe its validity as an academic source; (2) summary: briefly explain the content of the source in your own words; and (3) connection: consider and explain how the content of the source is relevant to your specific research topic. Each entry should be three to five sentences long.

Use Diana Hacker's *A Writer's Reference* for guidelines on MLA style.

For each research question, list and explain your sources while keeping the following guidelines in mind:

1. **Category:** (a) Identify the type of source. Is it an academic journal article or a journal article in an online journal in a database? A book of a specific type, such as a monograph, collection of essays, work of popular nonfiction, or anthology? Is it a video? A newspaper article? (b) Describe the source's credibility and validity as an academic source. Note whether the

source is from a peer-reviewed journal, or is written for researchers and professionals. Does the source include a bibliography of the sources used as support, or does it at least mention where the author got the information?

2. **Summary:** Explain the content of the source in your own words. Describe the larger concepts and ideas rather than focusing on numbers and statistics.

3. **Connection:** Consider and explain how the source is relevant to your research question. Note whether the source provides an overview, specific examples, or in-depth research that supports or answers your research question.

LITTLE BIG HISTORY ESSAY

Dominican University of California Big History Faculty

The "Little Big History" is a form invented by Big Historian Esther Quaedackers of the University of Amsterdam, Netherlands.

A Little Big History traces the origin of any entity—an object, person, place, or concept—from the Big Bang, and through all of Big History. It is, essentially, a causal investigative argument in which the author finds the conditions for—and the remote, proximate, and precipitating causes of—the existence of the subject in the chronological events that unfold across the Big History story. The result is that the chronospatial structure of the Big History story serves as the argumentative structure for the Little Big History essay.

In our program, we have used a version of the Little Big History form as the culminating project in our first-semester Big History course for first-year students (the results have included "A Little Big History of Beer," "A Little Big History of My Boyfriend," and "A Little Big History of my iPhone").

Below you will find the current version of the assignment prompt (developed by J. Daniel May).

In the following chapter, Dominican librarians explain how they have designed materials, exercises, and activities to support students in pursuit of the Little Big History as research paper.

And to end part 3, Cynthia Stokes Brown demonstrates the form by telling a Little Big History of Big History. It's a remarkable meta-history—a history of history.

A "Little Big History"

J. Daniel May

Overview of Assignment: This assignment is an essay in the form of a Little Big History—a history of a single topic from the beginning of the universe up to the present day and consideration of its possible future. The choice of topic is yours. However, you will make your task easier by selecting a topic that is specific, tangible, well defined, and genuinely interesting or meaningful to you.

Examples of such topics would be: a baseball, your grandmother, pencils, the Eiffel Tower, water, apples, the car you drive, the girlfriend or boyfriend you love, the cellphone you use, a piece of art you admire, a language you study, or the family you miss.

Avoid topics that are intangible (such as "propaganda" or "freedom"), abstract (such as "dignity" or "purity"), or hard to define precisely (such as "righteousness," "honor," or "love"). Although it is not impossible to write about such topics, it is unnecessarily difficult. Consult with your instructor if you are uncertain about the suitability of a particular topic.

Style and Audience: Imagine that your essay is going to be published in a magazine specializing in popular science and history, such as *Discovery, National Geographic,* or *Smithsonian,* or even a more general news magazine such as *Newsweek* or *Time.* Your reader is intelligent, curious, well read, and may have some college education (but is not necessarily a college graduate). Your reader is not familiar with this assignment, nor with the terms "Big History" and "thresholds," so your essay needs to be self-explanatory and self-contained.

Structure and Content:

Introduction: Your first paragraph should clearly identify your topic, explain why this topic interests you, and offer your reader an overview of what you plan to do in this essay. In addition, provide any necessary background or definitions. This might be easy and straightforward for a tangible, well-known topic such as the moon or plastic water bottles, but may require more detail and elaboration for a topic less familiar to the reader (e.g., "my cat Frisco," "Tierra del Fuego," or "Trajan's Marketplace"). You may also need to clarify a word with multiple meanings (e.g., is "Sagittarius" a constellation of stars, an astrological symbol, or a Roman archer?).

Body paragraphs: Follow your topic through every threshold of Big History, from the Big Bang to the present, plus your topic's possible future, connecting your topic to the themes discussed in every threshold in a meaningful way.

Topic sentences in every body paragraph should keep your reader aware of the relevance of your topic to each threshold. Transitional remarks should connect paragraphs so that the essay flows smoothly from beginning to end.

Conclusion: It is easy for a reader to become immersed in the details of an essay and temporarily lose track of why those details matter. Now that you have taken your reader on this journey through time and space, remind her or him of the purpose of the journey. What new insight, perspective, or appreciation might you have gained? What final emotion, thought, idea, or question do you want resonating with the reader at the very end?

Paragraph Outline for Completed Essay:

1. Introduction
2. Your topic and Threshold 1
3. Your topic and Thresholds 2 and 3
4. Your topic and Threshold 4
5. Your topic and Threshold 5
6. Your topic and Threshold 6
7. Your topic and Threshold 7
8. Your topic and Threshold 8
9. Your topic in the future
10. Conclusion

Additional Requirements:

Library research: Your textbook will be your major source of information, but you must have a minimum of three library-based research sources in addition to the textbook.

Citation and documentation: Your sources of information must be documented: use internal citations and include an end-page listing sources according to MLA, APA, or CMS specifications.

Length: 1,500–2,100 words (i.e., 5—7 pages), plus the end-page.

Format: Double-spaced, 1-inch margins all around, 12-point standard font such as Times or Courier.

Brainstorming Questions and Suggestions for Your Little Big History Essay   The questions below are intended to help you start brainstorming ideas and suggest areas for your research. (Note: This is *not* a checklist—you are not expected to answer each and every question!)

Threshold 1 (13.8 billion years ago): The Emergence of the Universe

All the matter / energy that exists now began in the Big Bang. What forms of energy (motion, heat, light, and so on) are relevant to your subject? What is your subject made of? What subatomic particles is it composed of? Are helium and hydrogen part of its composition? (Hint: Probably!) Does your subject have mass, and therefore gravity?

Threshold 2 (13.4 billion years ago): The Emergence of Stars and Galaxies

On what does your subject exert gravitational attraction (and vice versa)? How do stars form? What do they do? Can you relate your subject to star formation? Does your subject have anything to do with light? With heat?

Threshold 3 (13.4 to 4.6 billion years ago): The Emergence of Chemical Complexity

What elements heavier than hydrogen and helium does your subject contain? Oxygen? Nitrogen? Zinc? Gold? Lead? How and where did those elements emerge? What radioactive elements might be part of your subject's composition? What effect do heavy elements or radioactive elements have on your subject?

Threshold 4 (4.6 to 3.8 billion years ago): The Emergence of Our Solar System

What molecules is your subject composed of? When and where did these molecules emerge? Where on planet Earth (or beyond) is your subject located? Is your subject affected by celestial phenomena such as solar flares, ultraviolet radiation from the sun, or lunar cycles? Has your subject been affected by impact strikes of comets, asteroids, or meteors? What geological processes (continental drift, tides, weather, seasons, and so on) are relevant to your subject?

Threshold 5 (3.8 billion to 8 million years ago): The Emergence of Life on Earth

Is your subject living or nonliving? How does your subject affect the biosphere? How does the biosphere affect your subject? For *living* subjects: What evolutionary paths has your subject followed? What selection pressures does your subject experience, and what selection pressures does it exert on other living entities? For *nonliving* subjects: Are organic molecules part of your subject's composition? What influences or effects does this nonliving subject have on living entities?

Threshold 6 (8 million to 10,000 years ago): The Emergence of *Homo sapiens*

How was your subject relevant to our proto-human ancestors? How is it relevant to modern humans? Is it a resource that can be exploited? Is it a source of food, a means of obtaining food, part of the process of preparing or cooking food? Is it useful for making tools, shelters, ornamental objects (e.g., gemstones, precious metals, or seashells), or any other human artifact?

Threshold 7 (10,000 to 200 years ago): The Emergence of Agriculture

Is your subject affected by human farming? Does your subject have an effect on human farming? How does agriculture change the natural landscape? Forests? Pasture land? Water tables? Irrigation? Animal and plant habitats? Is your subject relevant to the emergence of city-states? Is your subject part of an urbanized lifeway?

Threshold 8 (200 years ago to the present): The Emergence of Modernity and Industrialization

How are fossil fuels (shale, coal, oil, and natural gas) relevant to your subject? How does the use of such fuels affect your subject? Local environmental effects? Global effects? Climate change? How has the recent exponential growth of human populations affected your subject?

Possible Futures

How long has your subject existed, how stable is its existence, and will it continue into the future? For how long? What will be the state or condition of your subject ten years from now? Fifty years from now? One thousand years? One million years? One billion years?

EIGHTEEN · Igniting Critical Curiosity

*Fostering Information Literacy through
Big History*

Ethan Annis, Amy Gilbert, Anne Reid,
Suzanne Roybal, and Alan Schut

The poet Thich Nhat Hanh explains interdependence thus:

> If you are a poet, you will see clearly that there is a cloud floating in your apple
> juice. . . . You cannot point out one thing that is not in the apple juice: time,
> space, the Earth, the rain, the minerals in the soil, the sunshine, the cloud, the
> river, the heat. Everything co-exists in a single glass of apple juice. In fact, one
> could say that it took almost the entire universe to produce a single glass of apple
> juice or a single human being. If any whole category, such as humans, is
> removed from the cycle, the entire cycle is truncated and we have no apple juice.[1]

Stories provide us with a past, a present, and a future—in short, with a path
through time. Within a library you can follow stories or paths almost anywhere
humans have ventured and left behind a trail of ink, paint, or pixels. The path we
follow through the Big History narrative is, by necessity, narrow, because we wend
our way through so much time in so little time.

In libraries, any point on the narrow path can be widened to incorporate almost
all known knowledge. Unlike in the past, almost all libraries are now interconnected,
essentially making almost all worldwide holdings available through almost any
library, even the smallest. The internet further expands information availability, but
the internet lacks filters. In contrast, information available in libraries, be it in print
or in databases, is usually screened by a publisher and then further screened by an
information specialist, so it is more likely to be reliable.

Academic librarians can teach students to discern what is relevant, accurate, and timely—to distinguish between information that is helpful and information that misleads, even if the information is unfiltered. More excitingly, they can teach students to uncover the stories of information itself, even the stories of something as seemingly mundane as a bottle of apple juice.

A COMMITMENT TO THE LIBERAL ART OF INFORMATION LITERACY

To be information literate is a skill, but it also is a state of mind. Beyond the simple mechanics of finding an answer, it is a habit of critical curiosity, of seeking out the best information and making connections across disciplines to enrich a subject or strengthen an argument. It is the ability to adapt to constantly changing technology, along with a "critical reflection on the nature of information itself."[2] To be sure, just trying to communicate 13.8 billion years of history in sixteen weeks is a Herculean task. But we are not just asking students to passively take notes and dutifully read the textbook. The more students are engaged in learning, the more they can be inspired to seek out new, reliable, and relevant information. We want to ignite this spark of critical curiosity that is at the heart of information literacy, a guiding light in the Big History program and the foundation of a twenty-first-century liberal education.

Dominican University is deeply committed to information literacy and has embedded it in general education and embraced it in disciplinary courses. The adoption of Big History provided the inspiration and the impetus to reimagine our approach. We realized that we could build an information literacy program that was flexible enough to respond to constant changes in technology and information tools and to students' continually evolving search behaviors. The program has evolved from a stand-alone course to a scaffolded, cross-disciplinary approach that builds increasingly complex skills, ethical awareness, and critical thinking across the curriculum and at all levels.

A PERFECT PARTNERSHIP: EMBEDDING INFORMATION LITERACY IN BIG HISTORY AND BEYOND

Information literacy is a key component of the "Essential Learning Outcomes" developed by the Association of American Colleges and Universities (AAC&U), on which the goals of the Big History program are based. The outcome relating to "Intellectual and Practical Skills" lists information literacy specifically, along with critical and creative thinking.[3] To achieve this outcome, we librarians and the first-year faculty have formed an ongoing partnership that includes collaborating on specific assignments geared to advancing information literacy, teaching class sessions, providing research support, and dedicating a faculty liaison to continue the dialogue at summer institutes and routine faculty development workshops. Drawing on the Association of College and Research Libraries (ACRL) Information Literacy Competency Standards for Higher Education, we create experiences and exercises that help students learn to locate and access information; critically evaluate and select relevant, credible, and high-quality information; and integrate and use that information ethically.[4]

A fascinating aspect of Big History is that it affords us the opportunity to imagine human experience at pivotal moments. In our mind's eye, we can see a human of the Paleolithic era hammering various stone types against each other to produce a basic tool in a more desired shape and form. And when she returns to the task a few days later, drawing on her experience, she will use only the stone types that served her best. She will bring refinement to her work, moving from hammering to chipping and flaking. Information literacy requires that same cycle of observation, experience, and refinement. In order to sustain our students' critical curiosity and reinforce learning, we have extended our efforts beyond the first year. While information literacy is a component of First-Year Experience "Big History," it is at this point that the Big History program becomes a component of our overall information literacy program. Students need to practice at each step in their education in order to move from acquiring basic skills to mastering university-level research. Librarians facilitate this transition by serving as liaisons to all academic departments, allowing the formation of meaningful working relationships with both faculty and students. We incorporate guided exercises to build basic skills, guest lectures to introduce discipline-specific resources, workshops to provide technological support, and one-on-one appointments to mentor deep critical thinking about information. In this way, we help our students move from hammering to chipping and flaking, and we put them on the road to lifelong success.

LIBRARY COLLECTION SUPPORT

Print and electronic library collections in all formats are essential to comprehensive syllabus design, to creative assignments, and to the provision of rich resources for supplementary reading and research. Collaboration between academic librarians and disciplinary faculty is critical to the identification of specific resources based on subject expertise and the development of focused selection criteria. Equally important is the recognition that library resources should mirror, whenever possible, the interdisciplinary and integrative learning models so characteristic of Big History pedagogy. Multiple perspectives on any Big History theme should include representative works from the humanities, social sciences, sciences, and professional disciplines—all with the goal of encouraging dialogue, critical thinking, and a holistic vision of the human condition in its cosmic setting.

In undertaking collection development for both foundational Big History courses and subsequent disciplinary "Through the Lens" courses, librarians acquire titles recommended by faculty, cited in established syllabi, or listed in bibliographies relevant to each of the eight overarching thresholds that structure the content of the Big History curriculum or to discussions of the future. With these core titles as a guideline, appropriate Library of Congress subject headings can be readily identified, and titles classified under these headings can then be assessed for individual acquisition. Alternatively, if the task of resource selection has been outsourced to a book vendor, the library can design detailed selection profiles to guide the vendor in the acquisitions process. In either case, the standard selection criteria of currency, accuracy, publisher reputation, bias assessment, and positive reviews, as further refined by faculty preferences, increase the scope, quality, and availability of significant resources over time. Ideally, as items are added to the collection, catalog records might note the relevance of a title to a given threshold, or annotated bibliographies could alert students and faculty to worthwhile library resources through the medium of online research guides. Library collections should work in tandem with library instruction to invigorate intellectual curiosity and instill appreciation for the culture of scholarship, evidence, and clear thinking.

INFORMATION LITERACY ASSIGNMENTS

Guided by the AAC&U Essential Learning Outcomes and the ACRL's Information Literacy Competency Standards for Higher Education, the library faculty, in collaboration with Big History faculty, design information literacy assignments that appear in both semesters of the Big History program. Each assignment is intended to build on the skills learned in the previous assignment, creating a foundation for information literacy throughout a student's college career. The assignments are resource-based to promote use of library tools and to teach students the fundamentals of using the resources available in the Dominican library.

The paragraphs below delineate our experience with three research assignments: (1) Mapping the Physical and Digital Library; (2) Annotated Bibliography; and (3) Little Big History. The first assignment is an introductory exercise supporting the next two assignments and should be embedded at the beginning of a course or program. The other two may be taught in interchangeable order based on the needs of the course. Descriptions and models for annotated bibliographies abound, but our assignment provides a well-defined template that is certain of meeting the learning objectives. In addition, the template is now used across programs and levels at Dominican and thus consistently reinforces the learned skills. Our third research assignment prepares students to write a Little Big History paper. A Little Big History paper contains a history of a single subject, from the beginning of the universe up to the present day, and ends with consideration of that subject's possible future. The choice of subject is the student's, but the requirement is that the subject be followed through every threshold of Big History, connecting in a meaningful way to the themes discussed in each threshold. (See chapter 17, "Activities for Multiple Thresholds," for the Little Big History paper assignment.)

BIG HISTORY LESSONS: MAPPING THE PHYSICAL AND DIGITAL LIBRARY

We launch the information literacy portion of the Big History program by helping students become effective observers of their surroundings. The students begin by exploring, and then mapping, the library's physical and electronic resources. Our intense focus on such fundamentals trains students to see and probe that which they might otherwise overlook in their everyday interactions. In completing this exercise, students examine the physical structure of a library and the various needs that must be met through that structure, familiarize themselves with collections of books and journals, and grasp the connection between the physical and digital worlds of the library as they use the online catalog for discovering hard copy books as well as titles we have purchased electronically.

And through the integration of information literacy into the Big History program, students learn—as they move through our eBrary database of scholarly e-books, articles available in our print and electronic journal collections, and the offerings of various world libraries accessible through WorldCat—that they are part of an interconnected world of data, libraries, and information. These Big History students eventually realize that today's interconnected world of information is the twenty-first-century equivalent of the ancient Silk Roads they study in class, and that they themselves are twenty-first-century practitioners of collective learning.

Working from Big History's interdisciplinary premise, we also support the teaching of other competencies connected with information literacy, the main one being writing. One of the significant sources we use in our instruction is Diana Hacker's *A Writer's Reference*, a seminal reference for students in their Big History and writing courses. This resource offers strategies for creating correct citations and bibliographies and for arranging a paper in MLA, CMS, or APA formats. We teach the anatomy of a citation, as it explains each part of the citation needed to correctly format a bibliography. Learning to avoid unintended plagiarism and to incorporate the correct use of citation formats is an essential aspect of being a student and a citizen of today's informationally vast world.

BIG HISTORY LESSONS: "ANNOTATED BIBLIOGRAPHY"

Big History is expansive in the knowledge and disciplines that it encompasses, yet it becomes accessible and useful through its thoughtful organization of that information. In the same vein, our annotated bibliography assignment is intended to familiarize students with the concept of organization of information—through categorization and connection—in addition to teaching them the skill of documentation. Students are asked to *annotate* their sources according to the same criteria that they use to *evaluate* the sources. A properly formatted citation is followed by short paragraphs that categorize, summarize, and connect each source to the student's research question. (See page 281 for an explanation of this assignment and the annotated bibliography at the end of this book, which models the form.)

This assignment is designed to help students think critically about their selected paper topic and about how the sources they have selected connect to and can shape their understanding of that topic. Students are asked to consider and categorize the type of source being used. How does a book differ from a journal article? What kind of information is contained in a book or a journal article or a reference article? And why does the type of source make a difference? Also, having the students summarize each source requires them to read each source more carefully and develop the skill of accurate summarization. Finally, connecting each source to the topic requires students to reflect on and evaluate the relevance and usefulness of each source for the paper.

BIG HISTORY LESSONS: THE LITTLE BIG HISTORY ASSIGNMENT AND ITS CHALLENGES

At this latter stage, teaching the use of Dominican's subscription databases introduces students to the abundant scholarly resources accessible to students 24/7, including primary, secondary, and tertiary resources that have been vetted for reliability and accuracy. By exploring different topics of their choosing in the databases, students hone the skills required for mapping a topic, identifying its different aspects, and ferreting out in-depth research sources. Librarians use the reference databases on the library website to assist students in finding keywords, subject terms, and anything related to their topic that may be helpful in forming a thesis and learning about the thresholds of Big History for their topic.

During our class visits, librarians instruct students on the use of library tools for locating and evaluating scholarly sources for their Little Big History papers. Most students are able to find books, e-books, and online articles in the library catalog or the library databases, respectively. However, one challenge in writing a Little Big History paper is that scholarly sources, especially journal articles, are not well-suited to a first-year research paper covering the expanse of 13.8 billion years. Textbooks, nonfiction general audience books, or articles in popular science magazines might be better choices. The first-year faculty have redefined the style and audience for this paper accordingly, asking students to "imagine that your essay is going to be published in a magazine specializing in popular science and history, such as *Discovery, National Geographic,* or *Smithsonian*" and that the "reader is intelligent, curious, and well-read . . . but not familiar with this assignment, . . . so your essay needs to be self-explanatory and self-contained."

Our Big History assessment efforts, specifically with the Little Big History paper, revealed some key challenges in teaching information literacy to twenty-first-century students. Students may lack the ability to critically evaluate and make the necessary connections between found information and the concepts of Big History. Today's students have grown up within a culture of abundance as regards access to information. The millennial generation is hardwired to expect instantaneous access to the information they seek. On the surface, this might make it seem as if these students lack critical curiosity when seeking out and processing sources for their papers. However, it is only a matter of properly training students to make a clear distinction between scanning and scrolling through vast amounts of information and gleaning true knowledge from consulted sources. Through our Big History program, we have rolled up our sleeves and engaged with this work.

Ultimately it is the "aha" moments that highlight the purpose of creating this type of program. One semester, a Big History student sought out a librarian at the reference desk. She wanted a librarian's opinion on the sources she had included in her Works Cited. Most of the sources were popular websites, some from a community-based Q & A website—and thus not a credible source for academic work. The conversation revealed that she had entered her precise paper topic as a question into a search engine text box. This act is not a rarity but the norm with the millennial student. This generation has grown up with Google and its algorithms and, thus, replacements of focused search keywords with broad search questions. The teaching of information literacy skills by librarians and first-year faculty enables students to use other search tools and discover more reliable or scholarly sources. This particular student learned to use popular public sources to generate distinctive keywords for searching in the library's tools. She laughed and admitted that she felt more confident about her Little Big History paper, which was on one of her favorite topics, gold jewelry. The librarian commented on the connections between the life cycle of the stars and modern man's cultural symbols. The student smiled. "I know," she said. "Isn't it cool?"

Acknowledging the challenges of a specific assignment is as important as meeting students where they are. Developing the skills required to navigate, use, and document popular sources properly is just as valuable as—and is valuable training for—being able to do that with scholarly sources. This serves as a good reminder of our larger goals for the program, which include teaching a specific form of literacy because doing so leads to a specific way of being in the world.

ASSESSMENT

Ongoing assessment and continuous quality improvement are key components of our Big History learning community. Librarians have participated in various assessment undertakings by Big History faculty. For example, in the spring 2012 semester, Big History faculty assessed a research paper according to an analogous standardized grading rubric. And in fall 2012, an assessment of the Big History survey course included both a perception-based student survey and a faculty-led evaluation of the Little Big History research paper using a standardized grading rubric. Each of these assessment tools included categories related to information literacy. The assessment was not intended to be comprehensive as regards the information literacy portion of the Big History program, but it proved valuable in providing specific insights about how familiar students were with the library, how they evaluated their own research skills, and whether the existing student artifacts showed improvement in their information literacy skills. In terms of information literacy, the results were very positive. It was clear that students' ability to find, evaluate, and properly cite sources was much improved from the previous semester.

Overall, our efforts have paid off. We've seen an increase in the number of student and faculty contacts. Traffic at the reference desk increased 31 percent from August 2011 to August 2012, and student appointments increased 62 percent. More importantly, our engagement with students and faculty has improved. Many times, students have approached the reference desk, saying, "You taught my class last week, can you help me?" They know we are here for them, and they are more likely to ask when they see a familiar face. In the Big History program, students have reported much greater satisfaction with and awareness of library resources. This has turned into one of the more successful partnerships on campus, not only from the librarians' perspective, but from those of the faculty and students as well.

Big History is exciting for many reasons, but its multidisciplinary inclusiveness is a significant one. One is able to find all kinds of connections among the stars, the Earth, how modern humans exist, and so on. And much like the glass of apple juice, our Big History program is made up of various disciplines and competencies, which, like elements, come together in an increasingly complex manner, yet function together organically and meaningfully.

NOTES

1. Adapted from Thich Nhat Hahn, *The Heart of Understanding: Commentaries on the Prajnaparamita Heart Sutra*. Berkeley: Parallax, 1989. Print.

2. Jeremy J. Shapiro and Shelley K. Hughes, "Information Literacy as a Liberal Art: Enlightenment Proposals for a New Curriculum." *Educom Review* 31.2 (1996): 3. Web. 13 September 2013.

3. Association of American Colleges and Universities, "Essential Learning Outcomes." *Aacu.org*. Association of American Colleges and Universities, 2013. Web. 13 September 2013.

4. These can be found at www.ala.org/acrl/standards/informationliteracycompetency.

NINETEEN

· A Little Big History of Big History

Cynthia Stokes Brown

Cynthia Stokes Brown demonstrates the form, in a meta-metanarrative.

Only human beings can tell stories. No other animals can, so far as we know. It took a universe to produce story-telling human beings. So, to tell the story of how people began to tell the Big History story, we need to go all the way back to how the universe began.

IN THE BEGINNING

In the beginning, some 13.8 billion years ago, out of a mysterious something or nothing that we know nothing about, all the energy and matter in our entire universe burst forth from one tiny spot. Temperatures were so high that matter could not form; there was only energy.

As temperatures cooled, the atoms of the simplest matter formed—hydrogen and helium. Gravity pulled these atoms into clumps, within which atoms began to fuse, forming galaxies and stars. Stars have a life cycle; they are born, exist, change, and die. As they burn, their atoms fuse into more complex elements. When stars run out of fuel, they die, scattering their remains around them into space. Sometimes, when a star is large enough, it collapses in an explosion called a supernova. The heat of this explosion fuses atoms into even more complex elements. In this way, the ninety-two naturally existing elements were formed and scattered into space to be absorbed into new stars and planets—all of which are chemically much more complex than the first stars.

Our home planet, Earth, formed around a fairly ordinary star, the sun, about 4.6 billion years ago. Earth, together with its sister planets, emerged out of gas and dust that the sun did not absorb. In this gas and dust were small amounts of all the elements that had been created in a nearby supernova. Out of Earth's particular combination of elements, living organisms emerged by chemical evolution; precisely how has yet to be worked out.

The first living organisms appeared about 3.8 billion years ago. They were microscopic, single-celled creatures, a few of which evolved about 600 million years ago into multicellular organisms. Multicellular animals began as sponges in the sea and gradually evolved into fish, some of which left the water as amphibians. One line of amphibians became reptiles; a line of those reptiles became mammals. After the extinction of the dinosaurs by a meteorite impact 65 million years ago, the mammals thrived. They eventually branched into many lines, including the great apes.

Then, about 5 to 6 million years ago, one line of apes diverged from the others. By about 250,000 years ago, this line had evolved into *Homo sapiens*—modern human beings.

Now, our story can begin to deal with storytelling itself; just when was it, over the 5-million-year development from our common ancestor with chimpanzees, that humans began to tell stories?

STORYTELLING BEGINS

No one knows what our common ancestor with chimpanzees was like, but, because chimpanzees cannot tell a story (though they can sign up to three hundred words if trained), we can deduce that the common ancestor was also unable to tell a story. The ability to think abstractly and symbolically, to speak in grammatical sentences using words to convey causality, must have developed gradually over at least the last quarter million years of hominine evolution, based on bipedalism, bigger brains, changed vocal cords, prolonged infancy, and social development.

No one can yet describe the development of the ability to speak symbolically with any certainty; it is another significant gap in the Big History story. Paleoanthropologists used to think that some genetic change occurred about 50,000 years ago that made full human speech possible, but more recent evidence from Africa suggests that this ability may have developed gradually starting at least 250,000 years ago, along with the other genetic and cultural changes that came to characterize *Homo sapiens*.

Humans lived as hunter-gatherers for more than 99 percent of their history, in a time usually called the Stone Age or the Paleolithic era. During this long period, humans preserved their knowledge in stories, songs, rituals, dances, and paintings. What we know about their lifeways comes from observing the very few people who still live as hunter-gatherers. From the stories of present hunter-gatherers we deduce that Paleolithic humans tried to explain where they came from and how everything in the world got here. The stories they told often involved other animals or powerful earthly elements, like sky, water, sun, or wind. These creation myths, or origin stories, gave explanations—usually natural ones—for what people didn't understand, and they often provided instructions for how people ought to behave.

ORIGIN STORIES DURING AGRICULTURE AND EARLY CIVILIZATION

By about ten thousand years ago, Earth's climate had entered a warm interglacial period coinciding with the full retreat of the last ice age, which had peaked about ten thousand years earlier. The warming climate made agriculture possible and necessary, and humans crossed a threshold into a new way of living by domesticating plants and animals. This enabled people to produce and store excess energy in the form of surplus food, and to expand significantly their population for the first time.

With agriculture also came a transformation in the stories that people told. Humans began to imagine and portray powerful beings—goddesses and gods—who influenced what happened on Earth, and who could be influenced by human rituals and prayers. Sky gods may have been part of a pastoral, nomadic existence. Goddesses of fertility may have predominated in people's origin stories during the transition to agriculture, as well as during the agricultural period itself, because people's survival depended on the fertility of crops and animals. The scale of origin stories remained localized.

Once sufficient agricultural surplus had been achieved in some places—often by coercion—some people could shift to city life, with many specializing in occupations other than farming and depending on others for food. In cities, the stories people told to explain their existence again underwent a transformation: while these stories had once featured beings in and of the Earth, now they emphasized a god or gods who resided in the sky, overseeing larger domains. Because life in cities proved difficult, people needed more complex moral codes to navigate the density of living close together. Philosophers and historians of religion often refer to the middle part of this era of agrarian civilizations as the Axial Age (c. 800 B.C.E.–200 C.E.), because it is when most of the world's current religions and basic philosophies arose (Islam developed later, out of Judaic and Christian traditions.) These religions became embedded in empires, which increased their institutionalization and scale.

THE SCIENTIFIC FOUNDATION

Humans crossed another threshold into the modern era when they interconnected the various parts of Earth with the development of the Silk Roads, with commerce in the Indian Ocean, and with the voyages of Zheng He and of the European sailors. When people began burning fossil fuel in the mid-eighteenth century, they brought about the onset of the industrial era. With modernity, humans' attempt to explain themselves and their universe saw another transformation, this time to the scientific form of explanation, based on physical and mathematical evidence. The Polish astronomer Nicolaus Copernicus launched the field of modern astronomy in 1543 with the publication of his book, *On the Revolutions of the Celestial Spheres*, in which he described a sun-centered cosmology that challenged the prevailing Earth-centered one. In 1610, the Italian Galileo Galilei provided preponderant—but not yet conclusive—evidence for this idea, with his improved telescope. In 1687, Isaac Newton formulated his law of universal gravitation, which explained why planets would orbit a larger sun with more gravity. Finally, scholars accepted that the Earth, too, was a planet.

Scientific research accumulated and accelerated, achieving great breakthroughs—in 1857, with the work of Alfred Wallace and Charles Darwin on the evolution of life; in 1907, with Einstein's theory of relativity; and in the late 1920s, with Georges Lemaitre and Edwin Hubble's understanding of an expanding universe of many galaxies. The story that science told differed from earlier stories in that it excluded supernatural elements; neither did it attempt to prescribe norms for behavior, leaving that sphere instead to states, religious organizations, and individuals.

The earliest attempts to tell the modern scientific story of the universe predated the scientific breakthroughs that finally made it possible. The human imagination leapt ahead of its observational powers. The two earliest-known pioneers were the Prussian naturalist and explorer Alexander von Humboldt and the Scottish entrepreneur Robert Chambers. Von Humboldt published the first volume of his *Cosmos* in 1845; he finished four volumes of his mainly descriptive account before his death. Chambers published his more speculative book, *Vestiges of the Natural History of*

Creation, in 1844; his authorship was revealed posthumously in 1884. His book sold well and helped prepare the way for Darwin's ideas.

During the late nineteenth century, most historians were busy specializing and becoming more nearly scientific in their historical work. In 1920, clearly wanting to foster global identity after the disastrous effects of World War I, the English author H. G. Wells published his *Outline of History: Being a Plain History of Life and Mankind*. In his account, he devoted fifty pages out of almost five hundred to describing Earth in space and time up to the development of *Homo sapiens*. He had no description of the beginning of the universe or its age, because this information did not exist at that time.

During World War II, the French paleontologist and Catholic priest Pierre Teilhard de Chardin wrote a book that became influential when it was published after his death in 1955. Called *The Human Phenomenon*, this book may be considered the first attempt at telling the modern Big History story. Teilhard accepted biological and cosmic evolution, but he saw them as taking place within spiritual evolution. He tried to formulate the implications of the evolving universe. He took seriously the uniqueness and the future of humanity; he saw a pattern of increasing consciousness and complexity. He believed that cosmic evolution is directed toward a goal he called the "Omega Point," a teleological idea not accepted generally by scientists. Teilhard's ideas significantly influenced Thomas Berry and Brian Swimme in the following generation.

During World War II another person began to formulate cosmic history—the Italian physician and educator Maria Montessori (1870–1952). Under house arrest by the British in India, she worked out ways to teach young children the story of how they were children of the universe, as the foundation for their education. Montessori's writing about this appeared in fragments in various places, rather than organized into a single narrative, but she clearly perceived the story line even before enough information was available to do so in a moderately rigorous way.

After the end of World War II in 1945, scientific research made several significant breakthroughs, which would make possible the construction of the modern Big History story. Perhaps most significant for the telling of Big History was what the historian David Christian calls "the chronometric revolution," based on the discovery that the age of rocks could be determined with the aid of the regularity of the radioactive decay of the nuclei of their atoms. This made possible the first accurate dating of the age of the Earth in 1953 by C. C. Patterson, a geologist at the California Institute of Technology. He measured the decay of uranium to lead in the meteorite that had created the Barringer Crater in Arizona, which had originated at the beginning of the solar system, thereby establishing the age of the Earth as 4.5 billion years. Other forms of more nearly accurate dating were also developed: counting the growth rings within tree trunks, counting genetic mutations in certain key species, and analyzing rock and ice samples, including the electromagnetic orientation of seafloor sea rock samples, which changes because the North Pole–South Pole magnetism has reversed, or flipped, many times in Earth's history.

Other scientific breakthroughs occurred in the 1950s and 1960s. The discovery of the structure of DNA in 1953 made it possible to track changes in the natural world with precision. In the 1960s, geologists began to understand plate tectonics, establishing that Earth's surface has changed over time as continental plates floated on the semimolten layers beneath them. In 1965, the discovery of the cosmic background radiation convinced astronomers of the Big Bang theory and persuaded them that the universe was evolving over time and had a history.

BIG HISTORY EMERGES

These scientific breakthroughs were accepted by the academic community by the 1970s and, together with photos taken by astronauts on the Apollo moon flights in the previous decade, inspired some to look at the whole story. One of the first modern Big History accounts is that of John Garraty and Peter Gay, who devoted forty-five of the one thousand pages in their edited *Columbia History of the World* (1972) to the period from the origin of the universe to the rise of agriculture. In 1978, Preston Cloud, a geologist at the University of Minnesota, published *Cosmos, Earth, and Man: A Short History of the Universe*. The story was beginning to take shape.

In the 1980s, scientists began to run with the idea of trying to tell the whole story. The most popular presenter was the astronomer Carl Sagan, with his book *Cosmos* (1980), written as a companion to his thirteen-part television series. (This influential series was updated in 2014 by a team that included astrophysicist Neil deGrasse Tyson.) But he was not alone. Also in 1980, the Austrian philosopher Erich Jantsch published *The Self-Organizing Universe*. In 1987, Siegfried Kutter, an astrophysicist at Evergreen State University in Washington State, published *The Universe and Life: Origins and Evolution*. In 1990, Larry Gonick, a cartoonist and mathematician turned historian, came out with volume 1 of *The Cartoon History of the Universe*. Two years later, the cosmologist Brian Swimme and the cultural historian and Catholic eco-theologist Thomas Berry published *The Universe Story*. In 2001, Eric Chaisson, a U.S. astrophysicist at Harvard and Tufts, contributed his synthesis, *Cosmic Evolution:*

The Rise of Complexity in Nature, which focuses on increasing complexity as the underlying pattern. (See also his earlier books, *Cosmic Dawn* and *The Life Era,* published in 1981 and 1987, respectively).

Published books present the material, but someone has to teach it to an audience beyond astrophysicists. Chaisson worked hard at Harvard and at Tufts to bring his work to students and to the general public. The pioneers among academic historians seem to have been John Mears, who at Southern Methodist University in the late 1980s instituted an introductory world history course that began with the Big Bang, and David Christian, who started a course in Big History at Macquarie University in Sydney, Australia, in 1989. Christian's course was followed by one at the University of Amsterdam in 1994, taught by biochemist and anthropologist Fred Spier and sociologist Johan Goudsblom, who had visited Christian in Sydney in 1992.

Mears took on the challenge of teaching his whole course by himself. He had always been attracted to sweeping treatments of the past and, in high school, had read Wells's *Outline of History* from cover to cover. By the mid-1970s, after teaching for a decade, he was increasingly frustrated by the intellectual fragmentation of the academy and went on to synthesize his own course on a big scale.

Christian, on the other hand, began his course by inviting experts from other disciplines to give the early lectures up to material familiar to him. He coined the term "Big History" when asked in 1991 to write an article about his course. Christian taught at Macquarie for twenty-five years before moving to San Diego State University in 2001. He published his seminal *Maps of Time* in 2004 and lectured in a series of Teaching Company videos released in 2008. He returned to Macquarie in 2009 and has continued as a prominent spokesperson for Big History, with many significant articles, books, and presentations to his name. Among his contributions are the use of thresholds of complexity as the structure of the story and the idea of collective learning as the unique attribute of humans.

Spier and Goudsblom, like Christian, began by inviting lecturers from several disciplines to their course. Spier has continued the course in this mode, now at several universities in Amsterdam. Spier's book *The Structure of Big History* (1997) helped to jell the structure of the story, using transformations in human ecological regimes—the control of fire, agriculture, and the burning of fossil fuel—as the major structuring principle in human history. In his recent book *Big History and the Future of Humanity* (2010), Spier has refined his account using the theme of increasing complexity and the Goldilocks conditions necessary for its increase.

Meanwhile, at Dominican University of California, educator and historian Cynthia Brown retired early from full-time teaching in order to write a Big History account. She had dreamed of this since reading Christian's description of his Big History course in Sydney. Brown taught several versions of a Big History course in the early twenty-first century, including a popular colloquium with biologist James Cunningham and philosopher of religion Philip Novak. In 2007, Brown's account appeared as *Big History: From the Big Bang to the Present*. Published for a general audience, the book emphasizes the reciprocal relationship between Earth and humanity and the urgent need for a transition to sustainability.

In 2010, the Dominican faculty became the first in the world to require all first-year students to take two Big History courses. Literary scholar Mojgan Behmand served as director of the First-Year Experience and general education program and soon began encouraging faculty to develop thematically linked upper-division courses informed by Big History.

Meanwhile, Spier contacted Walter Alvarez, the geologist renowned for leading the discovery of the global iridium layer, which is leftover material from the meteorite that doomed the dinosaurs. Both Spier and Christian encouraged Alvarez to start his own course at the University of California at Berkeley. Alvarez did so in 2006, as an elective in geology; soon, so many students clamored to take the course that Alvarez had to select among them based on essays they wrote in application. In his course and his lectures, Alvarez stresses the role of contingency in Big History.

In August 2010, Alvarez and his colleague Alessandro Montanari hosted a small group of Big Historians at the Geological Observatory of Coldigioco, Italy, and instructed them in reading history from the rocks. At Coldigioco, seven big historians—Alvarez, Christian, Lowell Gustafson (Villanova University), Barry Rodrigue (Southern Maine University), Spier, Brown, and Craig Benjamin (Grand Valley State University)—planned the formation of the International Big History Association (IBHA), which they formalized the following year, with the assistance of several spouses and a grant from Microsoft. Headquarters for IBHA were established at Benjamin's campus in Grand Rapids, Michigan.

Out of Alvarez's course (and Microsoft's research entity) emerged a project called ChronoZoom, an online, interactive timeline of all of history—the brainchild of Alvarez's student Roland Saekow. Microsoft developed the program with Alvarez and Saekow's guidance as a tool for creating interactive timelines over the 13.8 billion year history of the universe.

Microsoft's founder, Bill Gates, had taken a personal interest in Big History after watching Christian's Teaching Company video lectures. Gates called Christian in San Diego and asked to see him. Gates expressed the wish that he had been introduced to Big History in high school, where it would have given him a framework for learning about everything. Christian, who had always wanted to lay out a Big History curriculum for high school students, made a proposal, which Gates funded privately. Thus began the Big History Project, aimed at ninth grade, which in most public school curricula features an elective in social studies. Selected teachers and schools piloted the curriculum in 2011 and 2012. The project went live with free online courses in November 2013.

APPROACHES TO BIG HISTORY

Among academics, a remarkably high level of agreement exists on the story. Scientists tend to call it "the Epic of Evolution," while historians tend to call it "Big History," but they use the same basic structure to tell the story. E. O. Wilson, the biologist and ant specialist at Harvard, made popular the term "Epic of Evolution," and it is useful for listing the course in the science curriculum. Into this category fall works by Cloud, Chaisson, Spier, and one by the U.S. astronomer Russell Genet: *Humanity: The Chimpanzees Who Would Be Ants* (2007). Among the Big History accounts are those by Christian, Brown, and Spier—a scientist and anthropologist who seems to fit in both categories.

A third group of approaches admits of possible supernatural elements or addresses religious audiences to present the cosmic story to them in a persuasive way, while at the same time remaining empirically accurate. In this group one could place Swimme and Berry, with their book *The Universe Story* (1994) and follow-up film, *Journey of the Universe* (2011), produced by Swimme with Mary Evelyn Tucker (of both the Divinity School and the School of Forestry and Environmental Studies at Yale University). This film moves many people, whether religious or not, yet it subtly suggests that the universe is conscious in some way. A former United Church of Christ minister, Michael Dowd, published a book called *Thank God for Evolution* (2008), and he and his biologist wife, Connie Barlow, maintain a website, www.thegreatstory.org, that reaches both religious and nonreligious audiences. The sacredness of nature, without assuming the existence of God, is presented by biologist Ursula Goodenough (*The Sacred Depths of Nature*) and by philosopher Loyal Rue (*Nature is Enough*). Russell Genet and his wife, philosopher Cheryl Genet, edited a conference book, *Evolutionary Epic* (2009), which features essays from a wide range of perspectives.

In the first university-level Big History textbook, *Big History: Between Nothing and Everything* (2013), Christian, Brown, and Benjamin—all historians—combine their voices to tell the story organized around eight thresholds of increasing complexity, periods in which something new and transforming emerged. The authors hope that Big History courses will soon be taught globally, from South Africa to Russia, from Peru to Korea, helping to create a common understanding within which humanity can work together on the issues confronting it.

Additional accounts will be forthcoming, and the story will be tweaked with each retelling, as all good stories are. Scientific research will provide many new details in the coming years, and possibly major revisions. Perhaps, someday, a majority of people all over the world will be able to tell the universe story and build a personal connection to the cosmos.

In the meantime, we can rejoice in the astonishing capacity of humans to piece together and grasp the whole story. As Carl Sagan wrote in his concluding paragraph to *Cosmos:* "We [are] starstuff pondering the stars; organized assemblages of ten billion billion billion atoms considering the evolution of atoms."

Perhaps, in this pondering, humans everywhere will perceive our unity and understand our common fate, and with this new understanding, learn to conserve and to protect the environment that sustains us.

BIBLIOGRAPHY

Credit must go primarily to Fred Spier for doing early research into the history of Big History and for continuing this research. Much of what I have included I have learned from him.

Alvarez, Walter. "A Geological Perspective on Big History." *World History Connected* 6.3 (2009): n. pag. Web. 20 May 2014.

The Big History Project. The Big History Project. Web. 2 November 2013.

Brown, Cynthia Stokes. *Big History: From the Big Bang to the Present.* 2nd ed. New York: The New Press, 2012. Print.

Bryson, Bill. *A Short History of Nearly Everything.* New York: Broadway Books, 2003. Print.

Chaisson, Eric J. *Cosmic Dawn: The Origins of Matter and Life.* Bloomington, IN: iUniverse.com, Inc., 2000. Print.

———. *Cosmic Evolution: The Rise of Complexity in Nature.* Cambridge, Mass.: Harvard UP, 2001. Print.

————. *Epic of Evolution: Seven Ages of the Cosmos.* New York: Columbia UP, 2006. Print.

————. *The Life Era: Cosmic Selection and Conscious Evolution.* Bloomington, IN: iUniverse.com, Inc., 2000. Print.

Chaisson, Eric J., and Steve McMillan. *Astronomy: A Beginners Guide to the Universe.* 6th ed. San Francisco: Addison-Wesley, 2010. Print.

Chambers, Robert. "Vestiges of the Natural History of Creation." 1844. *Vestiges of the Natural History of Creation and Other Evolutionary Writings.* Ed. James A. Secord. Chicago: U of Chicago P, 1994. i–390. Print.

Christian, David. "Big History: The Big Bang, Life on Earth, and the Rise of Humanity" (course no. 8050). *The Great Courses.* The Teaching Company, 2008. Web. 23 Oct. 2012.

————. "The Case for Big History." *Journal of World History* 2.2 (1991): 223–238. Print.

————. "The Evolutionary Epic and the Chronometric Revolution." *The Evolutionary Epic: Science's Story and Humanity's Response.* Ed. Cheryl Genet, Russell Genet, Brian Swimme, Linda Palmer, and Linda Gibler. Santa Margarita, CA: Collins Foundation Press, 2009. 91–100. Print.

————. *Maps of Time: An Introduction to Big History.* 2nd ed. Berkeley: U of California P, 2011. Print.

————. "The Return of Universal History." *History and Theory* 49.4 (December 2010): 6–27. Print.

Cloud, Preston. *Cosmos, Earth and Man: A Short History of the Universe.* New Haven: Yale UP, 1978. Print.

Drees, Willem B. *Creation: From Nothing Until Now.* London: Routledge, 2002. Print.

Dowd, Michael. *Thank God for Evolution: How the Marriage of Science and Religion Will Transform Your Life.* New York: Viking, 2008. Print.

Duffy, Michael, and D'Neil Duffy. *Children of the Universe: Cosmic Education in the Montessori Elementary Classroom.* Hollidaysburg, PA: Parent Child Press, 2002. Print.

Garraty, John A., and Peter Gay, eds. *The Columbia History of The World.* New York: Harper and Row, 1972. Print.

Genet, Cheryl, Russell Genet, Brian Swimme, Linda Palmer, and Linda Gibler. *The Evolutionary Epic: Science's Story and Humanity's Response.* Santa Margarita, CA: Collins Foundation Press, 2009. Print.

Genet, Russell Merle. *Humanity: The Chimpanzees Who Would Be Ants.* Santa Margarita, CA: Collins Foundation Press, 2007. Print.

Gonick, Larry. *The Cartoon History of the Universe.* Vol. 1. New York: Doubleday, 1990. Print.

Goodenough, Ursula. *The Sacred Depths of Nature.* New York: Oxford UP, 1998. Print.

Hughes-Warrington, Marnie. "Big History." *Social Evolution & History* 4.1 (Spring 2005): 7–21.

Humboldt, Alexander von. *Cosmos: A Sketch of the Physical Description of the Universe.* 1845. Vol. 1. Hamburg: Tredition, 2006. Print.

Jantsch, Erich. *The Self-Organizing Universe: Scientific and Human Implications of the Emerging Paradigm of Evolution.* Oxford: Pergamon, 1983. Print.

Kutter, G. Siegfried. *The Universe and Life: Origins and Evolution.* Boston: Jones and Bartlett, 1987. Print.

Mears, John A. "Evolutionary Process: An Organizing Principle for General Education." *The Journal of General Education* 37.4 (1986): 315–325. Print.

———. "Implications of the Evolutionary Epic for the Study of Human History." *The Evolutionary Epic: Science's Story and Humanity's Response.* Eds. Cheryl Genet, Russell Genet, Brian Swimme, Linda Palmer, and Linda Gibler. Santa Margarita, CA: Collins Foundation Press, 2009. 135–146. Print.

Pius II, Pope. "Modern Science and the Existence of God." *The Catholic Mind* 49 (1952): 182–192. Print.

Ratzinger, Joseph Cardinal. *'In the Beginning . . .': A Catholic Understanding of the Story of Creation and the Fall.* Grand Rapids, MI: Wm. B. Eerdmans, 1995. Print.

Reeves, Hubert. *Hour of Our Delight: Cosmic Evolution, Order, and Complexity.* New York: W. H. Freeman, 1991. Print.

Reeves, Hubert, Joël de Rosnay, Yves Coppens, and Dominique Simonnet. *Origins: Cosmos, Earth and Mankind.* New York: Arcade Publishing, 1998. Print.

Rue, Loyal. *Everybody's Story: Wising Up to the Epic of Evolution.* Albany: State U of New York P, 2000. Print.

———. *Nature Is Enough: Religious Naturalism and the Meaning of Life*. New York: State U of New York P, 2011. Print.

Sagan, Carl. *Cosmos*. New York: Random House, 1980. Print.

Spier, Fred. "Big History." *A Companion to World History*. Ed. Douglas Northrop. Oxford: Wiley-Blackwell, 2012. 171–184. Print.

———. *Big History and the Future of Humanity*. Chichester, West Sussex, U.K.: Wiley-Blackwell, 2010. Print.

———. "Big History: The Emergence of an Interdisciplinary Science?" *World History Connected* 6.3 (2009): n. pag. Web. 20 May 2014.

———. "How Big History Works: Energy Flows and the Rise and Demise of Complexity." *Social Evolution and History* 4.1 (2005): 87–135. Print.

———. "Interpreting the History of Big History." *Teaching History* 146 (March 2012): 44–45. Print.

———. "The Small History of the Big History Course at the University of Amsterdam." *World History Connected* 2.2 (2005): n. pag. Web. 20 May 2014.

———. *The Structure of Big History: From the Big Bang until Today*. Amsterdam: Amsterdam University Press, 1996. Print.

Swimme, Brian, and Thomas Berry. *The Universe Story: From the Primordial Flaring Forth to the Ecozoic Era: A Celebration of the Unfolding of the Cosmos*. San Francisco: HarperCollins, 1992. Print.

Teilhard de Chardin, Pierre. *The Phenomenon of Man*. Trans. Bernard Wall. New York: Harper Perennial Modern Thought, 2008. Print.

Wells, H. G. *The Outline of History: Being a Plain History of Life and Mankind*. New York: Garden City Publishing Company, 1930. Print.

TWENTY

Big History at Dominican

An Origin Story

Philip Novak

> Our new sense of the universe is itself a revelatory experience. Presently we are moving into a meta-religious age that seems to be a new comprehensive context for all religions . . . [in which] the primary sacred community is the universe itself.
>
> THOMAS BERRY AND BRIAN SWIMME, *The Universe Story*[1]

Everyone who wants to tell a story is faced with the thorny problem of where to begin. Every "once upon a time" one can think of is already the result of other things that happened *before that*—unless of course our "once upon a time" is the Big Bang, from which our material universe really did begin, and which has no "before" that the human mind can comprehend. But to begin *every story* with the Big Bang, though technically correct, would be impractical—not to mention boring! So we usually begin our stories in a more recent, and thus somewhat arbitrary, way. This is what I shall do here.

The history of Big History at Dominican began in the summer of 1983, when I taught a course called "God and Evolution." It is true that had this not happened, Big History's history at Dominican would have been quite different, or might not have *been* at all! But what conditions—what Goldilocks conditions—gave rise to the course, to *that* little island of organized complexity?

There were three. First, there were the conditions pertaining to the instructor. I was fresh out of graduate school at Syracuse University, where I had earned a PhD—much to my surprise—in the history of religions. A decade earlier, nothing could have interested me less, but undergraduate readings in psychology and philosophy changed all that. How was it, I wondered, that intelligent people from all times and places should seek to harmonize their lives with a variously named, ultimately spiritual reality? What was going on there? Self-delusion? Or extraordinary perceptivity? I had gone to Syracuse for answers, but almost every answer raised new questions. Along the way my professors made me do lots of reading in modern science and, especially, evolution.

That brings us to the second set of conditions—the books. One that I taught that summer at Dominican was Loren Eiseley's *The Firmament of Time*. Eiseley was an anthropologist with a literary gift, once hailed as "the modern Thoreau," who, more than any other writer, opened my humanist's mind to the beauty of science. He had done so with a clutch of lyrical books including *The Invisible Pyramid*, *The Star Thrower*, *The Unexpected Universe*, *The Immense Journey*, and, of course, *The Firmament of Time*. A second book I taught in that course was Pierre Teilhard de Chardin's *The Phenomenon of Man*. Teilhard was a professional paleontologist but also a Jesuit Catholic priest. He believed biological evolution to be a phase in a more encompassing evolution of human consciousness that would eventually give rise to an enlightened and peaceful humanity. Teilhard's spiritual vision scandalized his fellow scientists, and his embrace of evolution scandalized his Catholic superiors— for this was 1955, decades before the Vatican would acknowledge evolution's facts. Rounding out the summer's reading were Theodosius Dobzhansky's *The Biology of Ultimate Concern* and Robert Francoeur's *Evolving World, Converging Man*.

The third set of favorable conditions was the students themselves. They were members of Elder Hostel, an organization of retirees intent on keeping their summer brains humming, and they were a delight. I suspected that many of them knew more than I did, but they kindly cheered me on anyway. One of those students was Faith James. And here we come to another crux of the story, for without Faith, the rest of the story would have been very different. But there she was, chatting me up at the end of the first couple of classes and teasing me about trying too hard to sound like Carl Sagan when I spoke about "billions and billions" of galaxies. At the end of the third class, Faith invited me to dinner at her home. Squeezing my hands between hers, she urged me to come because "there would be another guest whom you'd enjoy meeting." I don't remember if she told me his name, and if she had, I wouldn't have recognized it anyway. But I was a hungry young professor—so I went.

I imagined that it would be a Big Party at a Big Table. But, thank God, it turned out to be a small party at a small table where real conversation was possible. Of the four chairs, Faith and her husband occupied two. I parked myself in the third. And in the fourth sat Father Thomas Berry.

Berry was a Fordham University historian of culture in town on a lecture tour. I liked him immediately. He was sixty-nine at the time, and his eloquence was wrapped in a gravelly, gentle voice. There was a twinkle in his eyes and a bemusement around his mouth that testified to life's goodness. But its corners were tinged with melancholy: lifeboat Earth, he felt, was sinking shockingly fast.

Berry was steeped in modern cosmology and biology and well aware of their convergence in a powerful evolutionary metanarrative. He later called that narrative *The Universe Story* and considered it to be the key to a revolution in thought strong enough to defeat our ecological ignorance. "It's all a matter of story," he liked to say, and those words open "The New Story," one of his most important essays. Berry believed that our old stories—humankind's religions—having functioned upliftingly for so long, were now dysfunctional, because they were out of sync with hard-won modern knowledge. He believed that the story now being told by science, an empirically grounded panorama of cosmic, biological, and cultural evolution, though lacking a needed spiritual dimension, was the undeniable basis from which a new spiritual vision must spring.

Berry believed that by making the New Story a fundamental part of education we could gradually *"reinvent the human at the species level."* For him, the New Story meant that, for the first time, all humanity within earshot of an education could realize that it shared a *single, transcultural origin story.* As we passed the butter and the broccoli that night, Berry said: "You know, the universe is the only text without context. Everything else has to be seen in the context of the universe. The universe, the solar system, and planet Earth, in their evolutionary emergence, constitute for the human community the primary revelation of that ultimate mystery whence all things emerge into being."

Berry loved the wisdom embedded in humanity's great religions—he had written books on the religions of India and on Buddhism—but he also saw how easily religion could degenerate into folly. "I'm no longer a theologian," he smiled, "I'm a *geo*logian." We talked for a long time that evening. Berry's impression on me proved indelible.

Fatefully, right around that time, Dominican received from the Buck Fund a $1.2 million academic initiatives grant. One of its provisions was a fellowship program to bring leading educators to campus for a week of lectures, class visits, and informal conversations. I proposed Berry as a fellow and he arrived on campus in spring 1984.

It was a memorable week, in which all of us got to chew on the ideas I had first tasted at Faith's supper. In the question-and-answer period following one of Berry's public lectures, an incredulous listener asked, "Are you really a Catholic priest?" Without missing a beat, Berry replied, "Last time I checked."

The Dominican sisters *loved* Berry, and that was no small reason why the seeds of Big History continued to sprout here long after his departure. Sister Susannah, head of the Santa Sabina retreat center on campus, was so inspired by his ideas that she changed the name of her house cat to T-Berry. This admiration also meant that both Berry—and his equally illustrious younger collaborator, the physicist Brian Swimme—were frequent speakers at Santa Sabina over the next decade.

The Dominican curriculum began to absorb Big Historical concerns. Berry's visit made it clear that we needed a religion and ecology course in the curriculum. We created one and baptized it with the title of another of Berry's essays, "The Spirituality of the Earth." I then added to my course "The World's Religions" three context-setting Big History lectures, and experienced palpable gains in student comprehension of the course as a whole.

By far the most important curricular innovation in this regard was the Big History Colloquium itself. At Dominican, a colloquium is a group of courses designed for maximal thematic interplay. The idea for a Big History Colloquium came from my history colleague Cynthia Brown. Cynthia had attended Berry's lectures ten years before and had gone on to read so widely in the field that she was now writing her own book on it (published in 2007 as *Big History: From the Big Bang to the Present*). Cynthia proposed that our three-course colloquium be made up of an evolution course, taught by our biology colleague Jim Cunningham, her own Big History survey, and my Big History–slanted "World's Religions." We offered the colloquium successfully in 2003, 2005, and 2007.

That's how things stood in the 2007–2008 school year, when the Dominican faculty unexpectedly called for a revision of the general education program. No one could have guessed that this would eventually result in Dominican becoming the first university in the world to require a Big History course of all its first-year students.

Former Harvard president Derek Bok once said that it's as easy to change a curriculum as it is to move a cemetery, and anyone who's been through a general education revision knows how true that can be. The particular difficulty this time was the creation of a new First-Year Experience program for our entering students.

The committee devoted to this task debated options for two years, but nothing quite clicked. Its initial assumptions were that the first-year courses should provide creative freedom for faculty and freedom of choice for students. But, as the deadlocked months dragged on, some of us began to question those assumptions. We were, after all, in the midst of a renewed national conversation about the importance of truly *foundational* knowledge and a *common* first-year experience for entering students. The question became: *Is there any knowledge out there foundational enough to serve as a common and compelling educational core for an increasingly diverse student body in an increasingly transnational civilization?* The answer that bubbled to the surface was "Yes—Big History!" Not every committee member agreed, but enough did to start the ball rolling.

In the weeks ahead, as the proposal moved out of committee to faculty-wide discussion, its supporters caught three lucky breaks. First, Cynthia Brown was willing to come out of retirement to join the campus conversation. Her eloquent essays on this new academic field, not to mention her charm, softened resistance wherever it was found. Second, I was named dean of the new School of Arts, Humanities and Social Sciences and could now make a case for Big History directly to fellow administrators. Third, and most importantly, there was Mojgan Behmand, the committee's young English professor, to whom fell most of the leadership responsibility for the First-Year Experience. She had never heard of Big History and was initially skeptical. But the more she studied, the more academic promise she saw. Over the next few months her skills made the seemingly impossible possible: instead of another bloody general education war, Behmand commandeered a bloodless coup. She navigated the new First-Year Experience program and its Big History centerpiece past the tricky shoals of faculty and trustee approval.

Then came the hard part—finding the money to make it all go. We were clear from the outset that the program would fail unless we could train teachers willing to learn how to teach in it. Big History is, after all, the big mama of interdisciplinarity, and no one can really know all that teaching Big History requires one to know, or how to convey it to a group of eighteen-year-olds in the context of a course that builds basic academic skills. We knew we needed to create a summer institute for the teaching of Big History and that at least twenty of us would have to undertake an intensive weeklong training in order to achieve true lift-off. In tight budget times we asked our president and provost for $50,000 to launch. And they said yes.

For the first summer institute, in 2010, Brown and Behmand produced the syllabus and delivered key lectures. Physicist Brian Swimme accepted our invitation to deliver the keynote. Other experts spoke, but we also taught and learned from one another. By the end we were exhausted but satisfied: many of us felt that this was the most significant faculty development exercise we'd ever been through. University trustees began to take notice and offer support.

National media attention didn't hurt; nor did the discovery that Bill Gates has a particular interest in Big History. Dominican administrators found their way to supporting a second summer institute in 2011. We hosted guests from around the world at our summer institutes in 2012 and 2013, and at the second annual International Big History Association conference in 2014. And Dominican's new president,

Dr. Mary Marcy, has made Big History one of the pillars of her vision for the university's future.

The three hundred or so first-year students entering in fall 2014 were the fifth such group to begin their university education with a course in Big History. They have enjoyed a syllabus that has grown more elegantly structured and coherent every year; readings that have been more carefully weighed and selected than ever before; a beautifully illustrated, just-published textbook that, within its own genre of Big History, is the best of its kind; and a multimodal classroom experience that includes carefully honed hands-on activities, extra-classroom experiences in modern astronomy and California geology, a twenty-entry reading journal (one entry per class meeting), and a tightly crafted final writing assignment called "A Little Big History." Like my fellow program faculty, I am proud of our considerable pedagogical efforts and achievements.

But I also want to add a word of critique, which, fittingly, returns us to the heart of this piece—the ideas of Thomas Berry. When Dominican's Big History program was in its gestation, no influence was more seminal than Berry's. Now that influence is in eclipse. Today, for understandable reasons, the program is using a textbook—the best in its genre, as I have said above—whose historian authors work under the assumption of philosophical materialism, which is that matter is the only reality. Science's *methodological* materialism—that science will deal only with what registers to human senses or their technical extensions—is, of course, unassailable and a splendid path to knowledge. But from this, the truth of *philosophical* materialism does not follow. Knowing this, Berry (and many others) prefer to "do" Big History with an epistemic openness to other domains of reality. This openness means leaving room for—among other things—the possibility of a sentient universe.

There may, indeed, come a day when materialist Big History proves to be not just one way of telling the story, but the only way that is true beyond all rational doubt—but we seem far from that point. In the meantime, some of us, following Berry's lead, are bound to sense that materialist Big History needlessly forestalls some important spiritual possibilities.

With that in mind, there can be no better way to end this piece than with Berry's own words. Without them, it can be argued, the book you hold in your hands would not have come to be.

> The primary principle . . . is that the Universe—and in particular planet
> Earth—is a communion of subjects, not a collection of objects. If we don't
> learn that—nothing is going to work.[2]

Identification with the cosmic-earth-human process provides the context in which we now make our spiritual journey. This Journey is no longer the journey of Dante [or] . . . the journey of the Christian community through history to the heavenly Jerusalem. It is the journey of primordial matter through its marvelous sequence of transformations—in the stars, in the earth, in living beings, in human consciousness—toward an ever more complete spiritual-physical intercommunion of the parts with each other, with the whole, and with that numinous presence that has been manifested throughout this entire cosmic-earth-human process.[3]

This story [Big History] can fulfill its role only if the universe is understood as having a psychic-spiritual as well as a physical-material aspect from the beginning. This should not be difficult since we know what something is by what it does. Since the universe brings us into being with all our knowledge, our artistic, and our cultural achievements then the universe must be a knowledge-producing, an art-producing, and a communion-producing process.[4]

NOTES

1. Brian Swimme and Thomas Berry. *The Universe Story*. New York: HarperCollins, 1992. 255. Print.

2. Ibid.

3. Thomas Berry, "The Spirituality of the Earth." *Liberating Life: Contemporary Approaches in Ecological Theology*. Ed. Charles Birch, William Eaken, and Jay B. McDaniel. Maryknoll: Orbis, 1990. 158. Print.

4. Thomas Berry. *The Great Work*. New York: Random House, 1999. 81. Print.

TWENTY-ONE · Teaching Big History or
Teaching about Big History?
Big History and Religion
Harlan Stelmach

> You can't teach religion without taking a religious
> stance: if you think you are avoiding one, you just end
> up taking an anti-religious stance.
>
> ROBERT BELLAH

When "teaching" what is known as Big History, an answer to the question posed in the title of this essay is paramount. What appears to be a subtle distinction is at the heart of clarity about the subjects of Big History and religion. Teaching, rather than just teaching *about* a subject, assumes, if not agreement and commitment, at least deep sympathy for the subject. Some might even say it assumes a kind of "participation" in the subject. Not to understand this will lead to confusion on the part of our students and those of us teaching Big History. I was reminded about this distinction in a series of poignant conversations with my colleagues at Dominican University. I said, "We need to be clear about *teaching religion* in our curriculum." A literature professor challenged me, saying: "I hope we do not teach religion in our curriculum. Students should learn *about* religion but we should not *teach religion*." I asked two professors in the religion department if they "teach religion." Both suggested that in a university curriculum, unless one is teaching theology, one should only teach *about* religion.

This disagreement has two sources. First, there are differing assumptions about what we mean when we use the word "religion." The second is about the teaching itself. That is, what are we doing when we "teach" any subject about which we are passionate? This kind of teaching inspires others to appreciate and to explore something fundamental about their humanity as part of a larger community. A poet from our English faculty was clear that she was not just teaching *about* poetry; she was *teaching poetry*. My science colleagues all asserted, "I teach science." Of course they do. And of course they also teach *about* science. We do both.

The founder of modern Big History, David Christian, is clear in the assumption of "agreement and commitment" to one's subject in his book *Maps of Time*. In his introduction, subtitled "A Modern Creation Myth?" he forthrightly answers that question.[1] According to Christian, "this new myth" *based* in "modern science . . . can help us answer some of the deepest questions we can ask concerning our own existence, and that of the universe through which we travel."[2] The myth, he writes, answers the questions "Who am I? Where do I belong? What is the totality of which I am a part?"[3]

> By offering *memorable and authoritative* accounts of how everything began—
> from our own communities, to the animals, plants, and landscapes around us,
> to the earth, the Moon and skies, and even the universe itself—*creation myths*
> provide universal coordinates within which people can imagine their own
> existences and find a role in the larger scheme of things. *Creation myths* are
> powerful because they speak to our deep spiritual, psychic, and social need for
> a sense of place and a sense of belonging. (emphasis added)[4]

Christian is clear that he is offering, with his "Big History" story, just such a "memorable and authoritative" account of how everything began. He reiterates this position in his coauthored textbook, *Big History: Between Nothing and Everything*.[5]

This chapter will explore the implications of the following questions: How should we address the issues that emerge when we teach Big History and teach *about* Big History? And does the presentation of Big History as a "universal worldview" render it a modern "religion"? If there is even a remote chance of that, we need to articulate the issue clearly and deal with it explicitly. In addition, because practitioners of the academic study of religion have developed approaches both for teaching and for teaching *about* various worldviews and origin stories, the field may have something to offer for use in the Big History classroom.

When we teach "Big History" as the modern creation story, we often confront students with the issue of religion in multiple ways. For many, the Big History narrative challenges long-held personal beliefs about a religious creation story. For others, it is a welcome new story that appeals to a modern self-understanding, has little interest in "organized religion," and holds that humans are the makers—and sometimes destroyers—of their own destiny. Some students see religion as the pivotal force that, if not shaping the world for good or bad, has at least been a defining feature in how they live their lives. For them, Big History becomes problematic when many of its texts appear to assume a minor role for religion in the sweep of cosmic, biological, and cultural evolution. On the other hand, some students welcome our teaching a story that does not *explicitly* challenge their worldviews, whether they are religious or not.

For many, students and faculty alike, it is more comfortable to simply keep religious questions out of the classroom and let them remain private. Students often just want to understand the "facts" and get on with their lives. Faculty members are often not equipped to engage students on such deep questions and want to keep the focus on the content that the students should know. But there are some useful approaches that can help educators responsibly address religious questions that arise or fail to arise when we teach Big History. Just as important, we can learn to take advantage of a teaching opportunity that cuts to the grit of the meaning of liberal education: confronting the essential question *What does it mean to be human?*

I will address how our collective learning in implementing Dominican's Big History program has sought to deal with this issue. This journey should be instructive for others wanting to embark on a similar path of sorting out the many and profound issues around Big History and religion.

DEALING WITH DISORIENTATION

The first major issue to address is the general disorienting element of the Big History approach for both students and faculty. As the late sociologist Robert Bellah expressed in his magnum opus, *Religion in Human Evolution*, if Copernicus pointing out that the Earth was not the center of the universe was "unsettling to humans," then "how unsettling can you get" when David Christian states in *Maps of Time*: "Our Sun, it seems, is situated in an undistinguished suburb in [the Milky Way,] a second-rank galaxy (the Andromeda Galaxy is the largest in our local group) in a group of galaxies that lies toward the edge of Virgo Super cluster, which contains many galaxies."[6]

This story is certainly disorienting to many of our students, who come with either literal views of the creation stories they've learned in their religious training or a commitment to the idea that reality and meaning are not sufficiently informed by an evolutionary explanation of human life.

Some of the professors in the program have had to deal with their own disorientation. Most of us are discipline specialists now trying to teach a unified body of knowledge from astronomy, physics, and chemistry to biology, history, and anthropology. If you add teaching religion or dealing with issues of spirituality to the mix of disciplines, the task becomes even more daunting. Some professors do have a commitment to traditional creation myths in their ethical and symbolic meaning, if not their literal meaning. If a faculty member has an alternative meta-narrative to which she or he is committed, "teaching Big History" could become the most disorienting feature of the project.

Our program has addressed the complex set of problems surrounding Big History and religion directly, through an ongoing conversation. Accordingly, our solution to how the subject of religion should be addressed in both our first-semester survey Big History course and our second-semester "Through the Lens" courses has been an unfolding one.

Once we realized that we were not addressing the questions of meaning and religion well enough in the survey Big History course that we require all our students to take, we started conversations among ourselves and discussions in the classroom. We added a specific course on "Religion *through the Lens of* Big History," which the student may choose from a list of other discipline-based options, after having taken the Big History survey course in their first semester at the university. While in the first-semester survey course we essentially are providing our students with this disorienting and decentering myth, in the second semester, a self-selecting group of students, who may have been disquieted by the first-semester course, end up in "Religion through the Lens of Big History," which I teach.

David Christian, Cynthia Brown, and Craig Benjamin begin their book, *Big History: Between Nothing and Everything*, by asserting that the new story is different from origin stories of the past. "It is a literal account of the origin of everything," they write. "It expects to be taken seriously as a description of what actually happened 13.8 billion years ago."[7] The authors also state that it is the only origin story "accepted by scientists throughout the world." Yet they remind us that because it is a story based on evidence, "the same scientists also know that many of its details will change in the coming years. It is not a fixed or absolute story and does not claim to be perfect."[8]

In *Maps of Time*, Christian explains why there is a need for a new creation myth and what motivated him to suggest Big History as a viable alternative narrative. He has his own notion of what is disorienting in our world today: the lack of a coherent story of our origins using modern scientific knowledge has left us without a myth that "speaks to our deep spiritual, psychic and social need for a sense of place and a sense of belonging."[9] If the new story is now offered to play this role, it is not too surprising that some of our students find it disorienting.

We try to hit this issue head on. In the first two weeks of our survey course, we seek to get a sense of how disoriented or accepting our students are feeling about what we are presenting. We ask our students a variety of questions: How do you believe the universe began? Do you foresee any conflicts between the content of this course and any religious or spiritual beliefs you hold? Explain how you plan to handle these conflicts, if they arise. And if you do not foresee any conflicts, briefly describe your familiarity with the Big Bang theory and evolution.

Among the tenets of our program is that unsettling our students is an important goal of liberal education, one that will prepare them to cope with the uncertainty and ambiguity that are part of our complex, ever-changing world. But to leave our students unsettled without preparing them to explore the meaning of this disorientation would be to miss an opportunity for educational transformation. Mark William Roche, in his 2010 book *Why Choose the Liberal Arts?* affirms with Hegel that education brings a "self-alienated spirit." But Roche is just as insistent that we have the teaching skills to recognize and address the existential drama of this new alienated identity, if not identity *crisis*.[10] Much rides on our instructors' willingness and capability to follow up with discussions of the questions above.

BENJAMIN'S APPROACH: STUDENT-LED DISCUSSIONS

Textbook coauthor Craig Benjamin, in his reflective article "The Convergence of Logic, Faith and Values in the Modern Creation Myth," gives us the best starting point for addressing religion in a basic course on Big History that teaches this scientific creation story.[11]

Benjamin, a former student of Christian, embraces Christian's perspective that Big History is about "teaching a modern creation myth." He is also aware of how disorienting this new myth is, particularly for the new undergraduate "who walks into a classroom on the first day of her new life to be ambushed by an elegant, multi-disciplinary account of origins and futures—in essence a creation myth articulating an alternative explanation of origins and *purpose* for existence" (emphasis added). This observation leads Benjamin to further remark that what he "couldn't see at first was that when my students moved from their homes and high schools to a university dedicated to providing a secular liberal education"—he teaches at a state university in Michigan—"they found themselves confronted by a far more nuanced view of the world than the paradigm that had sustained their lives thus far."[12]

Benjamin says he understood that the "first crisis they had to face was a crisis of faith."[13] He said he then had an epiphany. He felt he was not fulfilling his obligation if he ignored the faith stance of his students and pretended that there was not an "elephant sitting in the room." Benjamin, a native of New Zealand, writes: "If I am genuinely committed to the task of fusing intellectual development with character and ethical development, surely a central goal of the great liberal education tradition of the United States, I need to concentrate on all the *functions* in my students and help them make connections. And a core *function* for most of them is their faith" (emphasis added).[14] Benjamin has given witness to what many instructors in our program have come to realize: we cannot avoid a conversation about religion in our Big History classes.

In the classroom, Benjamin explains, he puts this discussion in the hands of small teams of students who lead weekly discussions. With certain ground rules of civility to address "divisions along entrenched cultural lines," he encourages "frank and mutual consideration of the inner conflict" that competing worldviews produce. He is clear that Big History is a "worldview." The role of the professor, he says, is that of a "teacher-student in a classroom of student-teachers." By this approach he begins to let go of the chief role of "myth-teller."[15] This allows an "empowerment" in those hearing the myth. The assumption one would make from this approach is that not only would there be the possibility of the listener to be transformed by the myth but the conversation would also have an impact on the myth-maker. If this is so, then we really do have a classic example of transformative education, which may be mutually transformative for both the student and the instructor.

Yet we find it difficult to be influenced by our students on the issue of religion. We have our own private views of religion. We either have a bias against teaching religious issues—or even teaching *about* them—in class, or are just not equipped to have these conversations. One of the fundamental reasons no doubt rests on our personal view of religion. Quoting a fellow historian, Joel Tishken, Benjamin reminds us of the following: "Our academic definition of religion as an abstraction separate from the rest of life, is not one shared by most people past and present, but is *an invention of modernity.* . . . Religion does matter to most humans, and we must find ways to represent the power of religion by remembering that the academy's views on religion are not the global norm" (emphasis added).[16]

IDENTIFYING OUR OWN RELIGIOUS STARTING POINT

As Big History instructors, we must be clear about our religious—or nonreligious—starting point. It is not enough to say that we do not want to unduly influence our students with our point of view. By teaching Big History with our current texts we are already influencing students in a certain direction in respect to religion. We owe it to them to accompany them on that path.

In 1955, Paul Tillich, perhaps the most important American Christian theologian of the twentieth century, retired from Union Theological Seminary in New York to become University Professor at Harvard. Three years later, he gave a talk to the Board of Harvard Overseers entitled "Religion in the Intellectual Life of the University: The Faculty and Its Functions; the Dual Attitude; the Religious Dimension."[17] Tillich's talk illustrates both my approach to religion as a religious person and my approach to teaching religion, not just teaching *about* religion, in the university. Certainly, Harvard Overseers thought Tillich had a perspective that needed to be heard.

Tillich defines "the religious question" as the "the question of meaning of human existence and existence generally."[18] He reminds us that "the word 'meaning' points to the questions: What am I for, for what do I live? Is it worthwhile to live? Is being a whole big accident, or has it an ultimate inner aim? Religion gives symbols in which these questions are answered."[19] Tillich does not relegate these questions to realms "separate from the rest of life." Instead, Tillich says that these questions are "asked in every realm and on every level of human existence."[20] "The possibility of raising them makes man man [sic]. The answers belong to the religious traditions already discussed [such as the major historic world religions]. But the question belongs to the whole realm of human existence. Therefore, the question arises in every section of man's cognitive approach to reality."[21]

If we borrow the language of religion when we refer to Big History as "a new creation myth," then it is not surprising that Christian, and many of us, in trying to understand the profound meaning of Big History, exhibit what Tillich might call a religious sentiment. What Christian, Brown, and Benjamin have done, if we use Tillich's perspective, is to debunk current religious traditions as not expressing scientific rigor. What they have not done, and cannot do as humans, is to go against their own humanity. What makes us human is to ask, and to answer, these questions of meaning and coherence and eventually our destiny as humans. In one of the most poignant expressions of his humanity, David Christian, in a TED talk, explains why he is motivated to tell the Big History story: he wants his grandson to inherit a world worth living in.[22] Christian is hoping for a story that gives his grandson the knowledge and the meaning to face what is an uncertain future.

My view of religion is close to Tillich's: religion is not an archaic tradition that can be explained away with new knowledge. Rather, it is fundamental to who we are as humans; it is what makes us human. We may express this reality of our humanness through various traditions, but these traditions do not, and cannot, totally define us as the fundamentally meaning-seeking beings we are. Tillich explained this view of religion in his 1959 book, *Theology of Culture:* "Religion is not a special *function* of man's spiritual life, but it is the dimension of depth in all its *functions.* . . . [Religion] is home everywhere. . . . Religion is the dimension of depth in all of them. Religion is the aspect of depth in the totality of the human spirit. Religion, in the largest and most basic sense of the word, is ultimate concern" (emphasis added).[23]

I am aware that even this view of religion is disorienting to our students. What my students appreciate is a chance to engage on these issues. I am not trying to convince them of my view. I am disclosing my view as they disclose their view. I work to have empathy for their positions. I often long for the deep conviction that my students feel about their faith, knowing that my position springs from a largely intellectual understanding. My students are inviting me to do more than teach about religion in a detached way.

Robert Bellah, whose *Religion in Human History* is framed in a version of Big History and includes cosmological, biological, and cultural evolution on the topic of religion, spoke at Dominican, helping us to deepen our understanding of these matters. His book and the many talks he gave in the years after its publication have reinforced my position that religion is essential to what we as humans do to discover meaning in all realms of our lives. Religion is not just a separate part or function of our existence. In a talk recounting his journey as a sociologist of religion, as he moved from Harvard to Berkeley, Bellah concludes:

> In my teaching I treated religions in . . . the anthropological mode, that is, at arm's length, as specimens to be analyzed. I had some disturbing experiences in these early years of teaching. On more than one occasion I had a Bible-belt Protestant or a Jesuit-belt Catholic come to my office to tell me that my course had caused him to lose his faith. . . . I eventually came to realize that you can't teach religion without taking a religious stance: if you think you are avoiding one, you just end up taking an antireligious stance. . . . It was at that point that I realized that I had to teach religion before I could teach about it.[24]

My advice is that whichever stance we take, we need to be open and honest with our students. We are asking them to deal with deep identity questions; it behooves us to share our positions in a thoughtful way at the same time.

TEACHING BIG HISTORY OR TEACHING
ABOUT BIG HISTORY?

For his Harvard audience, Tillich identifies two essential components of university teaching, "detachment and participation."

On "detachment," Tillich says: "Without distance, no cognitive approach is possible. Otherwise the result is not knowledge, but emotional 'outcry.' And this detachment refers to the philological preciseness, to historical probability on the basis of documents, to a consistent use of concepts, to exact observation."[25] This is the so-called "rigor" that is required of all disciplines, including science.

Then he addresses "participation": "On the other hand, there is not understanding of any intellectual creation without participation. This participation is not identification with a concrete religious reality, but *empathy*, as it is demanded in all humanities" (emphasis added).[26]

It is clear that "participation" does not play second fiddle to "detachment." Though Tillich is making a formal distinction between the two, one without the other is not possible for our understanding of knowledge in the modern world. So he explains their unity as follows: "It is a unity of participation and detachment, which could be called 'existential hypothesis.' All detached knowledge remains hypothesis. It is preliminary; but participation brings the subject matter into us or us into it. Such participation produces the eros and the passion, which inspire the teaching without destroying the scientific soberness."[27]

Again, I think you can extend this requirement to the teaching of any discipline, including science, if it is truly "teaching" with a passion and appreciation rather than just teaching *about* something. One does not just teach about theories; a scientist participates in a community that also fundamentally teaches science. Tillich ends this essay trying to keep the scientific and the religious enterprise together: "Thus in many realms of the scholarly work of a university the religious dimension is revealed, independent of a concrete religious tradition. University education should comprise both strict scientific discipline and the opening up of vistas in which the questions of life are seen in the light that scholarly endeavor sheds upon them. This is what our best students are looking for; they should not be disappointed."[28]

A Big History course has the capacity to be the type of endeavor that Tillich is suggesting our "best students are looking for," if we are clear about what we are teaching. We need to recognize that we do not only teach *about* Big History, but that we teach Big History as, at least, the functional equivalent of a modern religious perspective. What kind of "religious sentiment" is our Big History teaching communicating? The answer to this question will continue to unfold.

DOMINICAN'S APPROACH: AN
ONGOING DIALOGUE

Thanks to an open process of exchange of ideas in our innovative weekly Big History faculty development sessions and annual summer institutes, we have taken on the vexing question of how to educate our students and ourselves about Big History and its implications. We have endeavored to address the question of teaching religion, from which secular university education in general most often backs away. It has taken us four years to arrive at our current level of conversation.

Cynthia Brown has been our resident Big Historian. She has engaged the Big History faculty on the topic of religion and the perceived limitations of her own *Big History*, and those of the textbook she wrote with Christian and Benjamin, with great sensitivity. All our positions have evolved as a result of this crucial dialogue. A yearlong exchange of views with key faculty members was an illustration of open communication. Our dialogue has culminated in a thoughtful critique of the materialist approach of the early founders of Big History.

In addition, we have opened our campus to the discussion. We have invited guest speakers to provide alternative perspectives, in order to expand and deepen our conversation. The speakers have included sociologist Robert Bellah, cosmologist Brian Swimme, theologian Ted Peters, Big Historian Fred Spier, and a representative from the Vatican Observatory, Father Paul Gabor. We have hosted student–faculty panels on religion to address the "elephant in the room." We organized a Big History faculty retreat on the topic to help faculty appreciate the variety of perspectives on religion that exist among them, but which, without a safe environment in which to express them, had remained inchoate. These perspectives ranged from "religious naturalism," to orthodox theistic positions on the self-described religious or "spiritual" side of the conversation, to materialist and atheistic positions on the so-called "nonreligious" side. The religion question has also had a serious hearing at the International Big History Association (IBHA) conference. Our colleague Scott Sinclair proposed an anthology on Big History and religion in 2012; and at the 2014 conference, several panels addressed religion or meaning.

The result of all these activities has not led to complete agreement on how best to address religion in Big History. Nor has it led to agreement on whether Big History as a worldview should be taught as a functional equivalent for religion. Nor has it concluded whether we should teach Big History or just teach *about* Big History. At this stage of our journey, it has at least led to better clarification of individual positions on religion. We are at a new stage of development that is allowing us to continue to deepen the conversation. It has certainly forced me to be more specific about who I am as a religious being and how I should "teach religion."

A FEW KEY OMISSIONS OF RELIGION IN BIG HISTORY CONTENT

For students taking my "Through the Lens" course, I spend time reframing the Christian, Brown, and Benjamin textbook. A couple of problems must be addressed to make this perspective as coherent and consistent as possible.

First, the story as it is told in *Big History: Between Nothing and Everything* does not incorporate the latest theories from neuroscience about religion. Theories are emerging that suggest that humans, after our long evolutionary development, are "hardwired" not just for morality but also for religion, certainly for meaning creation and belief. Whether this rises to a scientific consensus or is an adequate formulation will no doubt take time to sort out. But to not treat it in what is purported to be the modern "scientific" creation story—including cosmic, biological, and cultural evolution—is a lacuna. In an approach that seeks a unified knowledge, this missing ingredient needs to be included.

Second, because the subtitle of the Christian, Brown, and Benjamin textbook is *Between Nothing and Everything*, the slim discussion of religion in human history is also a mystery. Even if one has a negative view of religion because of the often violent role various religious traditions have played in history, one cannot deny that under the banner of religion humans have also conducted their affairs to help them reach to their highest ethical ideals. There are also theories about religion as being "adaptive" in our evolutionary struggle. David Sloan Wilson's book *Darwin's Cathedral: Evolution, Religion and the Nature of Society* is one excellent book that expresses this thesis.[29]

Religion is often accused of finding its reason for existence in the "gaps" in our scientific knowledge. This is a fair criticism, as far as it goes. But taken on its own terms, science often fails to acknowledge moments where such gaps are explained by pure speculation. Often we are told that we do not really know why certain physical events happen, but it is only a matter of time before we will be able to understand. This mantra is not science; it is belief. It is a faith. There have been and always will be deep mysteries that will never be totally explained by our knowledge of science. Our students deserve honesty and clarity on this issue.

What I am suggesting is a different view of religion that is missing in the work of many of the Big History authors with whom we are most associated in our program: religion is part and parcel of what defines our humanness and cannot be separated out from the deepest sense of our being. My position also assumes that we have not outgrown religion and become a new type of human being. In fact, even in science our religious nature is being affirmed.

APPROACHING THE ELEPHANT IN THE CLASSROOM

Where does this leave us? What is the best approach to educating our students and ourselves on the topics of Big History and religion? First, I believe we should acknowledge that we are teaching a *quasi*-religious stance when we teach Big History. This is the fact of our modern life, and we must deal with it. Let us be clear that we are at this moment in this intellectual and religious space in the modern world.

There are viable options that I call *content* approaches. As long as we are clear and open about which approach we privilege, all are equally viable as points of departure for a critical discussion in the classroom. We could also present all of them as options, making clear which one or ones we have the most sympathy for. Here are examples of these types of content approaches. There are, no doubt, other possibilities. For each option I suggest that the topic of religion be integrated into the content.

In summary form:

1. *Science:* This approach stays self-consciously within the domain of science. But the emphasis is the branch of neuroscience that is emerging as neurotheology. This branch seeks to explain the development of our species as being "hardwired" for meaning. Here we would have to address the notion that even the idea of being hardwired may be too reductionist. The key works for teaching this approach include those by Andrew Newberg. I have used his coauthored book *Why God Won't Go Away: Brain Science and the Biology of Belief* in my classroom.[30]

2. *Symbolic Representation:* This approach gives a modern understanding of religion as the discovery of and search for meaning through the symbols we create to express this meaning. It is the product of religious thinkers who have gone beyond the literalism of various faith traditions in their conversation with science. In a sense this approach understands modern science as a demythologizer of religious creation stories, but then uses myth to

reformulate our relation to truth. The previously mentioned works by Paul Tillich and Robert Bellah would fit this approach. This is my approach. Needless to say, it is not an approach my literalistic-oriented students find compelling.

3. *Creation Stories:* An approach that is much appreciated by my students is that espoused by Ted Peters and his colleague Martinez Hewlett in their book, *Can You Believe in God and Evolution?*[31] This book tackles all the various religions' responses to neo-Darwinism, from those proposing intelligent design to theistic evolutionists. The book is clear about the claims of science, philosophy, and theology. I have used portions of this book to really get at the elephant in the room.

4. *Religious Naturalism:* Religious naturalism, as I define it, is when one finds meaning in nature itself or just finds "awe and wonder" in nature and the splendor of our humanness. Much more can be said about this approach. I am most drawn to a more sophisticated version of this approach that appears in the work of Brian Swimme, although he does not identify with this label. His book and full-length film *Journey of the Universe* best exemplify this approach.[32]

5. *Philosophy:* Philosopher of religion John Hicks's work helps tease out the implicit philosophical and religious stances that scientists often gloss over. His book *The New Frontier of Religion and Science* is a must read for all wanting to be in conversation with religion and science.[33]

6. *Science Fiction:* The ideas emerging from what is called "singularity theory" are worth mentioning. The basic notion is that what we now understand as our current humanness will evolve, resulting in a new kind of human being; that without this evolution—or emergence—we will not be able to cope with our current level of complexity; and that, far from our world being random, we will be able to master our future in more predictable ways. This is becoming a serious discussion within science. It needs to be engaged and not left out of the discussion (as much as I personally think it is another version of the hubris that has been with us throughout human existence.) The work of Stanford University archeologist and Big Historian Ian Morris is representative of this approach. His book *Why the West Rules—For Now: The Patterns of History and What They Reveal about the Future* deserves to be read by all teachers of Big History.[34]

There is a seventh option, which is methodological, rather than focused on content. It is what I would recommend from my own experience as the most appropriate approach, given that we are treating deep identity issues for our students. This approach can complement any and all of the above content approaches.

The point is to start with where the students are. We need to give them a safe space in which they may reflect on their religious position. In a sense, this would be a Socratic and inductive approach. The goal is to encourage them to discover their deep "religious" self. This discovery would include, but be more than espousing, the given "beliefs" of their received traditions. How do we do this? It is not easy. What has worked for me is to get students to articulate moments of profound meaning in their lives. When they see that they have ritualized these moments in other parts of their life and that they can tell stories about these moments, they seem to get that religion is not a separate part of their lives but is part and parcel of the totality of it. Further, they then begin to understand how these rituals and stories become systematized in rational thought, and then to better understand how their faith traditions have evolved.

In order to establish this safe environment, the instructor needs to allow students to express very early in the course the key issues in their minds and hearts. These issues may range from the idea that Big History is "destroying my faith" and "I don't believe in evolution" to "I am not religious anymore, but I am spiritual" and "I want to believe, but I find it difficult." These discussions also allow students with nonreligious perspectives to become comfortable with expressing their positions, especially in the "Through the Lens" course, where their position is in the minority. They need to feel safe to say that Big History has given them the coherence that they find meaningful. Or that they find science more worthy of their attention to address the complex issues of our future than other fields.

Inspired by Benjamin, we at Dominican have formed student-led "exploration / discovery groups." These groups allow the students to take the lead in defining what is important to them. They get to think together on the issue, exploring the various positions within their peer group in some depth. Eventually, Benjamin suggests, they help educate the whole class as they share how their positions have evolved or deepened, often with greater conviction and well-constructed arguments. It is at these various moments that they become more authentic about themselves as "religious" beings—or, in other terms, about "what it means to be human." With this method we are able to keep our focus on being respectful of our students' evolving identity.

Big History is among the most identity-demanding courses of a college career, and it challenges our students at almost every turn. Our students are faced with this challenge in their first year! Teaching this course is a teacher's dream, but it is also a great responsibility. The method I have outlined also demands that we as instructors be open to expressing our own identity. This is a crucial standpoint from which the course needs to be taught. We have to be willing to explore existentially the meaning of the course for our own religious stance. This is more than just a statement about our rationalized position, such as one of the following: agnostic, atheist, religious naturalist, or from a received and considered tradition such as Christian, Buddhist, Jew, Muslim, Hindu, Pagan, or Humanist. At best, what these labels express is the final journey in our self-understanding. But, too often, they express a quick, unexamined position or a hiding behind a label. The student wants to know how we got there and why. It is the same thing we are asking of them. Where are you now? How did you get there? Where are you moving with your new knowledge? We are asking them to explore and express their deepest thoughts. This method demands that we return in kind.

Over the years of teaching Big History and its relationship to religion, I have stumbled on this approach with the help of many colleagues. The ingredients of this four-year process have been: (1) discussing the scope and content of our Big History text with the help of our resident Big History author; (2) numerous faculty retreats on every subject in Big History; (3) debates about the meaning of materialism and religion among faculty and students; (4) students' courage to express their concerns about these issues; (5) faculty assessment processes that include focus groups with students; (6) writing a book about Big History pedagogy, which really means reflecting on the content we teach and what this says about how we should teach it; (7) well-organized extracurricular lectures on most aspects of Big History; (8) specific retreats on the religious views of the Big History teaching faculty; (9) my own research on a "constructive religious ethic" that was essential to clarifying my position on religion; and (10) specifically for me, the influence of Robert Bellah's book *Religion in Human Evolution*, which walks a complex route through cosmic, biological, and cultural evolution on the topic of religion.[35] In short, my own stumbling has taken place in the midst of a profound intellectual community seeking answers to this complex set of questions. If it takes a village to raise a child, it takes a large intentional community to address how to teach religion in the academy. Like all knowledge, it takes participation, as Tillich pointed out to the Harvard Overseers. Without participation in a collective learning community of Big History, this journey would have been impossible and sterile.

Science debunks only certain historic religious practices and stories that claim a literal truth. Yet even on its own terms, science also is beginning to explain that, for humans, "religion" will not go away. Science can also help us understand the origin of religion in our evolutionary past as rooted in such theories as parental nurturing. Religion can be understood as compatible with, maybe even complementary to, science—for each, in its most narrow and formal sense, addresses different realities and uses different methods. Further, in the reality of a unified life that we live every day, these realities overlap, for both religion and science tell us a story about the deepest meaning of our humanness. Though science is most comfortable with materialism and reductionism in the so-called physical world, and though this approach has given us many new physical insights and technological breakthroughs, human consciousness continues to escape this kind of reductionism. Many of my students have come to see an origin myth such as the Abrahamic Genesis story to be more than a mere physical science tract, but rather a theological and anthropological perspective. They begin to see that these old myths speak to them in new ways. The world is big enough for different approaches and perspectives.

The goal of our instruction is to explore these many perspectives in an honest and open way. Sorting out which content approach is most grounded in the totality of our humanness will take a method that fosters this conversation. We must recognize that even when we are teaching *about* religion we have assumed a religious stance. A religious perspective has been unwittingly expressed. Assuming that we can only teach *about* something at first seems like the most tolerant thing to do, because we do not want to privilege one religious tradition over another. Yet addressing religion in this fashion actually becomes the most intolerant thing to do. We can do better. The more we examine in an open and honest way any meta-narrative, the better for the quality and the usefulness of that meta-narrative. I would agree with David Christian's challenge in the opening sections of *Maps of Time*, when he says that he is offering a story as coherently as possible with hope that together we can improve on it. Let this be our challenge.

NOTES

1. David Christian, *Maps of Time: An Introduction to Big History.* Berkeley: U of California P, 2005. 1. Print.
2. Ibid.
3. Ibid.
4. Ibid., 2.
5. David Christian, Cynthia Brown, and Craig Benjamin, *Big History, Between Nothing and Everything.* New York: McGraw-Hill, 2014. 4. Print.
6. Christian, *Maps of Time,* 52.

7. Christian, Brown, and Benjamin, *Big History*, 14.

8. Ibid.

9. Christian, *Maps of Time*, 2.

10. Mark William Roche, *Why Choose the Liberal Arts?* Notre Dame: U of Notre Dame P, 2012. 160–162. Print.

11. Craig G.R. Benjamin, "The Convergence of Logic, Faith and Values in the Modern Creation Myth." *World History Connected* 6.3 (2009): n.p. Web. Accessed May 2013.

12. Ibid.

13. Ibid.

14. Ibid.

15. Ibid.

16. Ibid.

17. Paul Tillich, "Religion in the Intellectual Life of the University: The Faculty and Its Functions; the Dual Attitude; the Religious Dimension." A talk to the Board of Harvard Overseers, November 24, 1958. *Harvard Alumni Bulletin*, 17 February 1959. 298–299. Print.

18. Ibid., 299.

19. Ibid.

20. Ibid.

21. Ibid.

22. David Christian, "The History of Our World in 18 Minutes." TED Talks. *Ted.com*, April 2011. Web. 14 October 2013.

23. Paul Tillich, *Theology of Culture*. New York: Oxford UP, 1959. 5–7. Print.

24. Robert N. Bellah, "C. S. Lewis Foundation Faculty Forum, UC Berkeley, Remarks for Luncheon Event," 11 October 2003. Presentation.

25. Tillich, "Religion in the Intellectual Life of the University," 299.

26. Ibid.

27. Ibid.

28. Ibid.

29. David Sloan Wilson, *Darwin's Cathedral: Evolution, Religion, and the Nature of Society*. Chicago: U of Chicago P, 2002. Print.

30. Andrew Newberg, Eugene D'Aquili, and Vince Rause, *Why God Won't Go Away: Brain Science and the Biology of Belief*. 2nd ed. New York: Ballantine, 2002. Print.

31. Ted Peters and Martinez Hewlett, *Can You Believe in God and Evolution? A Guide for the Perplexed*. Darwin 200th Anniversary Edition. Nashville: Abingdon Press, 2008. Print.

32. *Journey of the Universe: An Epic Story of Cosmic, Earth and Human Transformation.* Host Brian Swimme. Prod. Mary Evelyn Tucker and John Grim. InCA and Northcutt Productions, 2011. DVD.

33. John Hick, *The New Frontier of Religion and Science: Religious Experience, Neuroscience and the Transcendent.* 2nd ed. New York: Palgrave MacMillan, 2010. Print.

34. Ian Morris, *Why the West Rules—For Now: The Patterns of History and What They Reveal about the Future.* New York: Farrar, Straus and Giroux, 2011. Print.

35. Robert N. Bellah, *Religion in Human Evolution: From the Paleolithic to the Axial Age.* Cambridge, Mass.: Belknap Press of Harvard UP, 2011. Print.

TWENTY-TWO · The Case for Awe

Neal Wolfe

> When I heard the learn'd astronomer;
> When the proofs, the figures, were ranged in columns
> before me;
> When I was shown the charts and diagrams, to add,
> divide, and measure them;
> When I, sitting, heard the astronomer, where he lectured
> with much applause in the lecture-room,
> How soon, unaccountable, I became tired and sick;
> Till rising and gliding out, I wander'd off by myself,
> In the mystical moist night-air, and from time to time,
> Look'd up in perfect silence at the stars.
>
> WALT WHITMAN, "When I Heard the Learn'd
> Astronomer"

In the 1949–1970 radio and television series *Dragnet,* police sergeant Joe Friday famously admonishes the witnesses he interviews to stick to the facts. Inevitably, he succeeds in catching the bad guy; his no-nonsense approach is perfectly suitable for police work, at least of the popular culture variety.

The Big History narrative is the ultimate in factual approaches. After millennia of constructing creative, often rather fanciful stories to explain how the world began and how humans entered the picture, we finally are able to construct a science-based account of the origin and development of the universe, which we render as objectively as possible. It is rife with facts; indeed, for those of us who teach Big History, a rather daunting challenge is to whittle down the number of facts in order for it to not seem to take 13.8 billion years just to tell the story. But Big History involves so much more than "just the facts." We and our students are human beings, learning about and sharing in not only the telling but the ongoing construction of the narrative. As human beings, we are responsive—the story moves us, shapes us, and, at times, awes us.

So, should the teaching of Big History concern itself strictly with the scientific, fact-based account of universal and human development? Or should we also address—and cultivate—our human response to this most extraordinary account?

Is it not our subjective experience, about which science can say little, that provides the basis for our shared humanity? That we humans, flawed as we are, have reached a point in our evolution where we can comprehend the universe as a whole and are able to construct a reasonable explanation of its workings is truly profound. Reflecting on this realization alone can evoke a response unbound by the factual knowledge that fuels it.

I have used the word "awe" to describe the response that often is experienced when we contemplate the astounding features of the universe, such as the Big Bang, the nearly incomprehensibly vast scale of space and time, and our relationship to the life cycle of stars. It could also be called amazement, or wonderment. The philosopher Martin Buber writes of the awe-filled experience as the falling away of the distinction between I and thou, or between the observer and the observed. Buber uses the example of a tree: When we look at a tree we can see just a tree as a distinct object separate from ourselves, as is customary. But sometimes, if we are patient and quiet and open, the distinction between self and tree dissipates, and we are subtly drawn into relational unity with the tree.[1] When this happens as we contemplate, say, the overwhelming magnificence of the night sky, we lose our habitual sense of individuation and are drawn into the experience of being part of that which we contemplate.

The experience of awe to which I refer is not necessarily a religious one—at least insofar as religion is an organized set of beliefs. Instead, I am addressing an experience of amazement in the face of an overwhelming stimulus—the night sky, our understanding of the immensity of the universe, the realization that if all the strands of DNA in a human body were strung end to end, the resulting thread would stretch to the sun and back as many as seventy times! The human response to these stimuli is fueled by factual knowledge about them—such as that when we gaze upon the starry firmament we are looking back in time, millions or even billions of years! This realization does not require a religious interpretation, even though it may lead to one for many people who bring that conceptual framework to the experience.

Despite their differences in approach, both science and religion seek the same end, and both spring fundamentally from the perceived mystery of existence. What accounts for the existence of the universe? Both science and religion seek to answer this most problematic of questions. Ultimately, neither can succeed, in my view. No matter how many layers of mystery are removed by science, whatever remains is still stumped by the simple question, "Why this?" Why is there anything at all, rather than nothing? A religious belief in a creator as the first cause runs into a similar problem. How did this creator come into being? What is the evidence? In the end, what's left is the mystery—a wondrous, awe-inspiring mystery at the heart of all that exists. No matter which side of the balance one tends to favor—science or religion—our account of the universe's origins, development, and marvels leaves us in amazement. The Big History narrative, in addition to its many other significant educational features, provides a perfect opportunity to engage our students in that amazement. And we, as educators, can encourage their amazement by modeling our own.

Scholarly research into awe as an emotional state turns up some interesting observations. Social psychologists Dacher Keltner and Jonathan Haidt assert that two fundamental features are present in all cases of awe: perceived vastness and accommodation.[2] Vastness applies obviously to such things as the immensity of the scale of the universe, a central interest of Big History, but also to anything that appears larger than the self, or the self's ordinary frame of reference. This includes unequal social status, such as awe of royalty and fame, seen today in our cult of celebrity. The second characteristic of awe identified by Keltner and Haidt, accommodation, refers to the necessity created by "an inability to assimilate an experience into current mental structures."[3] When we gaze at the night sky with interest, we are easily dazzled by what we behold; when our gaze is informed with knowledge about what we are seeing—how stars are formed, how far away they are, that we are looking millions or billions of years into the past—such as students develop in our Big History classes, our amazement deepens. When we reflect that we owe our very existence to the life cycle of stars—that we are, in a real sense, "stardust"—a measure of awe arises quite naturally. How do we grasp this understanding in any ordinary sense? We are overwhelmed, and we must conceptually accommodate something so grand it shakes us from the relative lethargy of our habitual everyday sense of who and what and where we are.

Of particular interest to us as Big Historians, Keltner and Haidt assert that what we call awe is one of the "primordial emotions," the "hard-wired pre-cultural set of responses that were shaped by evolution and built into the central and peripheral nervous systems of the human species."[4] This particular emotion, they argue, was developed as a response of submission to powerful individuals, a survival trait also seen in other primates. This trait of submission to perceived power, Keltner and Haidt argue, applied even to the relatively egalitarian hunter-gatherer societies, and it can be generalized as the basis for our response of awe to other powerful stimuli to this day. So, they say, we are biologically predisposed to the emotional state of awe.[5]

In many cases—perhaps most—we experience awe in response to nature, whether at the ocean, in a forest of towering redwoods, or on a mountain peak. In his mid nineteenth-century essay "Nature," Ralph Waldo Emerson writes: "Standing on the bare ground,—my head bathed by the blithe air and uplifted into infinite space,—all mean egotism vanishes. I become a transparent eyeball; I am nothing; I see all; the currents of the Universal Being circulate through me; I am part or parcel of God. . . . I am the lover of uncontained and immortal beauty."[6]

Of course, not all people experience such a sublime reaction to natural beauty automatically. For many—perhaps all of us, in some fashion—such a response requires cultivation. So it is with our students. It is a matter of cultivating perspective. Perspective is illustrated by the following story about two bricklayers who are helping to build a church. "A visitor asked one worker what he was doing and he grumpily replied, 'I'm laying bricks. What does it look like I'm doing?' The visitor walked around to another part of the building and asked the second worker the same question. This worker stood, looked toward the heavens, and said, 'I'm building a cathedral to the glory of God.'"[7]

The renowned psychologist Abraham Maslow had a particular fascination with awe-inducing experiences, which he termed "peak experiences"—experiences of being in complete harmony with one's surroundings. Maslow felt that education should include what he termed "being cognition," which places learning on the largest possible scale. For him, that meant eternity, or that which transcends the particular. Maslow observed that a student can be taught to recognize the individual notes of a Beethoven quartet, but how does one teach the student to hear the *beauty* of those sounds?

As teachers of Big History, we tell the profound evidence-based story of how the universe as we know it began, developed, led to our emergence as a species, culminates in the present moment, and continues unfolding into the future. What ramifications does this suggest for our identity as products of the universe? As the poet Walt Whitman put it so compellingly, "Immense have been the preparations for me. I am an acme of things accomplished, and I am an encloser of things to be."[8]

Just consider, for a moment, that as a species that is still relatively young, and clearly has major flaws, we have achieved the extraordinary ability to take a reasonable, scientifically based stab at explaining how the universe began, how large it is, and how it works!

As if that isn't enough, consider that, if we accept the Big Bang as a reasonably accurate theory, then everything that presently exists (or ever will exist) was, implicitly at least, present in the very beginning. This perspective was advanced in ancient religious literature, such as the Bhagavad Gita, the great two-thousand-year-old Hindu text, in which Lord Krishna admonishes the warrior Arjuna, "Never was there a time when I did not exist, nor you . . . nor in the future shall any of us cease to be."[9]

While visionaries such as Walt Whitman and the author of the Bhagavad Gita understood these things conceptually, by the late twentieth century, science essentially confirmed their assertion of the non-individuated permanence of all phenomena. I don't think any adjective can adequately capture the profundity of such an achievement. We have placed ourselves at the heart of the universal narrative. Such awareness clearly goes well beyond "just the facts." We find that the universe is not just something "out there" that we may know about intellectually (from a student perspective, enough to answer questions on an exam, at least) but that it is what we are *actually* part and parcel of. One of my students put it simply in a written response to the stargazing event we held on campus last fall: "I realized that I am a part of something much bigger than myself, my life, and even my planet." It is important, obviously, for our students to learn the facts, but such awareness is just as important. If students experience themselves as part of a magnificent mystery, which exists on a scale nearly impossible to grasp, they will likely understand whatever facts they assimilate now and in the future in a much more expansive context.

So, how do we apply this perspective in the classroom? How do we engage our students in amazement, in awe, especially when we are not outside gazing into the vault of the night sky, when the experience might be a bit more readily accessible? We need to stop and have conversations with our students about their response to what they are learning and engage them in reflective writing. But more importantly, we need to allow ourselves to be in amazement ourselves and to model that amazement in the classroom. Our students are deeply influenced by us—by our tone of voice, our posture, our energy. We need to lay our factual bricks carefully and in proper order, but we must also remain aware that we are constructing a narrative cathedral that inspires wonder and awe. After all, we live on a planet orbiting a star in a spectacular luminous spiral galaxy that is one hundred thousand light years across!

NOTES

1. Martin Buber, *I and Thou*. New York: Scribner's, 1958. 19. Print.
2. Dacher Keltner and Jonathan Haidt. "Approaching Awe, a Moral, Spiritual, and Aesthetic Emotion." *Cognition and Emotion* 17.2 (2003): 297–314. Print.
3. Ibid., 297.
4. Ibid., 306.

5. Ibid.

6. Ralph Waldo Emerson, *Nature; Addresses and Lectures*. Boston: Munroe, 1849. 8. Print.

7. David N. Elkins, "Reflections on Mystery and Awe." *Psychotherapy Patient* 11.3–4 (2001): 164. Print.

8. Emerson, *Nature*, 77.

9. Lars Martin Fosse, trans. *The Bhagavad Gita: The Original Sanskrit and an English Translation*. Woodstock: YogaVidya.com, 2007. 14. Print.

Conclusion

On a hot Northern California afternoon in late May 2011, some thirty Dominican faculty members sat in the round, in a redwood-paneled ground floor classroom called Meadowlands Hall. We were deep into our second Big History summer institute, and we had broken into groups to brainstorm. What were our wishes for our young program? What did we want students to take away?

Each group of four or five had a sheet of big white paper and a few felt-tipped markers. It was late in the day, late in the session, and we were punchy and loose—full of new information and refreshed with new understanding of what we were doing. When we rejoined the larger group, we hung our posters on the old wooden walls of the low-ceilinged lecture hall. Afternoon light shone through the windows' stained-glass tops. We compared notes; many of our goals were the same. We categorized them: critical and creative thinking traits (ability to deal with nuance and ambiguity, openness to multiple perspectives, willingness to challenge one's own assumptions, ability to cope with cognitive dissonance, ability to evaluate arguments and relate information to larger, diverse concepts); meta-cognitive outcomes (ability to synthesize new ideas with one's understanding of one's place in the unfolding universe, intellectual and emotional capacity to grasp the concept of change); and synergistic outcomes (ability to distinguish knowledge from opinion by reference to evidence; large-scale context for future learning; broad familiarity with the story; understanding of themes, particularly thresholds and complexity; understanding of the place and impact of humans; ability to read, think, and research; a move toward thinking independently.)

From this long list, we synthesized, as a faculty, our student learning outcomes for the program:

THE GOALS OF THE FIRST-YEAR EXPERIENCE PROGRAM

The program is designed to promote

- critical and creative thinking in a manner that awakens curiosity, enhances openness to multiple perspectives, and increases willingness to challenge one's own assumptions;
- recognition of the personal, communal, and political implications of the Big History story, including insight into the interdependence between humans and the environment, and modes of positive empowerment to effect change; and
- development of reading, thinking, and research skills, enhancing one's ability to evaluate and articulate one's understanding of one's place in the unfolding universe.

It felt like a revelation, a dream agenda. It was ambitious, innovative, and exciting. It spoke to what we all wanted to accomplish as educators, what we wanted for our students.

Four years later, our assessment shows that it's working. Our students report that they understand the world differently. They perceive the connections among their various courses and the larger context in which their studies in their majors make sense. They are bringing their Big History understanding into their other classes—so that discussions and academic work throughout the university are informed by this larger context. And they are attuned to the future and to their own agency in shaping it.

So, to conclude, we'd like to reiterate what we think are some of the most important lessons for the Big History teacher. We'll revisit our approaches to teaching Big History in the classroom—what is the essential content and what are the essential approaches to teaching it? Then, we'll make a few recommendations for those who are interested in building a Big History program. Finally, we'll address a few challenges we've seen posed for this young field of study.

RECOMMENDATIONS FOR USING BIG HISTORY IN THE CLASSROOM

COMPLEXITY

It is crucial for students to understand complexity, because complexity is the conceptual underpinning beneath the thresholds model and beneath the Big History narrative. When students can comprehend a few concrete examples of what complexity is, and how its features manifest in familiar ways, they'll have an easier time visualizing and understanding the thresholds of complexity framework.

We can use what we think of as systems to illustrate what we mean by complexity and how it functions. Start by introducing the four features of complexity. Define them. Explain them at the atomic level. Start with a hydrogen atom. Once students begin to get their heads around the key features, the basic concepts, then you can ask them to think about the four features of complexity for the system in which they are the key components—the campus. What are its diverse components? What are the precise arrangements in which those components are arranged? How does energy flow through the campus? And what new emergent properties arise as a result? Start with the four features of complexity at the beginning of the Big History course, and then check back in with them as you cross each threshold (or, at the beginning and end of each unit).

When students recognize recurring patterns at every level of reality, at every new threshold or in each regime, they'll begin to understand that similar processes are at work throughout the universe, across time. It will certainly allow them to understand that the intellectual and material connections among what have long been treated as separate, unrelated, siloed academic disciplines are real, organic, and quite logical.

THRESHOLD I: THE BIG BANG

Begin your journey into the story by comparing it with other cosmologies. Students see that all origin stories fulfill common functions—to explain where the world came from, where people came from, and what our relationship is with plants and animals. They also see that the Big History story does a few things that other creation myths might not (such as quantify time).

Next, tell the story: the universe began some 13.8 billion years ago, when an atom-sized space, filled with all the matter and energy in today's universe, appeared out of nothingness and began very, very rapidly to expand—creating both time and space—and to cool, creating distinct matter and energy, as well as gravity, electromagnetism, and the strong and weak nuclear forces. The universe expanded to what may have been the size of a galaxy. Matter and energy stabilized into a plasma of protons, electrons, neutrons, and photons. The universe existed in this state for about 380,000 years, after which time it cooled enough that previously heat-jostled protons, neutrons, and electrons could succumb to electromagnetism and nuclear forces and begin to bind together and form atoms of hydrogen and helium. The resulting net neutral charge released the photons from the other particles' electromagnetic grip, resulting in a "huge flash" of light—as newly freed photons shot off in all directions.

Don't forget to lay out the key evidence: the cosmic background radiation that is evidence of the early universe's structure and formation; Hubble redshift, which proves the universe is expanding; the apparent age of all objects in the universe (13 billion years); change over time, observable using space-based telescopes; the abundance of hydrogen and helium, which appear to compose most of what we see, aligning with the theory that the universe formed so quickly that only the simplest elements would have had time to form.

Some students whose belief systems are informed by other ways of knowing—other cosmological epistemologies—may feel that their core beliefs are threatened by the scientific way of knowing about the universe. Be respectful of the traditions that often form the core of their identities. We want to challenge students, not to undermine them.

THRESHOLD 2: THE FORMATION OF STARS AND GALAXIES

The tangible quality of what we study in Threshold 2 and our intuitive understanding of gravity-driven processes present an opportunity to connect with those students not yet fully on board for the journey. Gravity is key in the evolution of complexity because it facilitates the first structures in the universe and is essential in creating energy gradients. The universe 380,000 years after the Big Bang was an expanding cloud of matter with only tiny variations in density. In areas of higher density the gravity was slightly more powerful, and the inward forces pulled these regions together. As density increased further, so did the force of gravity, and the process accelerated. After a few hundred million years, matter had clumped into protogalaxies and within these formed the first stars.

It is important to remind students that the universe is still expanding on the largest scales. Gravity can hold galaxies together, but galaxy superclusters are all moving away from one another; this is the expansion that was observed by Edwin Hubble. The cosmic background radiation data is a key piece of evidence in our understanding of how the universe developed its first structure. Show students the WMAP image and explain that the small differences in temperature correspond to differences in the density of the early universe.

Stars are formed by clouds of gas of sufficient density to undergo gravitational collapse. As matter falls inward the atoms smash together more and more violently and the gas begins to heat up until fusion begins. The continuous extreme heat and pressure inside stars is ideal for the creation of greater chemical complexity. The relevance of stars to the generation of chemical elements in Threshold 3 is key to this lesson. Students should also understand the significance of the immense energy sources now scattered throughout the universe. These are of central importance for the development of further levels of complexity—in particular, the Goldilocks conditions near one of these hot spots supported the evolution of life and human culture on planet Earth.

THRESHOLD 3: HEAVIER CHEMICAL ELEMENTS AND THE LIFE CYCLE OF STARS

Threshold 3 is all about understanding what stars are. When students understand stars as massive fusion reactors busily cranking out new elements, they'll view the night-time sky completely differently—and they'll understand where all the *stuff* comes from. Tell them (as David Christian suggests) that the gold or silver on their fingers or in their earlobes was forged in a supernova. This type of direct, palpable connection to the distant stars is a useful and unforgettable way to engage students in the material.

It's important in teaching Threshold 3 to refresh students' understanding of basic chemistry—atoms, protons, electrons, neutrons. Next, explain the processes at work inside a star—fusion, gravity, and stellar nucleosynthesis. A star is a mass of burning plasma caught in a tug-of-war between gravity and the ongoing fusion explosion (which, to begin with, converts hydrogen to helium) in its core. This explosive reaction releases photons—packets of pure energy, which travel slowly to the star's surface and then are freed into space, as light. When the star exhausts enough of its supply of hydrogen that fusion can no longer stand up against gravity—and its core is thus full of helium—the star collapses. When this happens, of course, such high pressure and temperature are generated in the core that the mass of helium begins to fuse into carbon—and the cycle begins anew.

The cycle continues on up the periodic table, fusing bigger and bigger atoms together until, in the largest stars, the core fills with newly minted iron. At this stage, when the star's fuel runs out, it collapses and explodes one last time, as a supernova, which is so hot that it creates not only neutrons, but all the rest of the naturally occurring elements up to uranium. In the force of the explosion, these new elements are all blasted off into space, someday perhaps to accrete into new stars, or planets of rock, or gas, or ice, or all of the above.

THRESHOLD 4: THE FORMATION OF OUR SOLAR SYSTEM AND EARTH

With Threshold 4, our grand narrative becomes more personal. The story has now arrived in our neighborhood; we are residents not just of Earth, but of a system of planets orbiting one of the stars that students learned about in studying Thresholds 2 and 3. To engage students, take them outside and help them situate themselves in relation to the sun and to the Earth. Our place on a planet in a solar system is more easily made apparent when one can see the sky and the sun. From there, one can discuss solstices and equinoxes as real events to be understood in relation to our orbit around the star. In addition, it is important to explore the history behind the collective learning that has led to our current understanding.

The content can be divided into discussion of the solar system, its formation, composition, and structures, and the Earth, its formation, composition, and structures. According to the prevailing nebular theory, just over 4.5 billion years ago, part of a massive molecular cloud collapsed as a result of gravity, possibly triggered by shock waves from a nearby supernova explosion. As the new star formed and planetary accretion took place, the components of the solar system—sun, planets, asteroids, comets—took shape, with the four smaller, terrestrial (denser) planets nearest the sun, and the four larger, gaseous (less dense) planets farther out. Similarly, within the newly forming Earth, intense heat caused the heavier materials to sink to the center, while the lighter materials rose to the surface, so that the newly forming Earth differentiated into separate layers—core, mantle, and crust. This structure drives plate tectonics; latent heat from the Earth's formation would go on to contribute to the origins of early organic compounds. Thus, with the formation of Earth and its privileged "Goldilocks" position relative to the sun—not too hot, not too cold, just right—the conditions were set for the emergence of life.

THRESHOLD 5: THE EVOLUTION OF LIFE ON EARTH

When teaching students about Threshold 5, the evolution of life on Earth, begin by getting them to think about the characteristics of living things. What makes living things different from nonliving things? They use energy to maintain themselves. They reproduce. They change; they evolve. A discussion of what it means to be alive helps students understand how unusual life is. The second law of thermodynamics states that everything in the universe is moving toward greater entropy, or randomness. But life is not like that at all. It is complex and getting more complex with time.

Next, discuss the process of evolution by natural selection. Go back and review the idea of increased complexity and what it means to say that something is complex. As life becomes more complex, it moves from heterotrophs to autotrophs, and then from prokaryotic to eukaryotic cells. Eukaryotic cells have a very important emergent property: they can clump together to form multicellular organisms and thus the vast diversity of species that we see today.

To conclude the teaching of Threshold 5, and to connect Threshold 5 with Threshold 6, summarize the history of life on Earth using the eight stages of life described in Christian, Brown, and Benjamin's book *Big History: Between Nothing and Everything*. Stage 1 consists of the first living organisms, which are thought to be similar to today's bacteria or prokaryotes. Stage 2 saw a new way of obtaining energy through the process of photosynthesis. It was at this stage that oxygen began to accumulate in the Earth's atmosphere. This abundance of oxygen led to Stage 3, in which organisms began to more efficiently break down food through the process of respiration.

During Stage 4, eukaryotic cells developed sexual reproduction. In Stage 5, these sexually reproducing eukaryotic cells became capable of clumping together to form multicellular organisms. The first vertebrates evolved during Stage 6. During Stage 7 we see the first organism venturing out onto land. Many organisms went extinct some 245 million years ago, resulting in the subsequent evolution of two important groups, dinosaurs and mammals, during Stage 8. A second large extinction occurred 65 million years ago, resulting in the elimination of the dinosaurs, which permitted mammals to evolve to fill many of the niches that had previously been occupied by dinosaurs. Some of these mammals eventually gave rise to our human ancestors.

THRESHOLD 6: THE RISE OF *HOMO SAPIENS*

The most significant part of teaching Threshold 6 is to get students to understand that the emergent property of this threshold is human culture; that because of humans' ability to learn collectively, human culture has now taken precedence over biological evolution; and that the next thresholds of agrarian and industrial civilizations were built on the foundations of this pre-ancient threshold.

The basic historical content of Threshold 6 is most approachable by framing the whole period as a narrative with three smaller turning points or mini-thresholds. The first stage is hominine evolution, occurring from 8 million to approximately 200,000–250,000 years ago. During this period, humans and chimpanzees evolved from a common ancestor and early forms of the genus *Homo* began appearing, roughly coinciding with the Pleistocene epoch. The appearance of a new species, *Homo sapiens*, in the Middle Paleolithic, sometime between 250,000 and 50,000 years ago, marks the second mini-threshold. Here we can explore the distinctiveness and uniqueness of humans in ways that open up classroom discussion. The third mini-threshold of Threshold 6 is the Paleolithic era, dating from approximately 200,000 years ago until 10,000 years ago, and the emergence of the Holocene, a new geologic epoch. The addition of this undertaught period of human history to the "world history narrative" is critical, because it is when humans realized their full potential physically, socially, technologically, and linguistically, developing powerful hunter-gatherer cultures and societies on all the continents except Antarctica. Not only is the Holocene the foundation of all subsequent world history; it constitutes 95 percent of all human history. Human history should not be separated from this earlier threshold, lest humans become disconnected from the larger story of their evolutionary past.

THRESHOLD 7: THE AGRARIAN REVOLUTION

Learning about agricultural activity and its implications is the most valuable lesson for students from Threshold 7. This is more manageable, pedagogically, than a survey of the achievements of the large number of civilizations that came and went between 2500 B.C.E. and 1800 C.E. The useful achievements of the agrarian civilizations can be quickly summarized in the advancements in astronomy, engineering, weapons, shipbuilding, and architecture; in each of these areas there was increasing sophistication during this period. But too much focus on the achievements of individual human civilizations diverts one's attention from the agricultural activity that underlay and supported all the agrarian civilizations.

The period between 2500 B.C.E. and 1800 C.E. should be seen as "humans with command of agriculture." The focus of the agrarian civilization portion of Threshold 7 should be on the forms and necessary infrastructures of agriculture, as well as its impacts and consequences, including increases in population; agriculture's impact on the biosphere (extinctions, forest loss, growth of cities); the lack of progress in real science and life expectancy; the continued emphasis for women on childbirth and rearing to the exclusion of other activities—but also the valuable advances in collective learning and the closing of the human web.

THRESHOLD 8: MODERNITY AND INDUSTRIALIZATION

Threshold 8 is where it all comes together for students. This is where the story is the most familiar, so they become more and more engaged—they will be seeing how the patterns we've observed throughout the 13.8-billion-year story have shaped the world in which they live right now, especially as they see how recent history fits into the larger scheme.

You might run through the transition from the age of agrarian civilizations to the very complex developments of the last 250 years or so. Networks of exchange developed in isolation across the three major world zones. After the Americas were "discovered," the globe circumnavigated, and the world connected for the first time in one massive, global network, collective learning and innovation spiked.

The central action of industrialization is that machines driven by fossil fuels produce wealth and leaps in human well-being. They also generate waste, and thus a host of environmental problems: toxic pollution of water, air, and soil, as well as carbon dioxide–forced global warming, which is already causing dramatic changes in Earth's climate, oceans, and biosphere. The industrialized warfare of the twentieth century led to unfathomable destruction but also to remarkable innovations: the splitting of the atom, the discovery of space, and the development of the internet, which is able to connect all the world's people in real time, via the mobile devices students have in their pockets or their pocketbooks.

A good way to approach Threshold 8 is to use the four features of complexity as an analytic tool. Allow students to talk about something they know well—their campus. Students will instinctively draw connections to the larger global civilization. Think about what the most important concepts are that you want to be sure to get across. Don't try to do everything. The most important thing is to leave ample time for them to reflect on how trends that we see across the massive scale of Big History—and in our own time—might shape the future.

THRESHOLD 9? POSSIBLE FUTURES

Students feel the greatest ownership of the content of a Big History course when it addresses matters of meaning or belief; when it stresses the interconnectedness of the disciplines; and when it foregrounds the future. Because there is no set of events or facts that can be ascribed with certainty to the future, think in terms of key concepts and approaches that allow you to address the problems of considering what the future could look like, and give students agency in thinking about the futures they'd like to see. Divide processes into those not influenced by humans (e.g., increasing expansion of the universe) and those influenced by humans (e.g., deforestation). Discussion of the two categories can be divided into three time frames: (1) the near future, about one hundred years out; (2) the middle future, one thousand to ten thousand years out; and (3) the remote future, one million to several billion years out. These can also be covered in reverse order so as to emphasize human agency in the near future.

Draw the important distinction between the future and a new threshold. Ask students to try to make sense of the patterns that progress through the thresholds, as well as those that recur. Have students attempt to extrapolate a possible future or a Threshold 9 in terms of the four features of complexity: more and more varied components, new arrangements, markedly higher energy flows, and new emergent properties.

For many students, the future is the most important part of the Big History story. Certainly, it is the most personal and the most relevant to them as individuals, and also potentially the most fun. Use science fiction texts—novels or films or even video games—to help them *see* some possible futures. Be careful to emphasize the positive so they are not overwhelmed by the challenges, which are easy to dwell on. Consider backloading your syllabus to foreground the future: build extra time into the end of the course, to ensure that if you run long in earlier units, you don't run out of time to consider the future deeply and thoughtfully. Consider teaching *each* threshold with an eye to its effect on the future. And leave ample time for students to apply their new Big History understandings to envision what might come next—and how to act accordingly. This is where they get to think about the sweep of the entire narrative and draw important conclusions that tie the course together.

OTHER KEY APPROACHES

REFLECTIVE WRITING

Throughout the course, it is important to get students talking about themselves. The practice of reflective writing enhances content retention, the achievement of learning outcomes, and understanding of "big picture" frameworks, and it enriches active and experiential learning or "real world" applications. It also can stimulate productive discussion and enhance students' understanding of the content's relevance to their own experience.

Embed reflective writing, and even specific questions, in the course syllabus. Identify specific learning objectives and develop reflective questions that prompt students to address them. Initiate reflective writing before a discussion or after a learning activity. Allow students time in class to "free write" in response to a clearly articulated prompt. Offer them additional time to reread their work and highlight their best ideas. Finally, request that students keep all their written reflections together for future reference. Free writing can produce valuable ideas for upcoming research papers or presentations. It helps students to internalize the writing process and enhances their ability to organize their thoughts on paper. And because writing is a solitary practice, it enables more reserved students, who might be reluctant to participate in class discussions, to express their ideas safely and at their own pace.

ACTIVE LEARNING AND ACTIVITIES

The active learning and associated activities that are described here have gone through rigorous vetting by our faculty and students. These activities are based in the assumptions that students have read, viewed, or listened to a common text in association with the activity, and that students have reflected (frequently in writing) on that text before experiencing the activity. Following the activities, process questions or prompts facilitate further connections and reinforce learning. Use reflective writing assignments in class or at home to help students connect the activities with the course content from textbooks. And, by all means, develop your own activities. You know your students and what will engage them. Be creative.

ADDRESSING BIG HISTORY AS A "WORLDVIEW"

When we teach "Big History" as the modern creation story—as a worldview—we confront students with the issue of religion in multiple ways. The first major issue to address is the general disorienting element of the Big History approach for both students and faculty. As Big History instructors, we must be clear about our own religious—or nonreligious—starting point. It is not enough to say that we do not want to unduly influence our students with our point of view. By teaching Big History with our current texts we are already influencing students in a certain direction in respect to worldview, and thus to religion.

Students appreciate a chance to engage on these issues. Create a safe space in which they may reflect on their religious position. Consider forming student-led discussion groups; allow them to take the lead in defining what is important to them, to think together on the issue, and to explore the various positions within their peer group in some depth. What emerges will help educate the whole class. With this method we are able to keep our focus on being respectful of our students' evolving identities.

Whichever stance we take, we need to be open and honest with our students. We are asking them to deal with deep identity questions; it behooves us to share our positions in a thoughtful way at the same time.

RECOMMENDATIONS FOR BUILDING A
BIG HISTORY PROGRAM

CURRICULUM REVISION

The story of Dominican's Big History program is the story of collective learning and community. Any institution will likely approach curricular revision with a mix of apprehension and a small dose of eagerness. Yet, if done right, curricular change can build community, strengthen cross-campus ties, and change institutional culture for years to come. The key is in involving all stakeholders, from the faculty to the staff and administration to the board of trustees. The level of involvement of course depends on the depth and expanse of revision, but being as inclusive as possible is always a goal to strive for in academia. And, where a school or university decides to offer a Big History course that lacks an obvious disciplinary home, this building of an academic community is a necessity.

When a group or committee has been empowered to undertake such a revision, encourage them to work in the open rather than isolating themselves. Have them invite colleagues and experts to provide new and creative ideas, to share practices previously unknown at the institution, and even to introduce ideas and models seemingly too ambitious. Have these guests present to the entire community rather than just the panel or committee. Solicit feedback and brainstorm together.

To build a Big History program, throw a wide conversational net and invite any faculty members who might be interested, regardless of discipline and status. Remember that, independent of the hierarchies and decision-making structures at a school, institution, or even a national curricular board, the actual teaching of Big History is a grassroots decision and begins with an interested teacher.

FACULTY DEVELOPMENT

The faculty development program, our practical application of the theory of collective learning, continues to evolve. The Big History faculty are comfortable knowing we need to continue to learn. The tiers of our process, which unfold over the course of the year, are

- annual summer institutes;
- weekly lunch meetings during the academic year;
- a daylong retreat at the end of each semester; and
- the development and implementation of quantitative, qualitative, and anecdotal assessments.

These steps illustrate the bottom-up faculty process that has grown from a genuine excitement about what we are doing. And they provide options for an institution planning to adopt Big History on a small or large scale.

At our summer institutes we use a combination of our own resources—ourselves—and lecturers invited from outside the campus. During weekly meetings we cover matters both practical and pedagogical. Daylong retreats at the semester's end involve some formal assessment and discussion of issues that have come to the fore. In summer, when preparing for the upcoming year, we use the results of formal assessments to change and improve courses.

While some of what is in this book is specific to our own institution, the lessons here about constant faculty engagement, transparency among colleagues, buy-in from the bottom up, our own collective learning, and our willingness to be adaptable, to experiment, and to not be experts can be instructive to most. Our experience can be adapted in pragmatic ways at other schools.

SOME CHALLENGES FOR BIG HISTORY

Certainly, Big History, as a young field, faces challenges and questions. Let's address a few of them in the context of teaching Big History.

COSMOLOGY AND EPISTEMOLOGY

Big History changes one's understanding of the universe, and so it challenges previous ways of understanding the nature of the universe—including traditional religions. Big History does not seek to supplant religions; religions are a different way of knowing and of finding meaning. Yet Big History, in explaining the origins of the universe and of humans in narrative form, does seem to compete with other worldviews in areas that have traditionally been their domain. This certainly causes some friction.

Big History will likely face resistance where the key pieces of observed, evidence-based knowledge that are its key components have met resistance throughout human history. Among these are, of course, a universe in which humans are not at the center; a solar system in which the Earth is not the center; Darwinian evolution via natural selection, in which humans are not necessarily the preordained apex; and anthropogenic global warming, in which humans have unwittingly seized control of the atmosphere, causing major disruption to Earth's climate systems. All these ideas have been opposed by existing political power structures, be they church, state, or corporation. All these ideas have challenged an existing accepted human understanding of how the world works. Yet all have survived and been largely accepted as material, provable fact—even, eventually, by those entities that had at first sought to suppress them.

Again, our approach, pedagogically, is to deal with such questions openly. These questions are among the most profound for students and among those they are most eager to discuss.

Interestingly enough, among students, hard scientists can be among the most resistant to the Big History framework. Some scientists worry that Big History could replace a solid science education, that it would lead to a loss of a granular understanding of the sciences. But Big History is not designed to replace science, in the same way that it is not intended to replace world history or national histories. It is a framework, a structural underpinning, that allows us to make sense of our disparate disciplines of human endeavor. What we study in a basic physics or chemistry class makes sense in the context of this framework. And a PhD-level understanding of astrophysics or biochemistry makes sense in the context of this framework.

Big History should never replace an education in science, or in history. It should enhance it and strengthen it, by making clear the very real threads that run through the material reality in which scholars in all disciplines are expert.

QUESTIONS ABOUT COMPLEXITY

Some scholars find the complexity model to be problematic. In complexity, as in the earlier thresholds, concepts developed by hard scientists are being interpreted by historians or other nonscientists. Some detail is lost in translation, and some concepts may be oversimplified.

In addition, the thresholds are not bright-line physical boundaries in time / space. They have been purposefully chosen for emphasis by Big Historians as the most important emergences in the single thread of the story of the universe that leads to contemporary, industrial *Homo sapiens sapiens*. These are the developments that changed everything, in ways that ultimately gave rise to us.

Even the concept of a threshold is an imprecise metaphor. A threshold is a doorway between rooms, yet the way Big Historians use the term "threshold" implies both the doorway and the room itself. And some of the big, threshold-making changes emerge over millions or billions of years. Terms such as "regime" or "transition" are used by some Big Historians instead—but they can be too expansive or imprecise to zoom in on some key moments ("regime" seems to indicate space, while "transition" connotes time; "threshold" at least balances the two).

Again, there is the fundamental question of whether the trend toward increasing complexity is a real phenomenon in the universe or merely a human-imposed framework that allows us to connect and make sense of this story. There is also the question of whether, then, the patterns that Big History illuminates are actual patterns in the fabric of the universe or imagined recurrences imposed by our human way of thinking. Was there nothing, and then a Big Bang that lit the fires of the universe? Or does even science's perception of the origin of the universe bear the imprint of the Genesis story?

We find that the very consideration of such questions is fundamental to the enterprise of human inquiry, and that the practice builds critical thinking skills.

In addition, the four features of complexity model is a useful analytic tool for understanding complex systems of any sort, from stars to ecosystems to college campuses to mobile devices such as smartphones and tablet computers. What are the key components? How are they arranged? How does energy flow through them? And what new properties emerge? Asking these questions can lead not only to rich and energetic classroom discussion but also to intuitive and nonideological answers to such otherwise complex and challenging questions as: How will humans use energy in the future? Or, what is political power, and how does it work? And, for that matter, what are the relationships among those two questions and their answers?

ANTHROPOCENTRISM

Another interesting problem of Big History is that it is not, in its entirety, an objective history of the entire universe but is, rather, necessarily anthropocentric—and, at this point, terracentric. Big Historians readily point out that Big History does not *really* tell the story of the entire universe—its bias is that it tells the 13.8-billion-year story of the universe as the story of contemporary *Homo sapiens*. It has that, at least, in common with old-fashioned state histories, which privileged the tellers.

There may be other life in the universe, but we can only tell the story of life on Earth. Surely, there are long, unknown histories even within our own solar system, but we have not yet uncovered them. We don't know the history of the future into which we are headed. The story of humans on Mars has only barely begun. Ultimately, we can't even view our own galaxy objectively—we can see it only from within and extrapolate through observation of other galaxies, which we believe to be similar to ours, how the Milky Way looks from outside (at this writing, the farthest-flung manmade object, NASA's *Voyager 1* space probe, launched in 1977, was at the outer boundary of our solar system, representing humanity on the verge of exploring interstellar space for the first time).

Nor does Big History, yet, tell the story of life on Earth from the perspective of other species who populate it alongside our own—dolphins and whales, for example, who are quite likely comparably sentient to humans, and who maintain transgenerational collective learning through oral tradition; or bacteria, Earth's most plentiful organism, whose sheer mass and pervasiveness, as sociologist Robert Bellah asserts, makes them Earth's true dominant species.

Still, one imagines that these threads, too, will be opened by scholars peering into the endless possibilities that working with a Big History understanding can provide for innovative study.

FINAL THOUGHTS

Our program is young, but the journey we have taken to reach this point in the development of our pedagogy has been long. Our colleague Philip Novak first brought Thomas Berry, the author of the "new story," to Dominican in 1984. The Dominican sisters embraced Berry's vision, and his ideas began to permeate the curriculum. Between 2003 and 2007, our colleagues Cynthia Brown, James Cunningham, and Novak—a historian, a biologist, and a theologian—taught Big History as a "colloquium" of three linked courses. And at the end of that first decade of the twenty-first century, the conditions were just right for Dominican to embrace Big History and make it the core of the university's general education program.

In 2010, we had our first summer institute, in which we began to build this curriculum and trained ourselves to implement it. In the years since, we have unpacked and streamlined the course content, enhanced our focus on the complexity framework, and developed active learning-based activities and reflective writing assignments to help students incorporate this complex course material and its high binding concepts.

In addition, we have designed and run numerous discipline-specific courses to approach this material from new directions. As of this writing, we have run some sixty-seven sections of our Big History survey course and sixty-six sections of "Through the Lens" courses, in business, history, political science, philosophy, religion, art history, studio art, creative writing, literature, music, health, women's and gender studies, and more. We have trained eighty faculty members and administrators in our summer institutes, hosted world experts in Big History, and brought scholars from around the world to our table to share in our collective learning.

Not only each year or each semester, but each week, in each class session, at each weekly faculty lunch, we learn something new that helps us to improve our courses, our approaches to this material, and our methods for teaching it in the classroom. Even the writing of this book has helped us to further crystallize our understanding of how we do this. That may be lesson number one: stay open, keep learning, and always be improving.

On a practical level, Big History provides our students with foundational knowledge across the disciplines. It frames students' four-year liberal arts education, and it can be used to complement critical thinking track courses and strengthen key academic skills. On a larger scale, our incorporation of Big History into our university's curriculum and our work to develop Big History pedagogy have proven transformative, for our faculty, for our students, and for our institution.

Building a Big History program is an opportunity to strengthen faculty unity across departments and to de-silo the institution academically, intellectually, and practically, giving educators in disparate disciplines a unifying, edifying experience that can bind the institution. It's also challenging for faculty—and fun. Every faculty member's work in her or his own home discipline has been informed by Big History. It is driving innovation in our teaching, and in our research.

Most importantly, Big History is a way to give students a tour of the wealth of human knowledge and, therefore, of the university. It shows them the infinite possibilities for their courses of study. It prepares them to understand, analyze, consider, and begin to solve the myriad complex problems that we face as a species in this still-young millennium. It empowers them with agency to shape their own individual futures, and our collective future. This is, indeed, what twenty-first-century students must know.

We believe that in the deep, intensive, and intensely engaging process of learning how to teach Big History, we have developed some important insight as to how to help our students get to know it, and how to use this material to engage eager young minds in the world of ideas.

We hope you find it useful.

ANNOTATED BIBLIOGRAPHY OF BIG HISTORY TEXTS AND RESOURCES

The annotated bibliography is an assignment in our discipline-specific "Through the Lens" courses. The assignment is designed to sharpen research skills, promote information literacy, and enhance critical thinking about source credibility and the relevance of source texts in various media. Students are asked to prepare and review a list of sources for a research paper, then write a brief entry on each, including a bibliographic citation, a categorization of the text, a summary of the text's contents, and a concise explanation drawing connections to the themes or topics that will be explored in the paper. Here, we meta-model the annotated bibliography as a resource for the Big History teacher.

KEY TO CONTRIBUTORS

AS   Alan Schut
CB   Cynthia Brown
CT   Cynthia Taylor
DD   Debbie Daunt
DM   J. Daniel May
EA   Ethan Annis
JC   Jim Cunningham
JH   Judith Halebsky
JLC   Jaime Castner

LD	Lindsey Dean
MA	Martin Anderson
MB	Mojgan Behmand
NW	Neal Wolfe
PCN	Philip Novak
RBS	Richard B. Simon

The 11th Hour: Turn Mankind's Darkest Hour into Its Finest. Dir. Leila Conners Petersen and Nadia Conners. Perf. Leonardo DiCaprio. Warner Brothers, 2008. DVD.

Category: Film. Documentary. 92 minutes.

Summary: Leonardo DiCaprio produces and narrates this documentary about the dire circumstance of our planet and realistic solutions for surviving our own progress.

Connection: Experts from a variety of fields—including Stephen Hawking and William McDonough—put humanity's ecological crisis in a Big History context by explaining how industrialization brought us where we are today. This film can be shared with students during an exploration of Threshold 8 or a unit on the future and used to provoke a discussion of their ideas for healing the planet. Be sure to also discuss the film's tone with students—do the content and delivery of information in the film inspire viewers to action or frighten unnecessarily? (JLC)

Allmon, Warren D. *Rock of Ages, Sands of Time.* Chicago: U of Chicago P, 2001. Print.

Category: Illustrations of painting series, science-based essays. 325 pages.

Summary: *Rock of Ages, Sands of Time* is a series of 544 contiguous painted panels, by artist Barbara Page, each representing a million years of the history of life on Earth since the Cambrian explosion, as represented by full-scale fossils and animals found within each geologic era. Paleontologist Warren Allmon's essays provide scientific context for the paintings.

Connection: Page notes that her extraordinary series of paintings "functions as a visual metaphor for an immense period of time—an eon," the period between the emergence of macroscopic life and the twenty-first century. Allmon's essays introduce each of the eleven geologic periods into which the timeline is divided, adding detailed scientific context behind the art. Of particular value for a course on Big History as understood through visual art, the project presents an unparalleled opportunity to grasp this vast timeline all at once, simultaneously offering the occasion to study the individual geologic periods it spans. (NW)

Alvarez, Walter. *T. Rex and the Crater of Doom*. New York: Vintage Books, 1998. Print.

Category: Nonfiction. 185 pages.

Summary: In this book written for the general public Alvarez describes his search for the cause of the Cretaceous extinction of the dinosaurs. Alvarez describes the evidence for the asteroid or comet that may have caused the extinction of dinosaurs.

Connection: Good description of one of several mass extinctions that altered the course of vertebrate evolution. With the extinction of dinosaurs, the era of birds, mammals, and eventually man could begin. (JC)

Asimov, Isaac. "The Judo Argument." *The Planet That Wasn't*. Ed. Isaac Asimov. New York: Avon Books, 1976. 225–237. Print.

Category: Essay. 12 pages.

Summary: This essay provides several arguments against the necessity of a divine creator for life to evolve in a universe of random processes. If conditions were different, then some other life form could be talking about the Goldilocks effect. *The Blind Watchmaker* by Richard Dawkins offers an expanded treatment of some of these arguments.

Connection: Asimov elegantly explains how the transition could have been made from the nonliving primordial soup to the beginnings of life, which we cover in Threshold 5. (EA)

Atwood, Margaret. *Oryx and Crake*. New York: Anchor Books, 2003. Print.

Category: Novel. Science fiction. Post-apocalyptic. 376 pages.

Summary: Renowned novelist Margaret Atwood's narrative creates a dystopian and post-apocalyptic vision of our near future.

Connection: Atwood's novel draws extensively on the achievements and failures of the twenty-first century in order to project into the future and create a possible scenario of our future. Her illustration of the possible outcomes of capitalism, corporate power, mass production, overpopulation, genetic engineering, and climate change render this text useful for discussions of the modern industrial era (Threshold 8) and the future. (MB)

Bacigalupi, Paolo. "The Calorie Man." *Pump Six and Other Stories*. San Francisco: Night Shade Books, 2008. 93–122. Print.

Category: Short story. Speculative fiction. 29 pages.

Summary: In this short story, a few corporations, using genetically engineered sterile seeds, control almost the entire planet's food supply. Blights have wiped out crops not produced by these corporations. The story can be found on the web.

Connection: This ties in to Threshold 8 or the future. Bacigalupi imagines what the world would be like if almost all seeds used in food production were sterile. (Michael Pollan's chapter on potatoes in *The Botany of Desire* provides an excellent overview of sterile seeds in general.) Because Monsanto now produces sterile seeds and blights already affect crops, Bacigalupi's vision seems possible. (EA)

Bacigalupi, Paolo. "The Tamarisk Hunter." *Pump Six and Other Stories.* San Francisco: Night Shade Books, 2008. 123–136. Print.

Category: Short story. Speculative fiction. 13 pages.

Summary: "The Tamarisk Hunter" is set in Colorado, in the near future. A multiyear drought has desertified much of the Western United States. The Colorado River has been channeled into a heavily guarded pipe ending in California. Upstream of the pipe, the main character, Lolo, poisons water-sucking tamarisk plants on the riverbanks. The story can be found on the web.

Connection: Currently, people use so much water from the Colorado River that it rarely reaches the ocean, and large world land areas are undergoing desertification due to droughts. Climatologists warn that this trend will get worse. This story can be tied in to Threshold 8, or to a discussion of the near future. (EA)

Becoming Human: Unearthing Our Earliest Ancestors. Dir. Graham Townsley. WBGH Educational Foundation, 2009. DVD.

Category: Documentary film. 180 minutes.

Summary: This engaging film from the PBS NOVA series presents the latest scientific theories on what makes humans human. Combining interviews with world-renowned scientists, computer-generated animation, and actual footage from archaeological sites of important fossil discoveries, *Becoming Human* provides valuable images of the course of human evolution. The film is divided into three one-hour segments: "First Steps" (part 1) explores the hominine split from other great apes; "Birth of Humanity" (part 2) examines Turkana Boy, an almost complete specimen of *Homo erectus*, as an example of how long-distance running might have been a critical factor in the survival of these early humans; and "Last Human Standing" (part 3) compares *Homo sapiens* with their now-extinct European cousins, the Neanderthals. Given the great variety of hominine species of the past, the filmmaker emphasizes the fact that modern-day humans, as the last of the hominine family, are a relatively new phenomenon.

Connection: This is a great film that can supplement reading on the evolutionary development of human beings and the complex history of the six to eight million years comprising Threshold 6. (CT)

Bellah, Robert N. *Religion in Human Evolution: From the Paleolithic to the Axial Age.* Cambridge, Mass: Belknap Press of Harvard UP, 2011. Print.

Category: Nonfiction. 746 pages.

Summary: Renowned sociologist of religion Robert N. Bellah provides readers with a unique Big History perspective in his interdisciplinary magnum opus, *Religion in Human Evolution.* After a concise summary of modern scientific cosmology and the impact of the Big History narrative on modern humanity's existential quest, Bellah explores the origins of religion and makes the case that many of our greatest cultural capacities have their primordial root in play. He traces the cognitive and cultural evolution of our species by looking to the Paleolithic era, the rise of tribal religion, the emergence of archaic religion in the city-state period, and the breakthrough of the world's major religious and philosophical traditions during the Axial Age.

Connection: This text is recommended to educators who want to incorporate the role of religion in human evolution as part of the Big History narrative. The following sections are recommended as supplemental reading: "Overlapping Realities" (which treats issues around Big History and meaning), "Play" (which can complement a discussion on human evolution and our creative and moral capacities), "Tribal Religion" (which would serve students exploring the Paleolithic era), and "The Axial Age" (one particular case could be discussed during a session on the agrarian era). (LD)

Bernstein, William J. *A Splendid Exchange: How Trade Shaped the World.* New York: Grove, 2008. Print.

Category: Nonfiction. 447 pages.

Summary: Bernstein, a financial theorist and historian, writes an accessible work illuminating the numerous ways trade has connected human beings, from prehistory to the present.

Connection: This book's value is in the many interesting examples of how uneven distribution of resources forces humans to trade in ever more complex ways. For example, Bernstein argues that trade is the only explanation for obsidian tools being found hundreds of miles from any possible volcanic source, and considers what exchanges must have been made along the way. (MA)

Berry, Wendell. "Manifesto: The Mad Farmer Liberation Front." *The Country of Marriage.* Berkeley: Counterpoint, 1973. Print.

Category: Poem. 2 pages.

Summary: This poem exposes the hollow values of materialism and advocates for a dedication to the natural world.

Connection: This accessible reading is useful for classrooms discussions on industrialization and environmental conservation. It is also useful in assignments that ask students to reflect on their personal values and future goals. (JH)

Berry, Wendell. *The Unsettling of America: Culture and Agriculture.* Revised ed. San Francisco: Sierra Club Books, 1996. Print.

Category: Nonfiction. 234 pages.

Summary: An exposé and critique of industrial agriculture, Wendell Berry's seminal jeremiad (1977) is now a classic in this genre of environmental writing and cultural criticism. Berry notes in his afterword to this edition that his argument is still relevant and that the continuing loss of sustainable farming practices, family farms, and our sense of attachment to the land is a national tragedy. Endnotes are included, but there is no bibliography.

Connection: Berry's narrative blends agricultural policy, philosophy, literature, spirituality, and American history into a rich and compelling work of art that offers students in Threshold 7 a model of humanistic reflection on a practical theme. Highly recommended. (AS)

Big History Project. bgC3. 2011. Web. 3 November 2013.

Category: High-school course resource site.

Summary: This course is not based on thresholds, but organized into ten units, which can be compressed into one semester or spread over two semesters of three weeks each. The ten units are: (1) What Is Big History? (2) The Big Bang, (3) The Stars Light up & New Chemical Elements, (4) Our Solar System & Earth, (5) Life, (6) Early Humans, (7) Agriculture and Civilization, (8) Expansion & Interconnection, (9) Acceleration, and (10) The Future.

Connection: This site brings together a wide array of engaging materials (short videos, essays, cartoons, and so on) for teaching and learning the whole Big History story. It also offers a parallel site for the public. (CB)

Big History: The Big Bang, Life on Earth, and the Rise of Humanity. Host David Christian. The Teaching Company, 2008. DVD.

Category: Series of forty-eight half-hour lectures. Available in DVD, CD, and digital download formats.

Summary: In these lectures, Christian uses the threshold model for the first time to structure the story. The introductory lectures include "What Is Big History?" "Simplicity and Complexity," and "Evidence and the Nature of Science." The final lectures include "The Near Future," "The Remote Future," and "Big History—Humans in the Cosmos."

Connection: Used sparingly, these 30-minute lectures are useful for preparing oneself or for presenting to students any topic in the Big History story that one feels unprepared to present oneself. (CB)

"The Black Death." *Ancient Mysteries.* **A&E Television Networks, 1997. DVD.**
Category: Film. Made-for-television episode. 50 minutes.
Summary: This *Ancient Mysteries* episode gives a riveting historical account of the Black Plague.
Connection: This DVD brings students into the time of the plague and presents an overview of the pandemic while giving insights into the lives of people, families, communities, and nations affected by the Black Death in Threshold 7. (DD)

Blade Runner. **Dir. Ridley Scott. Perf. Harrison Ford. Warner Brothers, 1982. Film.**
Category: Film. Science fiction. 116 minutes.
Summary: Many works of speculative fiction ask what it means to be human after humans have developed the means to manufacture superhumans. In *Blade Runner,* based on Philip K. Dick's 1968 novel *Do Androids Dream of Electric Sheep?,* the superhuman "replicants" are behaving in ways not desirable to their creators. The film asks, is it even realistic to expect to control such beings?
Connection: Today, we create thinking machines, such as Deep Blue and Watson, that exceed us in narrowly focused ways. Soon we will probably create beings with greater overall intelligence levels than those of humans. *Blade Runner* provides a good starting place for a discussion of this possibility. The film can be tied in to a discussion of Threshold 8 or the future. (EA)

The Book of Eli. **Dir. The Hughes Brothers. Perf. Denzel Washington, Gary Oldman. Warner Brothers, 2010. DVD.**
Category: Film. Post-apocalyptic fiction. 118 minutes.
Summary: Eli walks alone across a landscape scorched by "the Flash" (presumably nuclear war), carrying what may be the last copy of the Bible. A brutal gang boss who is, ironically, the only person creating any semblance of social order, wants to gain possession of the Bible so that he can use its words to reinforce his control over the illiterate postwar populace.
Connection: This film ties in with discussions of the future. The overtly religious components of the story obscure the more subtle fact that this is also a story for the Information Age. The struggle between the holy man and the gang boss is very much a struggle over who controls the flow of critical information. (DM)

Bowles, Samuel, and Herbert Gintis. *A Cooperative Species: Human Reciprocity and Its Evolution*. Princeton: Princeton UP, 2011. Print.

Category: Nonfiction. 249 pages.

Summary: Bowles and Gintis cull data from archaeology, genetics, and ethnography to provide a comprehensive history of the human species' relationship with morality and cooperation.

Connection: Valuable background reading for the teacher of Big History to inform study of Threshold 6 and beyond. The text is written at a high level and is likely a good fit only for advanced students, but its expansive scope covers a variety of Big History–relevant concepts: differential genetics, fitness, coevolution, power, war, and altruism. (JLC)

A Boy and His Dog. Dir. L. Q. Jones. Perf. Don Johnson, Jason Robards. Sling Shot Entertainment, 1974. Film.

Category: Film. Post-apocalyptic fiction. 118 minutes.

Summary: Based on the 1969 novella by Harlan Ellison. World War IV has ravaged the Earth with nuclear weapons. A young man scavenges for food and women with his remarkable companion: a telepathic dog who is more educated, erudite, and civil than any of the humans. Humanity has indeed sunk lower than the animals; human savagery and anarchy abound.

Connection: This film ties in with discussions of the future. For movie audiences who had experienced the social upheavals of the 1960s, future descent into anarchy seemed highly plausible. (DM)

Brown, Cynthia Stokes. *Big History: From the Big Bang to the Present*. New York: New Press. 2007. Print.

Category: Nonfiction. Big History. 320 pages.

Summary: Brown contextualizes human history within the broad scope of the origins of the universe, our solar system, the Earth, and life. As Brown is a historian, *Homo sapiens* appear early in her version of the story—on page 38. Brown's telling offers an understated wisdom, a subtly feminist perspective, and an environmentalist's bent. The future comes down to a series of resource problems—overpopulation, global warming, depletion of forests, soil, and water, and nuclear radiation—which she addresses in the final chapter.

Connection: Because the science is conceptual and explained simply, the text is particularly accessible for students without strong science backgrounds. Brown leaves room at the end of each chapter to address "unanswered questions"—gaps in human knowledge—which students appreciate. This book works well as a core text—for example, as a very readable grounding in Big History, in a class that might go on to focus on some aspect of the story (the future, for example). (RBS)

Cave of Forgotten Dreams. Dir. Werner Herzog. Creative Differences, 2010. Film.

Category: Film. Documentary. 89 minutes.

Summary: Werner Herzog's 3D documentary is about the paintings in the Chauvet Cave in southern France. The cave's 32,000-year-old images are among the oldest known paintings by humans. The film takes the viewers into these caves, which are physically accessible to only a few people.

Connection: Any student interested in the beginnings of human expression would glean a great deal from this documentary, which can be tied in to Threshold 6. (EA)

Christian, David. *Maps of Time: An Introduction to Big History.* Berkeley: U of California P, 2005.

Category: Nonfiction. Big History. 642 pages.

Summary: In many ways, *Maps of Time* is the seminal Big History text, as told from a historian's perspective. Christian points out the human need for a global, universal creation myth based in science and history—and then sets out to compile that myth through research that is broad and deep. Among the key concepts Christian emphasizes are collective learning, networks of exchange, the divergences and convergences of human civilizations that lead to increases in collective learning and innovation, and the environmental collapse that is the hallmark of human civilizations. Christian tells the story almost pointillistically—laying out all the details until the larger picture comes into focus.

Connection: This book can be used as the core text in any Big History course, or as a comprehensive primer in Big History in other contexts. It is dense and exhaustive—and probably best for advanced or upper-division students. A must-read for the Big History teacher. (RBS)

Christian, David. *This Fleeting World: A Short History of Humanity.* Great Barrington: Berkshire Publishing Group, 2011. Print.

Category: Nonfiction. 120 pages.

Summary: This book is divided into three parts: "Beginnings: The Era of Foragers," "Acceleration: The Agrarian Era," and "Our World: The Modern Era," told in 107 pages. In addition, it has a "Prequel" of eight pages, which gives a quick history of the universe before humans arrived.

Connection: This is the big picture of world history—useful for seeing the big outline of human history and for creating world history courses, but also for teaching Thresholds 6, 7, and 8 in Big History courses. Especially useful at the high school or college freshman level. (CB)

Contagion. Dir. Steven Soderbergh. Warner Home Video, 2012. DVD.
Category: Film. Fiction. 106 minutes.
Summary: A pandemic spreads throughout the world and public health departments race to combat the virus.
Connection: Students see a fictional account of what could be a reality. Although *Contagion* is set in the future, it highlights theories touched on in several thresholds and shows the reactions of people faced with a global healthcare crisis. (DD)

Crumb, Robert. *The Book of Genesis Illustrated by R. Crumb.* New York: W. W. Norton, 2009. Print.
Category: Graphic novel. 224 pages.
Summary: Legendary underground cartoonist Robert Crumb illustrates the biblical text, in a surprisingly straight reading.
Connection: Crumb's deeply researched depictions illuminate striking parallels between the Abrahamic creation story and the story science tells of the early universe; his portrayal of the agricultural civilizations in which the later Genesis stories unfold reveals Genesis as a key Big History text. As Genesis parallels most of the Big History story, this text will be useful throughout a Big History course—especially in helping students visualize human and power dynamics in agricultural civilizations. (RBS)

de Waal, Frans. *Our Inner Ape: A Leading Primatologist Explains Why We Are Who We Are.* New York: Riverhead Books, 2005. Print.
Category: Nonfiction. 304 pages.
Summary:: This book features chapters about power, sex, violence, and kindness; it concludes that humans have two inner apes—one competitive and selfish, the other cooperative and loving. We are born with a gamut of tendencies, from the basest to the noblest, and our morality "is a product of the same selection process that shaped our competitive and aggressive side."
Connection:: This book is connected to Threshold 6, human development and emergence. Students like it because they recognize themselves in the descriptions of primate behavior. See de Waal's TED talk for an excellent short oral and visual presentation of his ideas. (CB)

de Waal, Frans. *Primates and Philosophers: How Morality Evolved.* Ed. Stephen Macedo and Josiah Ober. Princeton: Princeton UP, 2006. Print.
Category: Nonfiction essay collection. 232 pages.

Summary: Is there a discontinuous line between emotion-motivated seemingly moral behavior in primates and the reason-based moral actions of humans? Or, as de Waal suggests, is this behavior continuous, with human moral behavior founded in primate behavior? Four commentators respond to de Waal's central essay, then de Waal responds.

Connection: This book can provide faculty background for a Big History survey course in discussions of Threshold 6, regarding unique human attributes and characteristics. Alternatively, it can be used in advanced courses or in follow-up discussions of science, religion, and morality. (CB)

Diamond, Jared M. *Collapse: How Societies Choose to Fail or Succeed.* New York: Penguin, 2005. Print.

Category: Nonfiction. 575 pages.

Summary: Diamond explores the factors that have caused societies in the past to collapse. He identifies five major factors that have played significant roles in these collapses: climate change, hostile neighbors, collapse of essential trading partners, environmental problems, and failure to adapt to environmental issues. He discusses factors resulting in the collapse of Easter Island, Pitcairn and Henderson Island, and the Anasazi, Maya, and Viking societies. Diamond also writes about the collapse of several modern societies. The book concludes by applying lessons learned from these earlier collapses to today's environmental and social problems.

Connection: This is a good resource for instructors when discussing the relationship between human societies and their environments. It is particularly relevant in discussions of Threshold 8 and the future. (JC)

Diamond, Jared M. *Guns, Germs, and Steel: The Fates of Human Societies.* New York: W. W. Norton, 1999. Print.

Category: Nonfiction. 494 pages.

Summary: Diamond attempts to explain why Eurasian civilizations have survived and conquered others. He argues against the idea that Eurasian domination is due to any form of Eurasian intellectual, moral or inherent genetic superiority, and posits that it is instead due to differences in geography and environment.

Connection: The book provides good background information for instructors teaching about human civilization and the influence of the environment on these civilizations. (JC)

Dolin, Eric Jay. *Fur, Fortune, and Empire: The Epic History of the Fur Trade in America.* New York: W. W. Norton, 2010. Print.

Category: Nonfiction. 442 pages.

Summary: Dolin, an environmental planner with a PhD in environmental policy from MIT, provides a detailed history of the two-hundred-year beaver fur trade from North America to Europe.

Connection: An impressive work that connects fashion and desire in Europe to economic activity through a stark discussion of how fur companies operated in North America. The book is a valuable cautionary tale of how the pursuit of wealth without concern for the consequences resulted in the near extermination of the beaver from North America in a relatively short period of time. (MA)

Dunbar, Robin. *The Human Story: A New History of Mankind's Evolution.* London: Faber and Faber, 2004. Print.

Category: Nonfiction. 216 pages.

Summary: Robin Dunbar, an evolutionary psychologist, explores *The Human Story* in seven separate imaginatively written essays, each exploring a different human role or occupation: creative artist, music-maker or poet. Ultimately, Dunbar concludes that religion is the universal trait among humans that separates them from their ape cousins; not only does it "bond communities," but it enables humans to "meet the challenges that the planet has thrown at them."

Connection: These beautifully written essays, all logically interlinked, provide a great deal of information on the evolution of humans (Threshold 6), especially from the development of human psychology, and are accessible enough for under-graduate students and nonscience majors. (CT)

Duncan, Todd. *Glimpses of Wonder.* Beaverton: Science Integration Institute, 2011. Print.

Category: Nonfiction. Book of essays. 99 pages.

Summary: Duncan's thirteen brief essays are meditations on the cosmos and interconnectedness of all things. Insightful and often funny, these light readings are grounded primarily in astronomy but touch on a range of subjects, including quantum reality, energy, extraterrestrial life, and even squirrels.

Connection: These essays support a Big History course with an emphasis on awe and could be used at any point in the curriculum. Each chapter is self-contained and offers an ideal reflective writing prompt. Essays also serve as useful models for students of creative writing in Big History. (JLC)

Eames, Ray, and Charles Eames. "The Powers of Ten: A Film Dealing with the Relative Size of Things in the Universe and the Effect of Adding Another Zero (Final Version)." Online video clip. YouTube. YouTube, 26 August 2010. Web. 9 June 2013.

Category: Short film. 9 minutes.

Summary: This short film expands the viewers' view by a power of ten every ten seconds until the viewer is observing the observable universe at 10^{24} meters. Then the film reverses the process until the screen is filled by a view 10^{-16} meters, an animation of quarks in the proton of an atom. Stephen Jay Gould's book *Time's Arrow* explores this theme, too. The short film can be found on YouTube.

Connection: "The Powers of Ten" illustrates the concept of orders of magnitude or logarithmic scale in the most easily graspable form I have encountered. In Big History we are covering time logarithmically, so getting students to understand logarithmic scales enhances their understanding of the entire course. (EA)

Fagan, Brian. *Cro-Magnon: How the Ice Age Gave Birth to the First Modern Humans.* New York: Bloomsbury, 2010. Print.

Category: Nonfiction. 295 pages.

Summary: Brian Fagan, emeritus professor of anthropology, provides many examples of daily life from Ice Age Europe and the various Cro-Magnon cultures (the Mousterian, Aurignacian, Gravettian, Solutrean, Magdalenian) who thrived in the harsh and cold climates of Europe before the onset of the warmer and more stable climate of the Holocene. Fagan uses the term "Cro-Magnon" in a generic sense, as interchangeable with term such as *Homo sapiens* or "modern and anatomically modern human (AMH)." By using this term, Fagan directs the reader to the region of prehistoric Europe, when modern humans existed side by side with Neanderthals.

Connection: This book complements other studies that explore the life of early humans during the Upper Paleolithic era (Threshold 6) and is especially useful in explaining difficult archeological terms. Throughout the text are "boxes" that introduce the nonspecialist reader to important kinds of evidence and terminology used by archeologists and paleoanthropologists. (CT)

Finlayson, Clive. *The Humans Who Went Extinct: Why Neanderthals Died Out and We Survived.* Oxford: Oxford UP, 2009. Print.

Category: Nonfiction. 273 pages.

Summary: Clive Finlayson, considered an outsider to the paleoanthropology profession, offers some fresh insights as to why *Homo sapiens* survived and Neanderthals did not. In his view, humans, in confrontation with their environment, were "innovators" when faced with harsh climatic conditions and therefore benefited by chance and luck in influencing their own evolution. Finlayson argues against the commonly held view that Neanderthals, the conservatives in this rendering of the story of our early ancestors, died out because of their encounter with a superior race of humans, *Homo sapiens.* Rather, the Neanderthals were in the wrong place at the wrong time. For Finlayson, the critical moment for the "triumph" of modern humans was not the Fertile Crescent of ten thousand years ago but the conquest of the Eurasian steppe-tundra thirty thousand years ago, which "marked a dramatic shift in the fortunes of a population that would swamp the whole of Eurasia and the Americas."

Connection: This well-written and accessible narrative provides another perspective on the early development of human beings, contemporaries of the now-extinct and ever-fascinating Neanderthals. This book, with its many graphic examples of life during the Upper Paleolithic, can be recommended to undergraduates to yield greater insights about Threshold 6. (CT)

Freyfogle, Eric T. *Agrarianism and the Good Society: Land, Culture, Conflict, and Hope.* Lexington: UP of Kentucky, 2007. Print.

Category: Nonfiction. 183 pages.

Summary: *Agrarianism and the Good Society* is an excellent overview of the many cultural, moral, and ecological challenges that confront the implementation of sound land-use policy in the United States. This work reminds us that agricultural issues must be understood within the larger context of property rights, the tension between individual and community claims, and the guiding principles that govern land-use decisions. Scholarly notes are included.

Connection: This provocative, accessible work is recommended as supplementary reading for students in Threshold 7 and as a guide to instructors in designing curricular content, especially in a more advanced or "Through the Lens" course. Freyfogle offers a helpful discussion of the novels *Cold Mountain,* by Charles Frazier, and *Jayber Crow,* by Wendell Berry. When a literary representation of these cultural themes is desired, we recommend including these fictional works in a syllabus. (AS)

Gattaca. Dir. Andrew Niccol. Perf. Ethar Hawke, Uma Thurman. Columbia Pictures, 1997. Film.

Category: Film. Science fiction. 106 minutes.

Summary: *Gattaca* takes place in a world where most babies have been genetically engineered to provide the best traits of both parents. Genetically engineered people have more opportunities in the film. *Gattaca* centers on a non–genetically engineered character who attempts to accomplish something only available to genetically engineered characters.

Connection: In the not-too-distant future it may be possible to alter the DNA sequences that cause undesirable congenital diseases or even to genetically enhance a baby. This film provides a good entry to discussions of the implications of this, tied in to Threshold 8 and the future. (EA)

Gilgamesh. Trans. Stephen Mitchell. New York: Free Press, 2004. Print.

Category: Verse epic. 290 pages.

Summary: Stephen Mitchell's engaging translation of the ancient epic depicts Gilgamesh's quest for renown and immortality, foregrounding the timelessness and universality of these human concerns and rendering the narrative relevant for today's students. Students love this translation.

Connection: An excellent portrayal of the agrarian era in Threshold 7. In the Fertile Crescent, the Sumerian city of Uruk shows the development of cities and states, professional specialization, and the ensuing social hierarchies and gender roles. Special points of interest: Shamhat as "priestess," Enkidu's process of becoming "human," Humbaba as the first environmentalist, and the depiction of the Flood. (MB)

Goodall, Jane. "Respect for Life." *Living Philosophies: The Reflections of Some Eminent Men and Women of our Time.* Ed. Clifton Fadiman. New York: Doubleday, 1990. 81–88. Print.

Category: Essay. 8 pages.

Summary: Goodall's essay is a sobering but hopeful outlook on the challenges faced by Earth and its inhabitants. Such themes as interconnectivity and interdependence, collaboration and compassion, support Goodall's argument that our planet's youngest generation is in a unique position to effect positive change.

Connection: This essay is best used as a summative or cumulative reading to excite students about their continuing education and active citizenship. Goodall is a significant figure in the movement for animal justice, but her message and her life's work explore a broad range of Big History–relevant topics. Students will benefit from a reflective writing or small group discussion exercise to follow this reading. (JLC)

Gore, Al. *The Future: Six Drivers of Global Change.* New York: Random House, 2013. Print.

Category: Nonfiction. 592 pages.

Summary: In a work based on solid research, Al Gore shows how six current trajectories could play out in the future. The six trajectories are an interconnected and interdependent global economy, a global mind linked by the web, politics that are increasingly influenced by money, explosive population growth, revolutions in medicine based on genetics and technology, and the effects of global warming.

Connection: Gore synthesizes a great deal of information about our present into an outstandingly readable book and shows what the near future might bring. This is a good book to introduce in Threshold 8 for students asking the question, "Where's it all going in the near future?" or pondering what forces might influence a Threshold 9. (EA)

Gould, Stephen Jay. *Wonderful Life: The Burgess Shale and the Nature of History.* New York: W. W. Norton, 1989. Print.

Category: Nonfiction. 347 pages.

Summary: Gould discusses the discovery and interpretation of the fauna of the Burgess Shale. The Burgess Shale of Canada represents one of the best-preserved examples of what is commonly referred to as the Cambrian explosion, that period of time during which the diversity of life rapidly increased. Gould's controversial interpretation of the Burgess Shale fauna is that many of the fossils found in this deposit represent groups of organisms that are not found today, and thus that random events play an important role in how evolution proceeds. If the story of evolution were to be replayed, the outcome would likely be quite different.

Connection: This book can be useful when discussing the origin and evolution of life in Threshold 5, particularly the Cambrian explosion. (JC)

The Great Dance: A Hunter's Story. Dir. Craig Foster and Damon Foster. Earthrise and Liquid Pictures, Off the Fence Productions, 2000. DVD.

Category: Documentary film. 74 minutes.

Summary: This documentary portrays the life of the San in the Kalahari Desert, one of the few contemporary hunter-gatherer societies. The film enables the viewer to see the world through the eyes of a San hunter. It climaxes in a running hunt, in which the hunter runs a kudu to death.

Connection: This film is perfect for teaching the Paleolithic, hunter-gatherer era of human life in Threshold 6. (CB)

Hazen, Robert M. *Genesis: The Scientific Quest for Life's Origins.* Washington: Joseph Henry Press, 2005. Print.

Category: Nonfiction. 368 pages.

Summary: This book describes scientists currently working on theories of the origin of life and narrows the search down to three hypotheses: (1) that metabolism came first, as an evolving chemical coating on rocks, (2) that self-replication came first, as a genetic polymer, or (3) that both of the above emerged together in some kind of cooperative chemistry.

Connection: This book provides an interdisciplinary approach to Threshold 5, the emergence of life, and offers a clear statement of what is known and what the frontiers are. Recommended for faculty and advanced students. (CB)

Hazen, Robert M. *The Story of Earth: The First 4.5 Billion Years, from Stardust to Living Planet.* New York: Viking, 2012. Print.

Category: Nonfiction. 320 pages.

Summary: Hazen presents the story of Earth from its emergence to the future, in nonacademic language that is accessible to most readers. The focus of this book is on the role of rocks in the emergence of life and the coevolution of rocks and life, the intertwined tale of Earth's living and nonliving spheres.

Connection: The book offers up-to-the-minute information about the emergence of Earth, the origin of the moon, the early planet, origins of life, extinctions, volcanism, and asteroids, plus predictions for the future. Particularly relevant in discussions of Thresholds 4 and 5, and to then contextualize developments in Thresholds 6, 7, 8, and the future. (CB)

Hazen, Robert M., and James Trefil. *Science Matters: Achieving Scientific Literacy.* New York: Anchor Books, 2009. Print.

Category: Nonfiction. 386 pages.

Summary: A manual not on how to achieve scientific literacy in society, but on how to achieve it by reading this book. Nineteen short, clear chapters cover the basic topics, including energy, the atom and chemical bonding, relativity, the code of life, evolution, and ecosystems. The authors are colleagues at George Mason University, each an outstanding writer with dozens of books and hundreds of articles.

Connection: Indispensable resource for non-scientist professors learning to teach Big History. Each chapter provides sufficient background for the scientific topic. (CB)

Higman, B. W. *How Food Made History.* Chichester, West Sussex, U.K.: Wiley-Blackwell, 2012. Print.

Category: Nonfiction. 265 pages.

Summary: *How Food Made History* is an extended exploration of how and why food choices and technologies of production have affected power relations, trade, transportation, and urbanization since the beginning of settled agriculture over five thousand years ago. An extensive, briefly annotated list of recommended reading is included.

Connection: Although marketed as appropriate for the general reader, this overview of global agricultural development is more suitable as an aid to instructors designing course syllabi for Threshold 7. The author elaborates central themes, chapter by chapter, and then draws conclusions, or "claims," that are useful in identifying important class discussion topics. (AS)

Hoban, Russell. *Riddley Walker*. Expanded ed. Bloomington: Indiana UP, 1998. Print.
   Category: Fiction. 254 pages.
   Summary: This 1980 novel imagines a future England ("Inland") that has been nuked back to the Iron Age. Agrarians ("formers") struggle for resources against pastoralists ("fentsers"). A remote central government seeks to rediscover man's lost technological glory in buried wreckage. It maintains social control with roaming Punch-style puppet shows that mythologize the misunderstood past. The story is told by newly initiated seer Riddley Walker in a post-apocalyptic cockney, in which familiar words retain hints of their original meanings, yet bear new meanings and resonances.
   Connection: In this vision, the outcome of industrialization is a terrifying loop back to the early agrarian age. Hoban shows us the evolution, from deep misconceptions about inexplicable events, of language, culture, and religion. (RBS)

***How the Universe Works*. Host Mike Rowe. Discovery Channel, 2011. DVD.**
   Category: Film. Two discs. Eight total episodes of approximately 43 minutes each.
   Summary: Cosmologists and physicists present the building blocks of the universe in this accessible series, which contains episodes on the following topics: "Big Bang," "Black Holes," "Alien Galaxies," "Extreme Stars," "Extreme Planets," "Supernovas," "Alien Solar Systems," and "Alien Moons."
   Connection: These self-contained episodes are a useful supplement to the curriculum in Thresholds 1 through 4. Definitions and explanations of abstract concepts are helpfully clarified. (JLC)

***Into Eternity*. Dir. Michael Madsen. Magic Hour Films, 2009. Film.**
   Category: Film. Documentary. 75 minutes.
   Summary: This documentary explores the challenge of hewing out an underground storage facility in Finland for storing nuclear waste. The builders hope the facility will last for one hundred thousand years, after which time the waste will no longer be hazardous. The filmmakers note that no large-scale building project made by humans has lasted more than a small fraction of that time span.
   Connection: Students wondering what goes into the long-term thinking necessary to effectively manage nuclear waste and what is ignored would learn a great deal from this film. It also provides a poetic meditation about a bigger time scale

than we are accustomed to thinking about. It can be tied in to Thresholds 7 and 8 and the future. (EA)

Into the Universe. Host Stephen Hawking. Discovery Channel, 2011. DVD.
Category: Film. Three episodes of varying length. Total of 180 minutes.
Summary: Stephen Hawking hosts three episodes in this fascinating series: "Aliens," "Time Travel," and "The Story of Everything." Hawking makes some imaginative inferences based on hard science, considering what extraterrestrial life forms might look like and how wormholes and time machines could conceivably transport us to the future.
Connection: This is a playful, optimistic series for the cosmic daydreamer. Try a chapter or two in class or as a take-home assignment toward the end of your curriculum. The content will no doubt spark an engaging discussion among your students and you may debate how realistic Hawking's predictions really are. (JLC)

Jane's Journey. Feat. Jane Goodall. Dir. Lorenz Knauer. Perf. Angelina Jolie, Pierce Brosnan, Jane Goodall, others. Bavaria Media, 2011. DVD.
Category: Film. 107 minutes.
Summary: Jane Goodall and a range of speakers who have worked with and for the illustrious conservationist and humanitarian narrate the story of her life. This journey begins at home in the English countryside and then moves to the jungles of Gombe and the numerous locations to which Jane has traveled for speaking engagements or with her program Roots and Shoots. The tone is serious but optimistic about our collective future.
Connection: Instructors of Big History might use the early chapters of this film to introduce the anthropological sciences. In a unit on prehuman ape species these chapters could assist in illustrating how chimpanzees behave socially, use tools, and offer us insights on how to work sustainably with our environment. For a unit on the future, the later chapters of the film discuss various ways in which young people may safeguard a healthy future for all living things and the planet. (JLC)

Johanson, Donald C., and Kate Wong. *Lucy's Legacy: The Quest for Human Origins*. New York: Harmony Books, 2009. Print.
Category: Nonfiction. 309 pages.
Summary: The renowned paleoanthropologist Donald Johanson, who discovered Lucy—a specimen of *Australopithecus afarensis*, a transitional creature between apes and humans and the most studied fossil hominid of the twentieth century— writes about the many new advances in paleoanthropological research since his initial discovery in 1974. Writing with Kate Wong, a senior editor at *Scientific*

American, Johanson first describes Lucy and the world of Australopithecines. The book's second and third sections explore the worlds of Lucy's ancestors and descendants. The authors conclude with several mysteries still needing scientific answers.

Connection: This book provides a great update on one of the great discoveries of the twentieth century. Of special interest to readers seeking to understand human origins and evolution, it captures the original excitement of Johanson's critical discovery and how it transformed both the academic and popular understanding of how and when humans evolved. (CT)

Journey of the Universe: An Epic Story of Cosmic, Earth and Human Transformation. Host Brian Swimme. Prod. Mary Evelyn Tucker and John Grim. InCA and Northcutt Productions, 2011. DVD.

Category: Film. Documentary. 60 minutes.

Summary: Filmed exploring the Greek island of Samos, host Brian Swimme tells the story of the unfolding universe and humanity's place in it with stunning visual images. A short (130 pages) accompanying book from Yale University Press (2011) amplifies the movie's text and provides an extensive bibliography. The focus of both film and book is on the creativity of the universe and on human creativity as the appropriate response to it. The authors call for people to bring forth a multiform, planetary civilization that will enable both life and humanity to flourish.

Connection: This packet of materials connects in many ways. The movie is an effective introduction to the whole Big History story. The book can be used as part of a course or study group. In the extended version, the first disc presents specialists on subtopics of the story (the beginning of the universe, galaxies, stars, the birth of the solar system, life's emergence, the evolution of the brain, the fossil record, early humans, and the role of humans). The second disc presents discussions of the emerging Earth community (transitions, breakthrough communities, eco-cities, ecological economics, permaculture, indigenous ways of knowing, sustainable energy, healing and re-visioning, the arts and justice, myths and metaphors, and teaching the story). (CB)

Judgment Day: Intelligent Design on Trial. Dir. Gary Johnstone and Joseph McMaster. *NOVA*, 2008. DVD.

Category: Documentary film. 112 minutes.

Summary: This documentary film puts a spotlight on a Dover, Pennsylvania public school classroom, and its battle over evolution, which landed in federal court in 2004. Included are interviews with Dover's key players, expert witness testimonies, and courtroom reenactments based on legal transcripts.

Connection: Revisiting *Kitzmiller v. Dover School District* raises persisting questions about religion and science relevant to Big History. (JLC)

Kean, Sam. *The Disappearing Spoon: And Other True Tales of Madness, Love, and the History of the World from the Periodic Table of Elements*. New York: Little, Brown, 2010. Print.

Category: Nonfiction. Science journalism. 416 pages.

Summary: Kean, a science journalist, presents nineteen entertaining stories connected with the periodic table. The book is a humorous, accessible, and informative history of the development of the periodic table, the discovery of each element, and the flamboyant, egomaniacal, genius personalities of the scientists behind it all. The blend of clearly explained science, history, and biography makes learning about the periodic table and the elements interesting.

Connection: While not written by a Big Historian, *The Disappearing Spoon* gives a clear explanation of the development of hydrogen and helium in the Big Bang and the creation of the other elements during fusion as stars die. Kean then gives the history of the idea of elements, slowly leading to the idea of organizing them into a table, and then filling out the table, all the while explaining the role of the particular elements in life and on the planet. The book is full of useful, interesting anecdotes about the elements and the scientists that will engage students in something we all otherwise think of as "something hanging on the wall in a high school classroom." (MA)

Knoll, Andrew H. *Life on a Young Planet: The First Three Billion Years of Evolution on Earth*. Princeton: Princeton UP, 2003. Print.

Category: Nonfiction. 277 pages.

Summary: Knoll discusses the "deep history" of life on Earth from its origin to the Cambrian explosion. The book is written for both specialists and nonspecialists and discusses the conditions of the early Earth that permitted the development of early life forms.

Connection: The book provides good background for faculty when teaching and discussing Threshold 5 and the conditions under which life developed and evolved. (JC)

Leach, Amy. "You Be the Moon." *The Best American Essays 2009*. Ed. Mary Oliver. New York: Houghton Mifflin Harcourt, 2009. 67–72. Print.

Category: Essay. 6 pages.

Summary: Leach's essay is a meditation on moons, their light, their oscillations, their gravity, and our own moon's implications for life on Earth. Embedded in the essay is a potentially effective active learning activity: invite students to represent Earth, the sun, and the moon and recite poems of prescribed length to one another demonstrating how long light takes to travel from one entity to the next.

Connection: Humans have been interpreting the celestial skyscape for millennia; thus, Leach is one in a long line of writers exploring the vastness of space. In a class combining literature and Big History, use Leach's essay as an example of literature for Threshold 4, the formation of our solar system and Earth. (JLC)

Lewin, Roger. *Human Evolution: An Illustrated Introduction.* 5th ed. Malden: Blackwell, 2005. Print.

Category: Nonfiction. 277 pages.

Summary: As paleoanthropologist Roger Lewin states in the preface of his latest edition of *Human Evolution*, "Clearly, hominin history is turning out to be much more complex than previously assumed." Yet Lewin's well-illustrated and well-organized text demonstrates that this is a good place for any teacher to start understanding the beginning of human origins. Divided into nine chapters, with photographs, maps, and useful graphs on every page, *Human Evolution* brings the reader up to date on the latest findings and theories of the fast-paced science of paleoanthropology.

Connection: This is a valuable source for anyone teaching Threshold 6, especially the development of early hominines. (CT)

Lorey, David, ed. *Global Environmental Challenges of the Twenty-First Century.* Wilmington: Scholarly Resources, 2003. Print.

Category: Nonfiction. Collection of twenty scholarly articles. 312 pages.

Summary: *Global Environmental Challenges* is a collection of articles on environmental challenges facing humankind. The book is organized into five sections: "Driving Forces," "Water," "Global Climate and Atmosphere," "Biodiversity," and "Sustainable Solutions." Topics covered include population growth, food production, water usage, fertilizer, and extinctions. Contributors are scientists, environmentalists, journalists, government officials, and academics.

Connection: This work is valuable as a quick study on various environmental problems, which can easily be linked to Big History themes. The article "Human Alteration of the Global Nitrogen Cycle: Causes and Consequences" is particularly valuable in introducing students to human dependence on fertilizer in food production. "The Real Impacts of Household Consumption" is helpful in providing students with a positive feeling that they can make a difference. (MA)

Lovelock, James. *The Ages of Gaia: A Biography of Our Living Earth*. New York: W. W. Norton, 1988.

Category: Nonfiction. Earth science. 252 pages.

Summary: Lovelock updates his Gaia hypothesis, which holds that the Earth's rocks, ocean, atmosphere, and biosphere—all life—are so closely intertwined as to constitute a self-regulating system, akin to an organism. Here, he explains how it works: once life seizes hold of the planet, it maintains conditions that are fit for life through various feedback mechanisms. In the beginning, photosynthetic cyanobacteria, perhaps the first life-forms, converted Earth's atmosphere from one rich in carbon dioxide to one rich in oxygen—which was their undoing. Obvious even in 1988 was that humans are doing the same in reverse, by pumping carbon dioxide back into the atmosphere.

Connection: Gaia is for the Earth sciences what Big History is for history: a unifying, holistic vision that binds the disparate disciplines and makes sense of their interrelation. Lovelock's book is an excellent resource for helping students understand how Earth's systems work and the tight relationship among those systems. It's complexity on the planetary scale, useful in illuminating the transition between Thresholds 4 and 5 (as well as the problems of Threshold 8 and perhaps 9), and the connection between the planet and life itself. (RBS)

Mad Max Beyond Thunderdome. Dir. George Miller. Perf. Mel Gibson, Tina Turner. Warner Brothers, 1985. DVD.

Category: Film. Post-apocalyptic fiction. 107 minutes.

Summary: In this third installment in the *Mad Max* trilogy, people, no longer fighting over gasoline, have begun to rebuild. Knowledge of renewable resources has become more important. This is neatly personified by an engineer who is producing energy from pig feces (methane); the film's climax is a struggle between two factions for possession of the engineer.

Connection: The film's dramatization of the newly popularized concern about shifting from fossil fuel to renewable energy can be tied in to discussions of the future. (DM)

Manning, Richard. *Against the Grain: How Agriculture Has Hijacked Civilization*. New York: North Point Press, 2004. Print.

Category: Nonfiction. Environmental journalism, agriculture. 232 pages.

Summary: *Against the Grain* describes the "devil's bargain" that Neolithic and modern humans have made with sedentism and agriculture, the biological and psychological advantages of hunter-gatherer societies, and the increasing threats to human and environmental welfare posed by industrialized agriculture. The author integrates reportage with personal reflection in a relaxed and appealing style. Notes and a selected bibliography are included.

Connection: This work questions our traditional assumptions about what we eat, why we eat, and how our food is provided. It is a provocative supplementary narrative for students in Threshold 7. (AS)

Marks, Robert. *The Origins of the Modern World: A Global and Ecological Narrative from the Fifteenth to the Twenty-First Century.* Lanham: Rowman and Littlefield, 2007. Print.

Category: Nonfiction. 220 pages.

Summary: Marks is concerned with the large structures of modern globalization dating back to the 1400s. He coined the phrase "biological old regimes" to distinguish societies based on agriculture from industrialized societies.

Connection: A useful overview of the development of global trade patterns and the impact of industrialization in creating a wealth disparity between Western Europe, Japan, and the United States, on the one hand, and the rest of the world, on the other. (MA)

Marx, Karl, and Friedrich Engels. *The Communist Manifesto.* New York: Oxford UP, 2008. Print.

Category: Primary document. 40 pages.

Summary: *The Communist Manifesto* is Marx and Engels's famous 1848 declaration of the intentions, objectives, and motivations of the newly formed Communist Party.

Connection: Marx and Engels's succinct declaration of the purpose and goals of a nineteenth-century anticapitalist faction serves as an excellent introduction to one of the most powerful political movements of the twentieth century (in Threshold 8), which stood in opposition to the dominant model of corporate capitalism. (CT)

McNeill, William H. *Plagues and Peoples.* Garden City: Anchor Press, 1998. Print.

Category: Nonfiction. History. 340 pages.

Summary: This book gives a historical account of the influence of disease on humankind from the era of early humans into the future, demonstrating that disease affects our activity and has shaped our history.

Connection: *Plagues and Peoples* connects to Thresholds 6 through 8 as well as to the future. The history of man is described in terms of the challenges evolving humans face in combating disease in the past, in the present, and into the future. (DD)

McNeill, William, ed. *Religion and Belief Systems in World History*. Great Barrington: Berkshire Publishing, 2012. Print.

Category: Textbook. 264 pages.

Summary: This book contains a ten-page introduction by Martin E. Marty, an American Lutheran religion scholar at the University of Chicago, followed by forty-four articles on specific religions, religious figures, and belief systems. The introduction summarizes the development of the world's religions and belief systems. Marty concludes that religions do as much as wars, earthquakes, famines, or catastrophes to alter the human landscape.

Connection: This book connects to Thresholds 6, 7, and 8—human history. It is a source of material to balance the meager treatment of religion in other textbooks. Yet the material is more a summary of the religions themselves than an analysis of their influence or role in Big History. (CB)

McNeill, J. R. *Something New under the Sun: An Environmental History of the Twentieth-Century World*. New York: W. W. Norton, 2000. Print.

Category: Nonfiction. 448 pages.

Summary: This is a valuable and comprehensive evaluation of the impact of human activities (industrialization, urbanization, population growth) on the environment during the twentieth century, providing a trajectory of those impacts from 1900 to 2000.

Connection: The work is divided into impacts on the lithosphere, pedosphere, hydrosphere, atmosphere, and biosphere. McNeill objectively evaluates the impacts, so proponents of near-term environment collapse will find little support here, as it appears Earth's carrying capacity will survive for some time into the future. McNeill concludes that "it is impossible to know whether humankind has entered a genuine ecological crisis" (358), but it is clear that current trends are unsustainable and the consequences are unknown if we persist in sustaining the trends. (MA)

Miles, Kathryn. "Dog Is Our Copilot." *The Best American Essay 2009*. Ed. Mary Oliver. New York: Houghton Mifflin Harcourt, 2009. 110–124. Print.

Category: Essay. 14 pages.

Summary: Miles's essay tells the story of Charles Darwin's beloved terrier, Polly. She gives us a new perspective on the often icily portrayed father of evolutionary theory and goes on to discuss the unique and enduring relationship between humans and dogs. Miles reveals a bizarre human obsession with straddling the divide between "wild" and "domestic" spheres as we yearn simultaneously for absolute control and utter abandonment of it.

Connection: Miles touches on such Big History themes as interconnectedness and power while she explores specific concepts like artificial and natural selection as well as evolutionary biology. This essay is useful during study of Threshold 5, life on Earth, particularly in a class combining literature and Big History. (JLC)

Milton, Giles. *Nathaniel's Nutmeg; or, The True and Incredible Adventures of the Spice Trader Who Changed the Course of History.* New York: Penguin, 1999. Print.

Category: Nonfiction. 388 pages.

Summary: This book tells the story of the seventeenth-century struggle between the English and Dutch to control Run Island in the Banda Islands of the Moluccas and thus obtain a monopoly over nutmeg. Famously, the English traded Run Island to the Dutch in exchange for Manhattan, at the time a poor exchange.

Connection: *Nathaniel's Nutmeg* provides an example of a human struggle over a desired plant at the source where it grew naturally. Connections can be made as to how the uneven distribution of plants by plate tectonics and evolution influences human behavior. (MA)

Mintz, Sidney. *Sweetness and Power: The Place of Sugar in Modern History.* New York: Penguin, 1985. Print.

Category: Nonfiction. 274 pages.

Summary: At the time of initial publication, Mintz was an anthropologist at John Hopkins University. This book was a groundbreaking work that took an interdisciplinary approach to analyzing a foodstuff and all the human activity surrounding it and the consequences thereof.

Connection: Of particular relevance to Threshold 7, this work makes us think of our connection to a plant in nature and our relationship to it—in this case, through our desire for a sweet taste. Mintz examines the economic structures built to obtain sugar, with their moral and labor implications, as well as the environmental destruction caused in the pursuit of growing it. (MA)

Newitz, Annalee. *Scatter, Adapt, and Remember: How Humans Will Survive a Mass Extinction.* New York: Doubleday, 2013. Print.

Category: Nonfiction. 305 pages.

Summary: Science writer Annalee Newitz reviews Earth's past mass extinctions, the ways in which survivors managed to endure and flourish under extreme circumstances, and how humans are set to fare during such an event in our collective future (a major human-caused extinction event appears to be under way).

Connection: All mass extinctions on planet Earth have had one thing in common: survivors. This text is remarkably accessible and has a built-in Big History perspective. Chapters on each of Earth's past extinctions can be used throughout the curriculum, accompanying units on natural disasters and epidemics. Still, Newitz's eye is on the future as she hypothesizes how future humans may realistically thrive in space or otherwise, based on a piece of sage advice: scatter, adapt, and remember. (JLC)

Niane, D.T. *Sundiata: An Epic of Old Mali.* Trans. G. D. Pickett. Harlow, U.K.: Pearson Longman, 2006. Print.

Category: Epic poem, oral history. 84 pages.

Summary: Modern-day griot and storyteller D. T. Niane depicts the great king Sundiata, who ruled the African kingdom of Mali from 1235 to 1255 C.E.

Connection: Half history and half legend, Niane's epic tale of the legendary Sundiata tells the gripping story of how a deformed and crippled child grew to be the powerful leader that united the kingdom of Mali in the thirteenth century. This text provides insights into a Threshold 7 civilization: the Mandingo peoples of West Africa. Structured as a long narrative poem, the text communicates the greatness of an ancient African civilization before it was conquered by Muslim invaders and later entrapped in the brutal European slave trade. (CT)

Nostalgia for the Light. Dir. Patricio Guzmán. Icarus Films, 2010. DVD.

Category: Documentary film. Spanish with English subtitles. 90 minutes.

Summary: Chilean filmmaker Patricio Guzmán explores the Atacama Desert, where the clear skies are a magnet for astronomers and arid heat preserves the remains of mummies, miners, explorers, and the executed political prisoners of Chile's 1973 military coup. A visually striking and moving film.

Connection: This film begs the enduring question, what does it mean to be human? It is a useful tool for helping students visualize such Big History themes as interconnectedness and contingency, while it draws valuable parallels between our contemporary human experience and the not-so-distant night sky. (JLC)

On The Beach. Dir. Stanley Kramer. Perf. Gregory Peck, Ava Gardner, Fred Astaire, Anthony Perkins. United Artists, 1959. DVD.

Category: Film. Post-apocalyptic fiction. 134 minutes.

Summary: Based on the 1957 novel by Nevil Shute, *On the Beach* established the convention of starting the story in the aftermath of nuclear war. Australia has been left untouched by direct warfare, but a cloud of radioactive fallout is gradually enveloping the entire planet. The film exemplifies Cold War nuclear fears.

Connection: This film can connect to the future, as imagined by a post–World War II audience that could identify with the sacrifice, discipline, and basic decency of the characters. (DM)

Origins: Fourteen Billion Years of Cosmic Evolution. Host Neil deGrasse Tyson. WGBH Boston Video, 2004. DVD.

Category: Documentary film. Two discs. Total 240 minutes.

Summary: The program is divided into four parts: "Earth Is Formed," "How Life Began," "Where Are the Aliens," and "Back to the Beginnings." Compelling graphics dramatically illustrate major events in Earth's history, starting with the Big Bang.

Connection: This film can be used to illustrate the early thresholds, from the Big Bang (Threshold 1) to the origin of life (Threshold 5). The documentarians also discuss the issue of extraterrestrial life and the search for life on other planets. (JC)

Overstreet, Wylie. *The History of the World According to Facebook*. New York: Harper Collins, 2011. Print.

Category: Parody. 153 pages.

Summary: Overstreet's contemporary farce imagines Facebook posts and conversations between cosmic entities and historical figures from the beginning of time.

Connection: *The History of the World According to Facebook* is a humorous spin on our scientific creation story and the emergence of life, humans, and modernity. Segments of history are distilled into brief conversations that could easily be extracted and used to illustrate an era or event in the Big History story. Have students craft their own Facebook conversation about a moment in time as a fun group activity or assign as a take-home exercise for a Big History creative writing class. (JLC)

Paarlberg, Robert L. *Food Politics: What Everyone Needs to Know*. Oxford: Oxford UP, 2010. Print.

Category: Nonfiction. U.S. and global food politics. 218 pages.

Summary: *Food Politics* is a balanced assessment of the current politics of food, including issues of hunger, food safety, the conflict between large-scale agribusiness and small organic farming, the controversy over genetically modified foods, and the impact of various agricultural practices on environmental degradation. The text offers forthright and accessible counterarguments to many of the claims made by advocates for alternative forms of food production and land management. A glossary of agricultural terms, a list of acronyms, and an extensive bibliography are included.

Connection: Paarlberg's informative, well-reasoned rebuttal to many of the arguments advanced on behalf of sustainability and organic agriculture provides a useful opposing viewpoint not frequently encountered in contemporary academic circles. The book is recommended as supplementary reading for students in Threshold 7 and as a source of lecture materials and discussion topics for instructors. (AS)

Packer, George. *The Unwinding: An Inner History of the New America.* New York: Farrar, Straus and Giroux, 2013. Print.

Category: Nonfiction. 434 pages.

Summary: Weaving together numerous narrative strands, George Packer provides a history of America from 1978 to 2012. The narratives come from different perspectives, ranging from a family living in poverty to a Silicon Valley billionaire and from a powerful member of a US president's cabinet to a mortgage holder dying of cancer who spends months trying get a mortgage modification.

Connection: This is a superbly written description of what has changed in the thirty-four years covered. Any student trying to understand the present would be well served by reading this book, which can be tied into Threshold 8. (EA)

Pinker, Stephen. *The Better Angels of Our Nature: Why Violence Has Declined.* New York: Viking, 2011. Print.

Category: Nonfiction. 832 pages.

Summary: *The Better Angels of Our Nature* lays out a compelling statistically based argument that we are becoming less violent as a species and then offers theories about why history may be headed toward a more peaceful future.

Connection: Anecdotal evidence appears to indicate that violence is not declining and that the means to perpetuate violence is accelerating. Pinker takes us a step back so that we can see patterns in history. The book may be tied into Thresholds 7 and 8, and to the future. (EA)

Rich, Simon. "Center of the Universe." *The New Yorker* 9 January 2012. Web.

Category: Short fiction. 2 pages.

Summary: Rich's humorous writing, published as a *Shouts and Murmurs* column, offers a twenty-first-century retelling of the Book of Genesis. The narrator is a single twentysomething in New York City struggling to keep his girlfriend happy while he works demanding hours creating the universe.

Connection: This lively reading demonstrates the continued relevance of creation stories. It is an excellent text to use in an assignment that asks students to write their own creation story. (JH)

Rimas, Andrew, and Evan D. G. Fraser. *Beef: The Untold Story of How Milk, Meat, and Muscle Shaped the World.* New York: Harper, 2008. Print.

Category: Nonfiction. 238 pages.

Summary: A journalist and a professor of Earth and environmental science team up to write an accessible history of the relationship between humans and cattle and the economic, environmental, and health consequences of the West's current cultural celebration of beef as food and source of protein.

Connection: An excellent work on how culture masks the impact of devoting vast resources to the cattle industry for the financial benefit of a few, with mostly negative consequences for individuals who consume beef—and for the environment. (MA)

The Road. Dir. John Hillcoat. Perf. Viggo Mortenson, Charlize Theron. Dimension Films, 2009. DVD.

Category: Film. Post-apocalyptic fiction. 111 minutes.

Summary: Based on the 2006 novel by Cormac McCarthy, this grimly realistic story follows a father and his young son trudging through a barren world that appears to be in the worst throes of nuclear winter. Savagery, anarchy, and even cannibalism are commonplace.

Connection: This film illustrates the most terrifying of possible futures. The modern audience knows many ways the world could end: nuclear war, bioterrorism, and global warming, among others. Therefore, the exact cause of the environmental collapse is never specified—the audience needs no explanation. (DM)

The Road Warrior. Dir. George Miller. Perf. Mel Gibson. Warner Brothers, 1981. DVD.

Category: Film. Post-apocalyptic fiction. 94 minutes.

Summary: The second and most highly regarded film in the *Mad Max* trilogy, *The Road Warrior* opens with a prologue establishing nuclear warfare and social decay, but ultimately emphasizes the scarcity of remaining oil and gasoline reserves. When the reluctant hero allies himself with a small group holding out in an oil refinery, his struggle is elevated to a fight to save the last remnants of civilization.

Connection: The film exploited the fears of movie audiences who had recently seen gasoline prices skyrocket, and who had heard a series of dire predictions about the world's oil supply being rapidly depleted. Useful in considering possible futures that might stem from Threshold 8. (DM)

Robinson, Kim Stanley. *2312.* New York: Orbit, 2012. Print.

Category: Novel. Science fiction. 640 pages.

Summary: Robinson projects current knowledge and technology out three hundred years to envision a world in which humans have terraformed and inhabited (or are in the process of doing so to) the other rocky planets and moons in our solar system: Mercury (the rolling city Terminator stays on the planet's dark side),

Venus (under construction), Mars (mostly Chinese), the moons of Jupiter and Saturn (thrill seekers surf the icy rings). Humans have also hollowed out asteroids and turned them into themed, terraformed space stations—orbiting arks in which a catastrophically degraded Earth's lost species are being repopulated.

Connection: What's more complex than a digitally connected planetary civilization? A whole solar system full of digitally connected planetary civilizations, populated by humans who have transcended the interface between technology and biology, are often ambi-sexed, are youthful well beyond one hundred years of age, and tap the sun's energy so effectively that energy issues are moot. A compelling vision, in other words, of Threshold 9. (RBS)

Rue, Loyal. *Nature Is Enough: Religious Naturalism and the Meaning of Life.* Albany: State U of New York P, 2012. Print.

Category: Nonfiction. 176 pages.

Summary: Rue, a professor emeritus of religion and philosophy, wrote an account of the new story, in down-to-earth philosophical language, called *Everybody's Story: The Epic of Evolution* (2000). In this 2012 book, he discusses more in-depth human goals, purposes, and values. He does so by logical argument and by analyzing various philosophical traditions and labels, thus helping the reader understand both the history of philosophy and how to sort out one's own beliefs. Rue defends the position of religious naturalist, defined by him as one who "is a naturalist, or seeks to be, religiously engaged with the natural order." He concludes that nature is enough to satisfy religious needs and that global culture is moving toward naturalizing religion and sanctifying nature.

Connection: *Nature Is Enough* can be recommended to students who are trying to sort out their own beliefs in regard to religion and Big History. (CB)

Sacks, Oliver, Jonathan Miller, Stephen Jay Gould, Daniel J. Kevles, and R. C. Lewontin. *Hidden Histories of Science.* Ed. Robert B. Silvers. New York: New York Review of Books, 1995. Print.

Category: Nonfiction essay collection. 193 pages.

Summary: Silvers has collected five essays written by noted scientists who together demonstrate that science has evolved through growth and changes in human perception and that growth will continue into the future.

Connection: These readings show scientists as creative thinkers who ponder the mysteries of life and question accepted theories regarding several thresholds. These authors serve as examples to students to explore, question, and think in order to discover their own views in an ever-changing world. (DD)

Sawyer, G. J., and Viktor Deak. *The Last Human: A Guide to Twenty-Two Species of Extinct Humans*. New Haven: Yale UP, 2007. Print.

Category: Nonfiction. Evolutionary biology. 256 pages.

Summary: Relatively up to date in a discipline with constant new discoveries and new interpretations, this guide features maps, as well as photos of both authentic fossils and lavishly sculpted reconstructions. The concise text is divided into orderly, discrete chapters on each species—making it easy to study only those necessary for immediate classroom use.

Connection: This book thoroughly covers hominine evolution from eight million to ten thousand years ago (Threshold 6). Each species is treated with a comprehensive account of climate, habitat, social and migratory behavior, diet, tool use, brain size and possible cognitive functions, and more, providing a "zoomed-in" portrait that complements the "wide-angle" view of some core Big History textbooks. The information presented in the book is also invaluable background information for any instructor conducting the "Hominoid Skull Lab" activity described in chapter 12 in class. (Note: the skull identified in many sources as *Australopithecus boisei* is referred to in this book as *Paranthropus boisei*. The difference in nomenclature reflects differing interpretations of where this particular fossil fits in the human family tree). (DM)

Scarre, Chris, ed. *The Human Past: World Prehistory and the Development of Human Societies*. 2nd ed. London: Thames and Hudson, 2005. Print.

Category: Nonfiction. 784 pages.

Summary: *The Human Past*, edited by Chris Scarre of Durham University, is an important textbook for anyone interested in connecting human evolution to the growth of early human societies and their development of cities, states, and empires. With the help of twenty-three scholars, contributing archeological and anthropological expertise, *The Human Past* is divided into two parts: "The Evolution of Humanity" (six million to 11,500 years ago) and "After the Ice" (11,500 years ago to the early civilizations). This text contains over 770 illustrations, many in color, and other helpful features, highlighting key discoveries, key controversies, and key sites. Each chapter includes a useful bibliography of further reading.

Connection: The strength of this classic textbook is how effectively it connects Threshold 6 (hunter-gatherer cultures) with Threshold 7 (agrarian cultures and civilizations). (CT)

Sinclair, Thomas R, and Carol J. Sinclair. *Bread, Beer, and the Seeds of Change: Agriculture's Impact on World History*. Wallingford: CABI, 2010. Print.

Category: Nonfiction. 193 pages.

Summary: *Bread, Beer, and the Seeds of Change* describes the impact of agricultural practices and crop selection on the development of ten different societies, both ancient and modern. The prose style is straightforward, if a bit dry, but the comparison of food production in very different contexts is unusual, rich in information, and valuable in providing a global perspective. Selected endnotes are included, but there is no bibliography.

Connection: This work is recommended as supplementary reading for students in Threshold 7, when crosscultural investigations are emphasized in the curriculum. It would also be useful to instructors in preparing lecture materials. (AS)

The Sleep Dealer. Dir. Alex Rivera. Perf. Leonor Varela, Jacob Vargas. Maya Entertainment, 2008. Film.

Category: Film. Science fiction. 90 minutes.

Summary: The plot of this movie revolves around a corporation damming a river and then selling water to local inhabitants. The corporation uses drone strikes to protect its interests and describes poor peasants as terrorists.

Connection: This movie explores issues that we read about daily and shows how they could resolve themselves in the near future. It ties into Threshold 8. (EA)

Southwood, T. Richard E. *The Story of Life.* Oxford: Oxford UP, 2003. Print.

Category: Nonfiction. 264 pages.

Summary: Southwood traces the story of life on Earth, from its origin to today's world. In the last chapter he also discusses the impact of humans on Earth's environment. The book is divided into twelve chapters, each covering several geological time periods. The book contains good illustrations of major groups of organisms found in the fossil records. Each chapter begins with an illustration that shows the position of the continents during that time period.

Connection: An excellent summary of the origin and evolution of life in Threshold 5. (JC)

Spier, Fred. *Big History and the Future of Humanity.* Chichester, West Sussex, U.K.: Wiley-Blackwell, 2011. Print.

Category: Nonfiction. 272 pages.

Summary: This eight-chapter Big History text was written by one of the leaders in the field, Fred Spier.

Connection: Spier's telling of Big History is notable for its relative brevity and its thematic unity. The brevity is a function of altitude: if Christian's *Maps of Time* flies at thirty thousand feet, Spier's book flies twice as high—he covers all of human history in two chapters totaling sixty pages—sacrificing detail for pattern recognition. The thematic unity of Spier's argument, which is advanced recursively in each chapter, is that at every level and at all scales, Big History is the story of the rise and demise of complexity as energy flows through material structures within certain boundary conditions. (PCN)

Standage, Tom. *An Edible History of Humanity.* New York: Walker, 2009. Print.
Category: Nonfiction. 269 pages.
Summary: *An Edible History of Humanity* recounts, in an accessible and entertaining style, the impact of plant and animal domestication on the transformation of all aspects of human society, from the invention of agriculture to the current industrial model. Special attention is given to the relationship of agriculture to social stratification, trade, war, and slavery. Scholarly notes and an extensive bibliography are included.
Connection: This panoramic overview of global agricultural development provides a host of engaging historical examples and is designed to capture the curiosity of the reader. It is recommended as supplementary reading for students in Threshold 7. (AS)

Stokstad, Marilyn, and Michael W. Cothren. *Art: A Brief History.* 4th ed. Upper Saddle River: Prentice Hall, 2010. Print.
Category: College-level art history textbook. 604 pages.
Summary: Art historians Stokstad and Cothren offer a comprehensive survey of the central role visual art plays in human history through its representation of cultural developments and universal human themes. Works of art, including architecture, often serve as our best means of understanding historical cultures and events.
Connection: While this book is targeted for a general art history survey course audience, it has features that can be used effectively for a study of Big History, beginning with the Paleolithic era. It offers a detailed examination of Paleolithic art, including cave paintings and female figurines. The book is divided into standard art historical categories, but each chapter begins with an informative section that places its category within a larger historical context, which relates well to the Big History narrative. These introductions are followed by images and discussion of many visual examples, which can be selected to illustrate Big History themes and developments. (NW)

Stoppard, Tom. *Arcadia, a Play in Two Acts.* London: Samuel French, 1993. Print.
Category: Play. 112 pages.
Summary: *Arcadia* is set in two periods: 1809–1812 and 1993. The characters in 1993 try to determine what happened in the earlier period. The audience sees what actually happened and how historians capture it.

Connection: *Arcadia* is an excellent way to begin a discussion about the certainty of knowledge and how our understanding of the past constantly evolves, useful across all thresholds, but perhaps most directly relevant to Thresholds 7 and 8. (EA)

Through the Wormhole. Host Morgan Freeman. Science Channel, 2011. DVD.

Category: Television series. Four discs containing sixteen episodes of 45–50 minutes each.

Summary: Is there a creator? Are we alone? Is there a sixth sense? Do parallel universes exist? Morgan Freeman explores these questions in a mind-bending television series featuring experts in the fields of astrophysics, astrobiology, quantum mechanics, and string theory, among others.

Connection: Season 1 loosely emphasizes universal origins and considers our collective past, while Season 2 offers a general focus on the future. Self-contained chapters are easy to use in class throughout a Big History curriculum, but especially during the study of Thresholds 1 and 2. (JLC)

Turner, Jack. *Spice: The History of a Temptation.* New York: Vintage Books, 2004. Print.

Category: Nonfiction. 352 pages.

Summary: Turner has written a thorough study of the spice trade, from Roman times through the seventeenth century.

Connection: Turner's rich work provides an opportunity to contemplate the consequences of humans trying to solve a couple of basic problems: (1) how to preserve food and (2) how to make basic foods, such as meat, bread, and porridge, palatable. In addition to describing the complex spice trade systems, Turner provides examples of how desire for and love of spices are found throughout literature and art. (MA)

Vonnegut, Kurt. *Galapagos.* New York: Dell, 1986.

Category: Novel. 304 pages.

Summary: The great satirist unfolds a Genesis story and a flood story for the Darwinian worldview, narrated by a ghost from one million years in the future. Through a series of random accidents during a 1986 financial meltdown, a group of hapless castaways includes the last human females untouched by a virus that destroys human ovaries. They colonize the island of Santa Rosalia, ensuring humanity's survival. The narrator says that humans' "big brains" used to cause a lot of trouble—but after a million years in the Galapagos, we're basically seals.

Connection: *Galapagos* is a creation myth about evolution, highlighting the randomness of the events that cause changes across the bigger time scale. Great for thinking about the logic of evolution and for pondering possible futures. (RBS)

Wade, Nicholas. *Before the Dawn: Recovering the Lost History of Our Ancestors.* New York: Penguin, 2006. Print.

Category: Nonfiction. 296 pages.

Summary: *New York Times* science writer Nicolas Wade convincingly demonstrates how recent scientific knowledge of the human genome, since 2003, has helped "create a new and far more detailed picture of human evolution, human nature and history" than ever before. In twelve chapters, each starting with a substantial quote from Charles Darwin's *The Descent of Man*, Wade explores big themes in human development, such as race, settlement, sociality, history, and language, to show how effective today's science of genetics, like yesterday's compilers of the Book of Genesis, can "frame a coherent account of human origins."

Connection: This book provides a great deal of information for students wanting to know more about early human development or Threshold 6, especially from the newest developments in genetics. Wade provides example after example of how genetics can fill in the gaps of 90 percent of human history that has been "irretrievably lost." (CT)

The War of the Worlds. Dir. Byron Haskin. Perf. Gene Barry, Ann Robinson. Prod. George Pal. Based on the novel by H. G. Wells. Universal Pictures, 1953. DVD.

Category: Film. Science fiction. 85 minutes.

Summary: This film is an adaptation of the H. G. Wells novel, set in 1950s California.

Connection: Substituting post–World War II America for the British empire, this film adaptation beautifully updates the story of a world superpower being humbled by technologically superior Martian invaders. When even the atomic bomb fails to stop the Martians, what hope can there be for humanity? The film's ending emphasizes the novel's suggestion that the process of natural selection is part of God's creation. (DM)

Weintraub, David A. *How Old Is the Universe?* Princeton: Princeton UP, 2011. Print.

Category: Nonfiction. 363 pages.

Summary: Vanderbilt University professor of astronomy Weintraub traces the history of scientific inquiry into determining the age of the universe. Beginning with calculating the age of objects in our solar system, and working outward to stars, to beyond the Milky Way galaxy, the evidence for concluding that the universe's age is 13.8 billion years is clearly explained.

Connection: While generally readable for nonscientists, this book does include technical explanations that some may have difficulty following. For teachers of Big History, especially, the book is useful for understanding a wide variety of astronomical phenomena, including expansion of the universe and dark matter, and is a good resource for examining how our knowledge has developed throughout history. (NW)

Weisman, Alan. *The World without Us.* New York: Thomas Dunne, 2007. Print.
Category: Nonfiction. 336 pages.
Summary: What would happen to humankind's great works of engineering should our species disappear in a flash? Weisman illustrates the decay, in time lapse of our own homes, the bridges and tunnels of Manhattan, the Panama Canal, nuclear power plants, and more. He visits sub-Saharan Africa and the Korean Demilitarized Zone—biodiversity hotbeds that would repopulate Afro-Eurasia with now-rare fauna.
Connection: Weisman's text connects geology, natural history, climate, archaeology, anthropology, and engineering to shed light on the transitory nature of human works. Useful in teaching industrialization and possible futures—and in contextualizing our place in time in the recurring pattern in human civilizations of resource overexploitation and collapse. The solution, Weisman makes clear, is to control population. He explores the matter in depth in a 2013 follow-up, *Countdown.* (RBS)

Wells, H. G. *The War of the Worlds.* 1898. New York: Bantam Dell, 2003. Print.
Category: Novel. Science fiction. 194 pages.
Summary: Writing at the end of the nineteenth century, H. G. Wells creates a trend-setting vision of Earth invaded by extraterrestrials.
Connection: Excellent portrayal of the anxieties that came with modernity (Threshold 8). The moral consequences of global imperialism and colonization are explored in a wry role reversal, as England finds itself under assault by technologically superior Martians, whose war machines anticipate the horrors of industrialized warfare. Cutting-edge nineteenth-century science is vividly discussed and integral to the plot, especially the ending, which depends on the theory of natural selection. (The book works well in combination with the 1953 film version; see separate entry). (DM)

Whitman, Walt. "When I Heard the Learn'd Astronomer." *Leaves of Grass.* Ed. Harold W. Blodgett and Sculley Bradley. New York: New York UP, 1965. 271. Print.
Category: Poem. 1 page.
Summary: Whitman's short (eight-line) poem expresses the need to go beyond "facts" about the universe in order to fully appreciate it through the immediacy of pondering the night sky.

Connection: Poems often express feelings and understanding that cannot be adequately captured by linear thought or explanation. Adding variety to the study of Big History, Whitman's poem takes us to a more intuitive and intimate comprehension of the mystery and magnificence of the universe through direct observation, rather than just intellectual accumulation of information about it. (NW)

William, Ian. *Rum: A Social and Sociable History.* New York: Nation Books, 2005. Print.

Category: Nonfiction. 340 pages.

Summary: Journalist Ian Williams has written an accessible history of the development of rum, the rum industry, and the consequences of the spirit'sconsumption.

Connection: Despite taking the opportunity to make many a play on words, like a chapter titled "The Spirit of 1776," this is a serious history that deals with human activity involving a truly addictive substance: alcohol. Connections can be made to economic behaviors that can lead to personally destructive results and how culture masks or justifies the consumption of alcohol, despite its negative social and personal impacts. (MA)

Wrangham, Richard. *Catching Fire: How Cooking Made Us Human.* New York: Back Books, 2009. Print.

Category: Nonfiction. 309 pages.

Summary: In this engaging study, anthropologist Richard Wrangham explains how "culture is the trump card that enables humans to adapt," especially because humans learned about the values and advantages of cooking. Wrangham believes that "the transformative moment that gave rise to the genus *Homo* . . . stemmed from the control of fire and the advent of cooked meals." He develops his cooking hypothesis in eight easy-to-read chapters that cover such topics as the human body's evolution, the growth of the human brain, and how cooking may have been a factor in the development of later patriarchal societies.

Connection: Wrangham's accessible study may be of interest to young college students learning about Threshold 6, which explores the evolution of modern-day human culture from primate cultures. (CT)

Zalasiewicz, Jan. *The Planet in a Pebble: A Journey into Earth's Deep History.* Oxford: Oxford UP, 2010. Print.

Category: Nonfiction. 251 pages.

Summary: Zalasiewicz presents a summary of universe history told through the story of one pebble of grey slate from a Welch beach or gravel bed, rendering technical information accessible and engaging.

Connection: This is a model Little Big History in book length, featuring wonderful accounts of the Big Bang, Earth's structure and currents within it, and the final end of our planet and star system. No geologic time scale is included; for this, refer to Zalasiewicz and Williams's *The Goldilocks Planet*. (CB)

Zalasiewicz, Jan, and Mark Williams. *The Goldilocks Planet: The Four Billion Year Story of Earth's Climate.* Oxford: Oxford UP, 2012. Print.

Category: Nonfiction. 256 pages.

Summary: The authors are geologists at the University of Leicester, England. They have compiled into one coherent account the story of climate change over the whole of Earth's history. They accept that the Anthropocene began about 1800 C.E., that the current warming is much more likely than not caused by humans, and that it is likely to continue for decades, centuries, and millennia. At the heart of the book is the mystery of what keeps the Goldilocks phase of our planet going.

Connection: This is an invaluable resource for teachers of Big History—it provides climate data for all periods of Earth's history, correlated with geologic time scales and geologic terms, complete with a useful discussion of our current situation. (CB)

Zuckerman, Larry. *The Potato: How the Humble Spud Rescued the Western World.* New York: North Point Press, 1998. Print.

Category: Nonfiction. 320 pages.

Summary: Zuckerman tells the story of the potato, from its cultivation by the Incas through its introduction into Europe and its eventual triumph as an accepted and central part of the European diet.

Connection: This work is valuable for its insights on the relationship between culture and food. The main part of the work is the long struggle for the potato to even gain acceptance as human food in Europe. Through his discussion of the potato famine in Ireland, Zuckerman provides a cautionary tale of how cultural dependence can lead to catastrophe. (MA)

CONTRIBUTORS

MARTIN ANDERSON (PhD, history, Stanford University, 2000) is Associate Professor of History and served as Associate Dean of Arts, Humanities and Social Sciences at Dominican University of California. His main field of study is modern British history, primarily the British empire. His publications include "Tourism and the Development of the Modern British Passport, 1814–1858," *Journal of British Studies* 49 (April 2010) and "The Development of British Tourism in Egypt, 1815–1850," *Journal of Tourism History* 4 (November 2012).

ETHAN ANNIS (MLS, University of North Carolina at Chapel Hill, 1990) is Head of Access and Technical Services Librarian at Dominican University of California.

MOJGAN BEHMAND (PhD, English and American studies, University of Dusseldorf, Germany, 1998) serves as Associate Provost and is Associate Professor of English at Dominican University of California, where she was the recipient of the Teacher of the Year award in 2010. She was cofounder and director of Dominican's First-Year Experience "Big History" program and has published and presented widely on Big History, in addition to articles on medieval Persian epics and contemporary Iranian novels. She serves on the board of the International Big History Association.

KIOWA BOWER (PhD, biochemistry and molecular biophysics, California Institute of Technology, 2009) is Assistant Professor of Biology at Dominican University of California.

CYNTHIA STOKES BROWN (PhD, history of education, Johns Hopkins University, 1964) taught history and education at Dominican University of California from 1981 to 2001. She serves on the board of the International Big History Association. She has written

Big History: From the Big Bang to the Present (New Press, 2007 and 2012) and, with David Christian and Craig Benjamin, *Big History: Between Nothing and Everything* (McGraw-Hill, 2014).

THOMAS BURKE (MFA, writing, University of San Francisco, 1996) is Assistant Professor of English and Social and Cultural Studies at Dominican University of California. He is the author of *Where Is Home and Other Stories* (Fithian, 2005) and numerous anthologized pieces of fiction and critical essays.

JAIME CASTNER served as Program Coordinator for the general education program at Dominican University of California, where she is working toward her MA in humanities. She is the author of "A Cosmic Journey from English Lit to Big History," *Metanexus* (2013) and coauthor with Mojgan Behmand of "Big History, Big Lesson," *Metanexus* (2012).

JAMES B. CUNNINGHAM (PhD, zoology, University of Canterbury, Christchurch, New Zealand, 1986) is an ornithologist and Associate Professor of Biology at Dominican University of California. He has performed extensive research on native birds of New Zealand and authored numerous publications on the subject.

ROBIN CUNNINGHAM (PhD, mathematics, University of Michigan, 1991) is a lecturer in Statistics and Operations Research at the University of North Carolina, Chapel Hill. He was awarded a fellowship in the Society of Actuaries and has taught mathematics of physics at every level from eighth grade to graduate school. He is the author, with Richard Herzog and Richard London, of *Models for Quantifying Risk*, 5th Ed. (Actex Publications, 2012.)

DEBBIE DAUNT (MSN, University of Missouri, Columbia, 2003) is a registered nurse and Assistant Professor of Nursing at Dominican University of California. Her graduate research area was lymphedema as a complication of breast cancer treatment. She is pursuing a doctorate of nursing science at the University of Hawaii, Manoa. Her current study focuses on how nursing students impact hospital safety and quality of patient care.

LINDSEY DEAN is currently working toward her PhD in interdisciplinary studies (ethics and religious studies) at the Graduate Theological Union in Berkeley, California. She has been an instructor at Dominican University of California since 2009. She presented "Big History: The Modern Creation Myth and Its Contribution to Global Ethics" at the inaugural International Big History Association conference, and published the work in Dominican's Big History journal, *Thresholds*.

AMY GILBERT (MLIS, San Jose State University, 2002) is Associate Director and Coordinator of Assessment and Information Literacy at Dominican University of California. She is the author, with Barbara Jean Ganley and Diane Rosario, of "Faculty and Student Perceptions and Behaviours Related to Information Literacy: A Pilot Study Using Triangulation," *Journal of Information Literacy* 7.2 (2013).

JUDITH HALEBSKY (PhD, performance studies, University of California, Davis, 2009; MFA, poetry, Mills College, 1998) is Assistant Professor of Literature and Language at Dominican University of California. She trained in Japanese literature at Hosei University in Tokyo, on a fellowship from the Japanese Ministry of Education, and has published articles and book reviews in the *Asian Theatre Journal, Theatre Research in Canada*, and *Canadian Literature*, among others. Her collection of poems, *Sky = Empty*, won the New Issues Poetry Prize.

SEOHYUNG KIM (PhD, American history, Ewha Womans University, Seoul, Korea, 2011), whose field is American history, is the first to teach Big History at different levels (university, middle school, high school) in South Korea. She translated David Christian's *This Fleeting World: A Short History of Humanity* into Korean and is now translating *Berkshire Essentials: Big History*. Her current project is a twenty-book Big History series in Korean for middle school and high school students. She has participated in a Korean Big History Pilot program with bgC3, a teachers' workshop for extracurricular Big History classes in Korean schools, and presented in October 2013 at TEDxBusan.

JENNIFER LUCKO (PhD, cultural anthropology, University of California, Berkeley, 2007) is an Associate Professor in the School of Education at Dominican University of California. Her recent publications include "Here Your Ambitions Are Illusions: Boundaries of Integration and Ethnicity among Ecuadorian Immigrant Teenagers in Madrid," *The Journal of the History of Childhood and Youth* 7.1 (Winter 2014) and "Quiero estar con mi gente: La negociación de la identidad étnica en la escuela," in *Negociaciones identitarias de la población migrante* (Identity Negotiation among the Migrant Population), edited by Matilde Fernández Montes (Commonground Publishing, 2013).

J. DANIEL MAY (MA, English with an ESL emphasis, San Francisco State University, 1995) has taught English as a second language, linguistics, writing, and literature at Dominican University of California since 1996. He currently serves as director of First-Year Experience "Big History."

MARTIN NICKELS (PhD, anthropology, University of Kansas, 1975) is Professor Emeritus of Anthropology at Illinois State University and Adjunct Professor of Anthropology at Illinois Wesleyan University.

PHILIP NOVAK (PhD, religion, Syracuse University, 1981) is Professor of Philosophy and Religion at Dominican University of California, where he has taught for thirty-three years. He is a recipient of Dominican's "Teacher of the Year" award and is one of its Sarlo Distinguished Professors. While Dean of the School of Arts, Humanities and Social Sciences he conceived of the requirement for freshmen in Big History, the first such program in the world. He is the author of *The World's Wisdom* (Harper, 1994), *The Vision of Nietzsche* (Element, 1996), and *The Inner Journey: Views from the*

Buddhist Tradition (Morning Light, 2005), and coauthor with Huston Smith of *Buddhism: A Concise Introduction* (Harper, 2003).

BILL PHILLIPS (PhD, psychology, University of Nebraska, Lincoln, 1997) is Professor of Psychology at Dominican University of California, where he conducts research examining cultural differences in perception.

MAIRI PILEGGI (PhD, media and communication, Temple University, 1998) has published in such journals as *Cultural Studies, New Media & Society,* and *The Electronic Journal of Communication.* She has contributed chapters to *The Power of Global Community Media* (MacMillan, 2011) and *Restoring and Sustaining Lands: Coordinating Science, Politics, and Action* (Springer, 2012).

ESTHER QUAEDACKERS (M.Sc., Eindhoven University of Technology, 2006) is a lecturer in Big History at the University of Amsterdam and a board member of the International Big History Association. She invented the "Little Big History" approach and is currently developing it as an educational and research tool.

ANNE REID (MLIS, San Jose State University, 2007), Certified Archivist, is Reference and Instruction Librarian at Dominican University of California. Reid is the First-Year Experience "Big History" library liaison as well as University Archivist.

SUZANNE ROYBAL (MLIS, San Jose State University, 2005) is Instruction, Reference and Outreach Librarian at Dominican University of California. She is the author, with Madalienne Peters, of "Faculty-Library Collaboration: Embedding Information Literacy in Educational Research Graduate Classes," Education Resources Information Center (eric.ed.gov), 12 July 2012.

ALAN SCHUT (MLIS, University of California, Berkeley, 1990; MA English, University of Oregon, Eugene, 1975) is Senior Librarian at Dominican University of California.

RICHARD B. SIMON (MFA, creative writing, fiction, San Francisco State University, 2002) is Adjunct Assistant Professor of Literature and Languages at Dominican University of California, and also teaches creative writing, composition, and critical thinking at City College of San Francisco. He is a contributing editor at *Relix* magazine. His writing on music, arts, culture, current affairs, and environmental issues has appeared in *Rolling Stone, Juxtapoz,* at MTV / VH1.com, on San Francisco's NPR affiliate KQED, and elsewhere.

LYNN SONDAG (MFA, painting, California College of the Arts, 1997) is Chair of the Art Department at Dominican University of California. An artist who exhibits nationally and internationally, she also has a strong and diverse background in arts education. In 2011, she received the Teacher of the Year award at Dominican.

HARLAN STELMACH (PhD, religion and society, Graduate Theological Union, Berkeley, 1977; MTS, Harvard Divinity School 1970) is Professor of Humanities at Dominican University of California. His specialty is the interdisciplinary field of religion and society. He is the former director of the Center for Ethics and Social Policy and is the coauthor of *Doing Ethics in a Diverse World* (Westview Press, 2007) with Robert Traer.

CYNTHIA TAYLOR (PhD, American religious history, Graduate Theological Union, Berkeley, 2003), an American historian, is Assistant Professor of Religion and History at Dominican University of California. Taylor is the author of *A. Philip Randolph: The Religious Journey of an African American Labor Leader* (NYU Press, 2005) and a contributor to the forthcoming anthology *Reframing Randolph*.

NEAL WOLFE (MA, humanities, Dominican University of California, 2001) was Assistant Professor and Director of the First-Year Experience program at Dominican University of California.

INDEX

AAC&U. *See* Association of American Colleges and Universities

absorption lines, redshift or blueshift, 94, 105–7, 123, 129, 346

accommodation, as component of awe, 338

accountability, assessment goal, 41

accretion: activities and exercises for teaching, 144, 146–47; defined, 138, 140

activities, for multiple thresholds, 275–286; Amateur Astronomers Star Party, 275, 279–280; Annotated Bibliography assignment, 275, 281–82; classroom recommendations, 352–53; Eight Thresholds Video Project, 275, 280–81; Little Big History Essay, 275–76, 282–86; Walking the Big History Timeline activity, 275, 277–78. See also *under specific thresholds*

actual luminosity, 124–25

adaptive systems, complex, 80, 240

adenosine triphosphate (ATP), 154

affective assessment tools, 46

Africa: behavioral innovations of Middle Stone Age, 192; human evolution in, 70, 166, 174, 195

Afro-Eurasian world zone, trade routes, 216–17

Age of Information, 222

Agrarian era, 166

agrarian revolution (Threshold 7), 3, 6, 14–15, 16, 201–14; animals susceptible to cultivation, 203–4; Big History Project, 59; challenges in teaching, 209–10; classroom recommendations, 350; collective learning activity, 211–14; complexity features, 205–8, 212; disciplines for studying, 17; ethical regimes developed during, 20; fertilization, 205; health and healing perspectives, 252–54; Holocene epoch, 166, 168–69; human and animal labor limits, 204, 208; irrigation, 205; land, water, and nutrient requirements, 204; learning outcomes and assessments, 208–9; in Little Big History, 286,

atom bomb, 220

atoms, defined, 121

Australia, Big History programs, 26n3, 58, 302, 303

australopithecine fossils, 165–66

Australopithecus afarensis ("Lucy"), 14, 165–66; hominoid skull comparison lab, 180–89; hominoid skull lab, 176–79

Australopithecus boisei: hominoid skull comparison lab, 180–89; hominoid skull lab, 176–79

autopoeisis, 159

autotrophs, evolution of, 150

awe, the case for, 336–341

Axial Age, 299

backward-design principles, 43, 55n11

Bacon, Francis, 218

bacteria, as endosymbionts, 252

Bain, Bob, 57–58, 62, 63

Baker, David, 64

baking with complexity activity, 86–87

Barlow, Connie, 304

Barringer Crater (Arizona), 300–301

beak types (birds), natural selection simulation activity, 157–59

Becoming Human (NOVA film), 166–67

Behmand, Mojgan, 30, 35, 303, 315

Beld, Jo, 55n11

Bellah, Robert, 327

Benjamin, Craig: "The Convergence of Logic, Faith and Values in the Modern Creation Myth," 322–23; curriculum development role, 34; IBHA role, 303. See also *Big History: Between Nothing and Everything*

Berry, Thomas, 300, 301, 304, 312–14, 316, 358

Bhagavad Gita, 339–340

Big Bang (Threshold 1), 3, 6, 14, 16, 91–108; Big History Project, 59;

challenges in teaching, 97–98; classroom recommendations, 345–46; complexity, 95–96; concept overview, 93–95; conclusion, 98; creation myth creative writing workshop, 99–100; creation myth tunnel book activity, 101–5; disciplines for studying, 17; evidence of Big Bang, 93–94; health and healing perspectives, 250–51; learning outcomes and assessments, 97; in Little Big History, 285, 296; origin story comparisons, 91–93; redshift demo activity, 105–7; reflective writing questions, 269; South Korean curriculum, 68, 69, 70

Big Big History, 57, 65–66

Big History and the Future of Humanity (F. Spier), 65, 302

Big History: Between Nothing and Everything (D. Christian, C. Stokes Brown, C. Benjamin), 29, 77, 327; collective learning, 27–28; complexity model, 15–17, 78, 304–5; eight stages of life, 349; modern origin story, 91–92, 319, 321; religious discussions, 328

Big History Colloquium, 314

Big History Education Program for Talented Students (South Korea), 69

Big History: From the Big Bang to the Present (C. Stokes Brown), 29, 43–44, 215, 303, 314

"Big History Movie Night," 77–80

Big History overview, 1–3; approaches, 304–5; challenges for teaching, 4–6, 18–19, 355–57; classroom recommendations, 344–353; cultivating a sense of awe, 336–341; goals, 8, 12, 343–44, 358; investigative tools, 12–13; liberal education goals and, 21–26; as modern creation story, 319, 322–23, 329–330, 333, 336, 353; origin of program, 301–4; program

Chambers, Robert, 299–300
Chaplin, Charlie, *Modern Times*, 77–80
chemical differentiation, 140
chemical elements, formation of
(Threshold 3), 3, 6, 14, 16, 120–134;
Big History Project, 59; challenges in
teaching, 129; classroom
recommendations, 347; complexity,
127–28; conclusion, 129; disciplines
for studying, 17; found poem activity,
132–33; gravity vs. fusion activity,
130–31; hydrogen atom illustration,
80–81; learning outcomes and
assessments, 128; in Little Big History,
285, 296; Netherlands' program, 67;
reflective writing questions, 269;
stellar nucleosynthesis concept,
121–27, 347; Threshold 2 emergent
properties giving rise to, 114–15, 347
chemical evolution concept, 152
chemistry, in thresholds of complexity
framework, 17
childbirth: agricultural revolution role,
210; human vs. chimpanzee, 168;
genetically engineered children, 246
chimpanzees: common ancestor with
humans, 165, 297; compared to
modern humans, 167–68, 297;
hominoid skull comparison lab,
176–89;
China: Industrial Revolution role, 206–7;
relationship with Britain in 1700s, 208
chloroplasts, in eukaryotic cells, 154
Christian, David: Australian Big History
curriculum, 302, 303; Big History
Project role, 57–63, 303–4;
chronometric revolution, 300–301;
curriculum development role, 56, 324;
Dutch Big History curriculum, 302;
South Korean Big History curriculum,
69. See also *Big History: Between
Nothing and Everything*; *Maps of
Time*

Christianity, 299, 324–25. *See also*
creation myths (origin stories)
chromosome numbers, human vs.
chimpanzee, 168
chronometric revolution, 91, 300
ChronoZoom, 303
Civil Wars: American, 216, 219; Spanish,
219
claim testing, Big History Project, 59, 60
climate, impact on agriculture, 204, 205,
210. *See also* global climate change
closing vs. opening the loop, assessment
goals, 41, 42, 54n2
coal: Anthropocene origin, 220–22;
Industrial Revolution requirements,
207, 218–19. *See also* fossil fuels
cognitive assessment tools, 46
Cold War, 220
collective learning: to address religious
questions, 320; agricultural
civilizations activity, 211–14;
agricultural revolution role, 206, 350;
Big History Project, 59; defined, 18,
27–28; exponential acceleration, 245,
246–47; innovation role, 15, 217–18; as
meta-education (in faculty
development) 18, 27–40; program
development as, 353; as Threshold 6
emergent property, 165, 168, 172–73;
at weekly assessment meetings, 32, 36
Columbia College's learning portfolio
model, 264–65
Columbia History of the World, 301
Common Core Standards for Literacy, 62
communication: collective learning and
improvements in, 217; exponential
acceleration through technology, 244,
246–47; Gutenberg's printing press,
218; through language, 167–68, 297.
See also internet
communism, 219; Marx and Engels on
the Anthropocene activity, 228–230
The Communist Manifesto, 228–231

complex adaptive vs. nonadaptive
systems, 80, 240

complexity model (eight thresholds), 3,
6, 15–17; baking activity, 33, 34,
86–87; Big History Project, 59, 60;
blocks activity, 88–90; challenges in
teaching, 18–19, 83; classroom
recommendations, 80–83, 344–45,
356–57; complexity in pedagogy,
83–85; learning outcomes and
assessments, 83; prehistory, 12–13;
teaching strategies, 77–90. *See also*
four features of complexity; *specific
thresholds*

compliance mentality, assessment goal, 41

comprehension, in Bloom's taxonomy, 264

computers, Moore's law, 245. *See also*
internet

consciousness, advances in, 247–49, 312,
330

consumption and consumerism:
capitalism, 217, 219, 227–230; waste
generated by, 242; worldwide patterns
of, 235–36, 351

content approaches to religious topics,
329–331

continental drift, 141, 237–38

convergence, of early human cultures,
217

convergence education, 68–69, 71, 72

"The Convergence of Logic, Faith and
Values in the Modern Creation Myth"
(C. Benjamin), 322–23

Copernicus, Nicolaus, 137, 299, 320

corporations, 219, 227–28

cosmic background radiation (CBR), 94,
110–11, 115; inflation balloons activity,
117–19

Cosmic Education curriculum,
elementary education, 26n3

*Cosmic Evolution: The Rise of Complexity
in Nature* (E. Chaisson), 241, 301–2

cosmogenesis, defined, 248

cosmology: challenges in teaching,
355–56; creation myth creative writing
workshop, 99–100; creation myth
tunnel book activity, 101–5; early
astronomers, 137, 299–300, 320;
evidence-based, 92–93; in thresholds
of complexity framework, 17; student
reactions to studying, 116. *See also*
creation myths (origin stories)

Cosmos (C. Sagan), 301, 305

*Cosmos, Earth, and Man: A Short History
of the Universe* (P. Cloud), 301

Cosmos television series, 301

creation myth creative writing workshop,
99–100

creation myths (origin stories): modern
creation story, 319, 322–23, 324,
329–330, 333, 336, 353; of students'
faith traditions, 91–93, 97–98, 143,
339–340; teaching challenges related
to, 97, 143, 320–22; Thomas Berry
on, 313; urbanization and changes in,
298–99; value of, 319. *See also*
religious beliefs, teaching Big
History and

creation myth tunnel book activity,
101–5

"Creative Writing and Big History," 22

crop rotation, 205

cultivation: land, water, and nutrient
requirements, 204; plants and animals
susceptible to, 203–4; preparing land
for, 204–5

cultural evolution, 240, 248

Cunningham, James, 30, 303

curiosity, igniting critical, 4, 287–295. *See
also* information literacy

curriculum development: backward-
design principles, 43, 55n11; Big
History Project, 58; complexity model
and, 83–85; first year at Dominican,
32–33, 42; second year at Dominican,
35–36; third year at Dominican,

36–40. *See also* summer institutes, Dominican

cyanobacteria, 151

Daring Kings Play Chess On Fine Grain Sand (mnemonic device), 169–170, 174

dark matter and dark energy, 93, 95, 113

Dart, Raymond, 166

Darwin, Charles, 150, 299, 300

Darwin's Cathedral: Evolution, Religion and the Nature of Society (D. Sloan), 328

Dean, Lindsey, 34

deGrasse Tyson, Neil, 301

density differences, concept overview, 110–11, 112

Descartes, René, 218

detachment, Paul Tillich on, 326

Dewey, John, 47

dinosaurs: evolution of, 152, 349; extinction of, 152, 166, 172, 221, 297, 303, 349

diseases: biotechnology advances to reduce/eliminate, 245–46; defined, 251; health and healing perspectives, 250–54; labor shortages due to, 204; Malthusian cycles, 217; plague simulation activities, 34, 252–53; susceptible vs. resistant populations, 216, 251–52

divergence, of early human cultures, 216

diverse components, complexity feature, 15, 78, 79; Big History classroom illustration, 81; hydrogen atom illustration, 80; Threshold 1, 95–96; Threshold 2, 113; Threshold 3, 127; Threshold 4, 142; Threshold 5, 154; Threshold 6, 171; Threshold 7, 205, 212; Threshold 8, 223; Threshold 9, 240–41, 248

DNA: discovery of structure, 301; evolution of, 152, 153; in humans vs. chimpanzees, 168

Dominican University of California, 6; approach to religious topics, 327–333; guest speakers, 327; incorporating Big History curriculum at, 21–26; model questions about Big History program, 266–67; origin of Big History at, 311–17. *See also* summer institutes, Dominican

Doppler effect, 94, 97, 105–6

$E = mc^2$, 113, 115

Earth, formation of (Threshold 4), 3, 6, 14, 16, 135–148, 320; accretion, 138, 140; accretion activity, 144, 146–47; age of Earth, 139, 300–301; Big History Project, 59; challenges in teaching, 143–44, 320–22, 355; chemical differentiation, 140; classroom recommendations, 347–48; complexity, 142; conclusion, 144–45; continental drift, 141; disciplines for studying, 17; early studies, 137, 299; Hadean eon, 140–41; learning outcomes and assessments, 143; in Little Big History, 285, 297; plate tectonics, 141; radiometric dating, 139, 300–301; reflective writing questions, 269–270; remote future projections, 238; solar nebular theory, 138–39, 143; teaching models and exercises, 144

Earth as center of universe, 100, 137, 320, 355

earthquakes, fracking and, 222

eBrary database, 291

economic systems: capitalism, 217, 219, 227–230; communism, 219, 228–230; consumerism, 235–36, 242, 351; fossil-fuels based, 215

Eight Thresholds Video Project, 275, 280–81

Eindhoven University of Technology (Netherlands), 64

Einstein, Albert, 113, 299

Eiseley, Loren, 312
Elder Hostel, 312
electricity, solar- and wind-powered generation of, 220, 246. *See also* fossil fuels
electrons, defined, 121
emergent properties, complexity feature, 16, 17, 78, 79; Big History classroom illustration, 82–83; hydrogen atom illustration, 81; Threshold 1, 95–96; Threshold 2, 114; Threshold 3, 128; Threshold 4, 142; Threshold 5, 154; Threshold 6, 172–73; Threshold 7, 206; Threshold 8, 224; Threshold 9, 241–42
Emerson, Ralph Waldo, 339
emotional intelligence, 246
empathy, Paul Tillich on, 326
encephalization, human evolution and, 167, 168
end-of-life decisions, 253–54
The End of Oil (P. Roberts), 218–19
endosymbiosis, 151, 153, 252
energy, 220–22; economic considerations, 215; Einstein's relativity theory, 113, 299; impact on global trade networks, 217; near future projections, 235; nuclear power, 220; solar power, 220, 246; wind power, 220. *See also* fossil fuels
engaged learning, 33–34
entropy, 150, 248
environmental damage: agricultural revolution role, 210; air pollution, 235–36, 351; habitat loss, 236; near future projections, 235; student awareness of, 234; water pollution, 236, 351; worldwide consumption patterns and, 235–36. *See also* global climate change
enzymes, evolution of, 153
epistemology, challenges in teaching, 355–56

Erasmus University College (Netherlands), 64
Erathosthenes, early Greek astronomer, 137
essay questions, curriculum development, 37–38
"Essential Learning Outcomes," AAC&U, 42, 288–89, 290
ethoi, defined, 248–49
eukaryotic cells: components of, 154; evolution of, 151, 154–55, 172, 349
evaluation, in Bloom's taxonomy, 264
Evolutionary Epic (R. and C. Genet), 304
evolution of life on earth (Threshold 5), 3, 6, 14, 16, 149–163; Big History Project, 59; cell membranes, 153; challenges in teaching, 155–56; chemical evolution concept, 152; classroom recommendations, 348–49; complexity, 154–55; conclusion, 156; cultural vs. biological evolution, 240; disciplines for studying, 17; early stages of life, 172; endosymbiosis, 153; genetic material, 153; health and healing perspectives, 251–52; learning outcomes and assessments, 155; in Little Big History, 285–86, 297; microorganism design principles activity, 159–162; natural selection, 153; natural selection simulation activity, 157–59; reflective writing questions, 270; remote future projections, 237–38; South Korean curriculum, 69, 70
Evolving World, Converging Man (R. Francoeur), 312
Ewha Womans University (South Korea), 57, 68–69, 71
expansion of perception, measuring, 47–48
experiential value, measuring, 47–48
extinctions: agricultural revolution role, 210; climate change role, 221, 236; of

dinosaurs, 152, 166, 172, 221, 297, 303, 349; habitat loss and, 236
extremophiles, survival techniques, 252

faculty development: assessment role, 41–42; assessment tools, 39; Big History Project, 61; collective learning in, 18, 27–40; semesterly retreats, 22, 39, 354; motivation strategies, 42; recommendations for, 354–55; retreat to discuss religious beliefs, 327, 331–32; weekly lunch meetings, 22, 28, 33, 38–39, 354. *See also* summer institutes, Dominican
faith, importance of addressing, 322–23, 325, 328, 331. *See also* religious beliefs
farmworkers, working conditions, 202
fascism, 219
female and male hierarchies, human vs. chimpanzee, 168
fermentation, energy obtained through, 151
fertilizing plants, 205, 235
fire, use of, 168
The Firmament of Time (L. Eiseley), 312
flooding: at end of ice age, 216; impact on agriculture, 205, 210
flows of energy, complexity feature, 17, 78, 79; Big History classroom illustration, 82; hydrogen atom illustration, 80–81; Threshold 2, 114; Threshold 3, 127; Threshold 4, 142; Threshold 5, 154; Threshold 6, 171–72; Threshold 7, 206; Threshold 8, 223; Threshold 9, 241, 248, 249
focus groups, student-led, 48–49, 55n19; First-Semester Little Big History Paper Rubric, 50–51; Second-Semester Big History Paper Rubric, 53; values and beliefs issues, 49
food scarcity and famines, 202, 210; labor shortages due to, 204; Malthusian

cycles, 217; near future projections, 234–35
forager era, South Korean curriculum, 69, 70
fossil fuels: Anthropocene origin, 220–22; coal, 207, 218–19, 220–22; discovery of, 15; economic systems based on, 215; environmental damage caused by, 220–21; global climate change role, 15, 236; industrialization and, 15, 216–17, 299, 351; natural gas, 219, 222; oil, 219; wealth produced by, 219, 220. *See also* global climate change
found poem activity, 132–33
four features of complexity, 15–16; activities and exercises for teaching, 78–79; baking activity, 86–87; blocks activity, 88–90; challenges in teaching, 83; complexity in pedagogy, 83–85; illustrating in classrooms, 80–83; learning outcomes and assessments, 83; *Modern Times* illustration, 79; Threshold 1, 95–96; Threshold 2, 113–15; Threshold 3, 127–28; Threshold 4, 142; Threshold 6, 170–73; Threshold 7, 205–8, 212; Threshold 8, 225; Threshold 8 emergent properties giving rise to, 223–24; Threshold 9, 240–42, 248. *See also* diverse components; emergent properties; flows of energy; specific arrangements
FOXP2, language-related gene, 168
fracking (hydrofracturing), 222
Free University (Netherlands), 64–65
fusion reactions, 121–22; gravity vs. fusion activity, 130–31; star and galaxy formation and, 113, 346–47
future or possible futures (Threshold 9), 16, 232–260; advances in consciousness, 247–49, 312, 330; Big History Project, 59; challenges in

future or possible futures *(continued)*
teaching, 242–44; classroom
recommendations, 351–52; complexity,
240–42, 248; focus group findings, 52;
foregrounding in every threshold, 233,
249–254; key concepts in teaching,
234–38; in Little Big History, 286;
meta-educational concepts, 18; middle
future, 237, 351; near future, 234–37,
351; opinion snake activity, 255;
post-apocalyptic film festival, 256–59;
reflective writing questions, 273;
remote future, 237–38, 351;
technological revolution, 244–47;
Threshold 8 emergent properties
giving rise to, 223; threshold concept
applied to, 238–240

Gabor, Father Paul, 327
galaxies, formation of. *See* stars and
galaxies, formation of
Galileo Galilei, 299
gametes, in sexual reproduction, 151
Garraty, John, 301
Gates, Bill, Big History Project, 26n3, 56,
57–63, 303–4, 315
Gay, Peter, 301
Genet, Russ and Cheryl, 304
genetically modified food, 235, 241
genetic material (RNA and DNA),
evolution of, 152, 153
genus, in scientific taxonomy, 170, 171,
174
geoengineering, to remove carbon
dioxide, 222
geography, in thresholds of complexity
framework, 17
Geological Observatory of Coldigioco,
Italy, 303
geology, in thresholds of complexity
framework, 17
global citizenship, importance of, 69
global civilization, self-awareness of, 15

global climate change: as anthropogenic,
355; fossil fuels role, 15, 236; as "a
hoax," 225; impacts of, 20, 221–22, 236,
351; middle future projections, 237;
near future projections, 235, 236–37
global era, South Korean curriculum, 69,
70–71
globalization, origin of, 217
gold, elemental, 120–21, 347
Goldilocks conditions, 67; for early
galaxies, 113; energy flows, 16, 61, 78,
248; for human evolution, 171–72;
increasing complexity, 59; for
Industrial Revolution, 208; for life on
Earth, 85, 115, 142, 347, 348
Goodenough, Ursula, 304
gorillas: hominoid skull lab, 176–89
Goudsblom, Joop, 63–67, 302
gravity: concept overview, 109–10, 296,
346; Newton's theory, 299; solar
nebular theory, 138
gravity vs. fusion activity, 122, 130–31,
346–47
greenhouse gases (methane and carbon
dioxide), 152, 221, 257. *See also* global
climate change
Gustafson, Lowell, 303
Gutenberg printing press, 218

habitat loss, 236
Hadean eon, 140–41
Haidt, Jonathan, 338
health and healing, defined, 251
"Health and Healing through the Lens of
Big History," 22, 249–250
health-care professionals: academic
demands on, 250; Threshold 1
applications, 250–51; Thresholds 5 and
6 applications, 251–52
helium, hydrogen fusion and, 93, 121–22,
127, 296, 346, 347
Hertzsprung-Russell Diagrams, 125, 126,
129

industrialization and modernity *(continued)*

teaching, 225; classroom recommendations, 350–51; collective learning and innovation, 217–18; complexity, 223–24, 225; conclusion, 225–26; cosmological paradigms resulting from, 91; disciplines for studying, 17; fossil fuels role, 216–17, 220–22, 299, 351; health and healing perspectives, 252–54; industrialization, 218–19; learning outcomes and assessments, 224; in Little Big History, 286, 299–301; Marx and Engels on the Anthropocene activity, 228–231; reflective writing questions, 272–73; stock market trading activity, 227–28; teaching strategies, 77–78; Threshold 6 emergent properties giving rise to, 173; Threshold 7 emergent properties giving rise to, 206–8; transition between Threshold 7 and, 216–17

industrialization vs. agrarian civilizations, 216, 219

Industrial Revolution, agricultural revolution role, 206–8

Indus Valley, early agricultural practices, 205

inflation (universe expansion), 93, 238, 346

inflation balloons activity, 117–19

Information Age, 245

information literacy, 6, 290–91; annotated bibliography assignment, 292; assessment of, 294–95; assignments, 290–91; Big History role, 288–290; Dominican's commitment to, 288; librarians' roles, 287–88; library collection support, 289–290; Little Big History assignment, 293–94; mapping physical/digital library assignment, 291–92

Information Literacy Competency Standards for Higher Education, ACRL, 289, 290

innovation: agricultural revolution role, 204, 206, 208; collective learning and, 217–18; commercial activity expansion and, 217; Malthusian cycles that stifle, 217. *See also* technology

inquiry, models for effective, 263

Institute of World and Global History, 57, 68–69

"Intellectual and Practical Skills," AAC&U, 288–89

intellectual thought, agricultural revolution role, 204, 208. *See also* collective learning; scientific empiricism

interconnected and multidisciplinary content, focus group findings, 52

internal combustion engine, 219–220

International Big History Association (IBHA), 26n3, 36–37, 303, 315, 327

internet, 220; access limitations, 62; fragmentation problems, 65; lack of filters, 287; library collections, 289–290; mapping physical/digital library assignment, 291–92; subscription databases, 291, 293; teaching tools, 144; thegreatstory.org, 304. *See also* Big History Project

inverse-square law, 123, 129

irrigation, 203, 205

Islam, 217, 299

Italy, Big History programs, 303

James, Faith, 312

Jefferson, Thomas, 23, 26n4

Journey of the Universe (B. Swimme and M. E. Tucker), 304, 330

Judaism, 98, 299

Kaufman, Stuart, 248

Kearns, David, 24

Keltner, Dacher, 338
Kim, Seohyung, 57
knowledge: in Bloom's taxonomy, 264; fragmentation of, 3–4
Korean Foundation for the Advancement of Science & Creativity, 69
Kurzweil, Ray, 246

land requirements for agriculture, 204, 205
language: human vs. chimpanzee, 168; FOXP2 gene, 168; human evolution of, 167–68, 297; symbolic, 297
language change simulation activity, 194–200
learning outcomes. *See* assessment, Big History and; student learning outcomes
Lemaitre, Georges, 299
liberal education: AAC&U liberal education initiative, 23, 24; Big History and goals of, 21–26, 322, 323; defined, 25n1; perceived impracticality of, 23–24; Thomas Jefferson's support for, 22, 26n4; university's educational philosophy, 44
libraries: assessing students' ability to use, 294–95; collection development and support, 289–290; as interconnected, 287–88, 289–290, 291–92; for Little Big History paper, 293–94; mapping physical/digital library assignment, 291–92. *See also* information literacy
life expectancy: agricultural revolution role, 210; biotechnology to increase, 241–42, 245–46; of contemporary students, 234
lipids, cell membranes and, 150
Little Big History essay, 44, 120, 275–76, 282–86; Big History Project, 61; first-semester paper assessment rubric, 50–51; information literacy challenges,

293–94; Netherlands' program, 65–66, 282
Little Big History of Big History, 282–83, 296–308
living things, basic characteristics of, 149
Lucko, Jennifer, 48, 54n3
luminosity, actual, 124–25

Macartney embassy to China, 208
Mad Max (film), 257
Mad Max Beyond Thunderdome (film), 257–58
male and female hierarchies, human vs. chimpanzee, 168
Malthusian cycles, 217, 235
mammals, evolution of, 152, 172, 349
manure, as fertilizer, 205
Maps of Time (D. Christian), 29, 302, 333; on chemical element formation, 120; on collective learning, 218; on modern creation story, 319, 321
Marcy, Dr. Mary, 315–16
Mars, unmanned spacecraft to explore, 139, 220
Marx and Engels on the Anthropocene activity, 228–231
Maslow, Abraham, 339
mass of stars, calculating, 125
May, J. Daniel, 265
McTighe, Jay, 43
Mears, John, 302
Mesopotamia, early agricultural practices, 205
metabolism, enzyme role, 153
meta-education, in Big History narrative, 18; collective learning and faculty development, 27–40
methane, 152, 221, 257
microorganism design principles activity, 159–162
migration, human: from Africa, 70, 166, 174; ocean exploration and, 207; Paleolithic era, 216–17, 350

military: atom bomb, 220; cities defended by, 206; industrialization of warefare, 80, 219, 351
millennial students, skill gaps, 46, 55n14
Millennium Simulation Project, 112
Miller, Stanley, 152
mining technologies, 218–19
Miocene epoch, 166
mitochondria, in eukaryotic cells, 154
Modern Times (film), 77–80
Mongols, 209
Montanari, Alessandro, 303
Montessori, Maria, 300
Montessori schools, Cosmic Education, 26n3
moon, manned spacecraft to explore, 220, 301
Moore's law, 245
moral intelligence, 248–49
motivated use, measuring, 47–48
multicellular organisms, evolution of, 151, 172, 349
multiregional hypothesis vs. Out of Africa hypothesis, 174
"Myth and Metaphor through the Lens of Big History," 261
"Myth and Ritual through the Lens of Big History," 22, 31

Nagel, Thomas, 249
nanobot "grey goo," 246
nanotechnology, 245, 246
NASA: Hubble Space Telescope, 111, 139; *Voyagers 1 and 2*, 139
natural gas, 219, 222. *See also* fossil fuels
natural selection: defined, 150, 153; evolution role, 150, 299; resistance to concept, 355; simulation activity, 157–59
nature, awe as response to, 338–39
Nature is Enough (L. Rue), 304
Nazi Germans, 219
Neanderthals *(Homo sapiens neanderthalensis)*, 14, 168, 174, 176–79

Netherlands, Big History program, 57, 302; challenges and conclusions, 66–67; how program works, 64–66; origin of program, 63–64
neutrons, 121
neutron stars, 123
Newcomen steam engine, 218–19
The New Frontier of Religion and Science (J. Hicks), 330
Newton, Isaac, 299
Nhat Hanh, Thich, on interdependence, 287
niches, ecological, 152, 349
Nickels, Martin, 265
Nile River, early agricultural practices, 205
Nisa: The Life and Words of a !Kung Woman (M. Shostak), 261–62
nonadaptive systems, complex, 80, 240
Novak, Philip, 21, 30, 303, 311–12, 358
nuclear power, 220
nuclei, membrane-enclosed, 154
nucleic acids, evolution of. *See* evolution of life on earth
nucleotides, evolution of, 152
nutrient requirements for agriculture, 204, 205

oceans: impact of climate change on, 221, 237; life moves to land from, 172; overfishing, 236; seafloor spreading, 141; water pollution, 236, 351
ocean voyages of discovery: British role 1700s-1800s, 207–8; challenges in teaching, 209; "discovery" of Americas, 217, 350; impact on migration and trade, 207; scientific empiricism and, 217–18, 299
oil, industrialization role, 219. *See also* fossil fuels
Oldowan cultures, 167
Omega Point (P. Teilhard de Chardin), 300

reflective writing *(continued)*
University's Big History model,
266–67; on memories, 261–62;
reflection questions activity, 268–273;
Thresholds 1-3, 269; Threshold 4,
269–270; Threshold 5, 270; Threshold
6, 270–71; Threshold 7, 271–-72;
Threshold 8, 272–73; The Future, 273
regimes, systems distinguished from, 80
relativity theory, 113, 299
Religion in Human Evolution (R. Bellah),
320, 325, 330, 332
"Religion Through the Lens of Big
History," 22, 321, 328
religious beliefs: agricultural revolution
and changes in, 298–99; Christianity,
299, 324–26; differing assumptions
about meaning of, 318; historical
violence due to, 328;
institutionalization of, 299; Islam, 217,
299; Judaism, 299; neuroscience
theories on, 328, 329; questions about
mystery of existence, 328, 337–38;
science as challenge to, 217–18,
299–300, 331–33; spiritual evolution,
300, 312
religious beliefs, teaching Big History
and, 318–335; acknowledging what we
don't know, 95, 143; creation myth
creative writing workshop, 99–100;
creation stories approach, 330;
Dominican's approach, 327–333;
faculty retreats to discuss, 327, 331–32;
focus group findings, 49, 52;
identifying religious starting points,
323–25; omissions of religion in
content, 328–29; perceived as
challenge to religion, 19, 319, 320–22,
325, 345; philosophical approach, 330;
religious naturalism approach, 330;
resistance to religious references, 318,
319–320, 323, 328; resources and
inspiration, 304, 312–14; science

approach, 329; science fiction
approach, 330; student discussions, 38,
322–23, 327, 331, 353; symbolic
representation approach, 329–330; vs.
teaching *about* Big History, 318–19,
326; vs. teaching religion, 318, 319,
320–21, 323, 326, 353. *See also* creation
myths (origin stories)
religious naturalism, 330
respiration: ATP, 154; evolution of,
151, 349
retreats, end-of-semester, 22, 39, 354
RNA, evolution of, 152, 153
The Road (film), 258
The Road Warrior (film), 257
robotics, 245, 246
Rodrigue, Barry, 303
rubric construction resources, 55n12
Rue, Loyal, 304

The Sacred Depths of Nature (U.
Goodenough), 304
Saekow, Roland, 303
Sagan, Carl, 301, 305
satellites, 220
scales, differing in time and space, 59, 60
Scarre, Chris, 167
science fiction: approach to religious
topics, 330; and the future, 233, 237,
242, 352; post-apocalyptic film
festival, 256–59; as teaching tool, 352
scientific empiricism: 19th and 20th
centuries, 300–301; as approach to
religious topics, 329; Industrial
Revolution role, 207, 208, 217–18;
Little Big History of, 299–301;
religion to fill "gaps" in, 328; religious
resistance to, 19, 38, 95, 97–98,
217–18, 331–33, 345; replaced by Big
History framework, 355–56; teaching
strategies, 318
seafloor spreading, 141
second law of thermodynamics, 149–150